Fuels and Petrochemicals Division 2014

Core Programming Area at the 2014 AIChE Annual Meeting

Atlanta, Georgia, USA
16-21 November 2014

ISBN: 978-1-5108-1261-1

Printed from e-media with permission by:

Curran Associates, Inc.
57 Morehouse Lane
Red Hook, NY 12571

Printed by Curran Associates, Inc. (2015)

For permission requests, please contact AIChE
at the address below.

AIChE
120 Wall Street, FL 23
New York, NY 10005-4020

Phone: (800) 242-4363
Fax: (203) 775-5177

www.aiche.org

Additional copies of this publication are available from:

Curran Associates, Inc.
57 Morehouse Lane
Red Hook, NY 12571 USA
Phone: 845-758-0400
Fax: 845-758-2634
Email: curran@proceedings.com
Web: www.proceedings.com

TABLE OF CONTENTS

376445 Production of Sugars, Lignin and Lignin Monomers from Biomass Using Gamma-Valerolactone As a Solvent

Abstract:

Widespread production of biomass-derived fuels and chemicals requires cost-effective processes for breaking down cellulose, hemicellulose and lignin into their constituent monomers. Soluble carbohydrates can be produced at high yields (70 to 90%) from corn stover, hardwood and softwood in a solvent mixture of gamma-valerolactone (GVL), water, and dilute acid (0.05 wt% H_2SO_4). When used as a solvent, GVL promotes thermocatalytic saccharification of biomass by complete solubilization of the biomass, including the lignin fraction, and can be produced at high-yields from biomass itself (about 60% yield from cellulose). Rates of acid-catalyzed reactions in GVL appear to be linked to the increased catalytic activity of the proton due to solvent effects.

The carbohydrates can be recovered and concentrated (up to 127 g/L) in an aqueous solution by addition of CO_2 to form a CO_2-expanded GVL phase no longer miscible with water. This strategy is well suited for catalytic upgrading to furans or fermentative upgrading to ethanol at high titers and near theoretical yield. In parallel, the solubilized lignin stream can be recovered at high yields (>70% of the original lignin) by precipitation in water. Analysis of the isolated lignin by NMR demonstrated that its chemical properties remained remarkably similar to native lignin or enzymatically-isolated lignin despite the high temperatures and acid used during processing. Because of this, we show that this stream can be converted catalytically converted to lignin monomers at similar yields to those obtained from unprocessed biomass.

389390 Synthesis of Sulfonated 3DOm Carbon Catalysts for Hydrolysis of β(1-4) Glucan

Abstract:

Selective hydrolysis of cellulose into glucose is the key process for the utilization of biomass. It is the first step in the reaction network for converting biomass into fuels and high-value chemicals[1-3]. Conventional processes for the hydrolysis of cellulose using mineral acids, cellulase enzymes, and supercritical water have several drawbacks, such as the high cost for separation of products and catalysts, corrosion hazards, severe controls of enzymes, low selectivity and waste fluids. The increasing need for environmentally sustainable chemical processes has stimulated the use of recyclable heterogeneous acid catalysts as replacements for the unrecyclable homogeneous acid catalysts and enzyme-based techniques. In particular, sulfonated carbon catalysts (C-SO_3H) have shown promising activity for the selective hydrolysis of cellulose in water under moderate conditions. However, due to the limited understanding of the interaction between the solid catalyst and water-insoluble cellulose, the catalytic property of carbon catalysts for this reaction is still elusive.

The aim of this study is to design a bifunctional heterogeneous catalyst with both acid and adsorption sites for the selective hydrolysis of β(1-4) glucan to glucose. Three dimensionally ordered mesoporous (3DOm) carbon with a cage size of 35 nm was successfully functionalized with sulfonic acid through covalent attachment of sulfonic acid-containing aryl radicals produced by homogeneous reduction of diazonium salt with hypophosphorous acid. It is found that the adsorption of saccharides on this carbon catalyst could be dominated by the hydrophobic interaction between the π-domains on the carbon surface and the CH groups on the saccharides.

The carbon catalyst was used to hydrolyze acidulated ball-milled cellulose (ABMC) to produce glucose. ABMC was prepared by incipient wetness of microcrystalline cellulose with dilute sulfuric acid and ball-milled, producing short β(1-4) chains (DP < 10) with α(1-6) branching. This ABMC has excellent solubility (10% in water) compared to ordinary β(1-4) glucan. The turn over frequency (TOF, (mole glucose)(mole $H^+)^{-1}h^{-1}$) for ABMC reacted with the carbon catalyst and HCl was 1.5 and 0.18, respectively. A max glucose yield of 65% was obtained by reacting ABMC with the carbon catalyst. NMR studies revealed that β(1-4) glycosidic bonds break down rapidly while the α(1-6) bonds remain mostly unreacted using the carbon catalyst. The high reaction rate of the β(1-4) glycosidic bonds in ABMC when using the carbon catalyst is likely due to the interaction between α(1-6) branched β(1-4) glucan chains and adsorption on the carbon surface.

References:

1. Dornath, P.; Fan, W., "Dehydration of fructose into furans over zeolite catalyst using carbon black as adsorbent". *Microporous Mesoporous Mater.* **2014,** *191* (0), 10-17.

2. Chang, C.-C.; Green, S. K.; Williams, C. L.; Dauenhauer, P. J.; Fan, W., "Ultra-selective cycloaddition of dimethylfuran for renewable p-xylene with H-BEA". *Green Chem.* **2014,** *16* (2), 585-588.

3. Williams, C. L.; Chang, C.-C.; Do, P.; Nikbin, N.; Caratzoulas, S.; Vlachos, D. G.; Lobo, R. F.; Fan, W.; Dauenhauer, P. J., "Cycloaddition of Biomass-Derived Furans for Catalytic Production of Renewable p-Xylene". *ACS Catal.* **2012,** *2* (6), 935-939.

378810 Synergistic Effect Between Defect Sites and Functional Groups in the Hydrolysis of Cellulose Using Activated Carbon

Abstract:

Carbon with Brønsted acid sites can function as a stable and effective solid acid catalyst for reactions in hot liquid water. It is able to hydrolyze cellulose, an abundant and renewable biomass, into glucose, which is a versatile platform molecule. However, there are limited reports on the effect of different chemical treatments and the nature of active sites. In this study, the effect of chemical oxidation by H_2O_2 and H_2SO_4 on activated carbon was investigated, and the materials were used in the hydrolysis of cellulose for comparison. The oxidation procedure was characterized by XRD, FTIR, ^{13}C DP MAS NMR, Boehm titration, N_2 physisorption, CO_2 physisorption and Raman spectroscopy. Oxidation by H_2O_2 imparts acidic functional groups such as phenol, lactone and carboxylic acid, while treatment with H_2SO_4 imparts all of those oxygen functional groups along with sulfonic acid groups. Characterization reveals that H_2O_2 is capable of generating and enlarging defect sites, while H_2SO_4 creates more edges in the carbon material, and this effect is more pronounced with higher temperature.

Adsorption isotherms demonstrated that sulfonic groups enhance the adsorption of glucose monomers more than phenols, lactones and carboxylic acids by hydrogen bonding, and the influence of this enhanced-adsorption effect decreases as the length of glucan chain increases. This can be explained by the increasing influence of van der Waals interactions between the CH groups and the polyaromatic rings. All of the chemically treated materials exhibit catalytic activities in the hydrolysis of cellulose despite the presence of weakly acidic functional groups. Carbon treated by H_2SO_4 showed higher reactivity even though it has less functional groups and lower pore volume distribution. This can be attributed to the synergistic effect between the edge/defect sites and functional groups in the carbon material, where the glucan chains are immobilized and the exposed glycosidic bonds are forced to interact with the in-plane functional groups.

Fossil fuels provide approximately three quarters of today's fuel energy, but current fossil fuel world reserves will not meet future energy demands. The burning of these energy sources is one of the main causes of the increase of CO_2 concentration in the atmosphere that has been associated with Climate Change. Consequently a transition from nonrenewable (fossil fuels) energy sources to renewable ones is desirable. Biomass is the main renewable source that can substitute fossil fuels, but to achieve this goal some technical challenges must be overcome. One of them is the design of catalysts with good thermo solvent stability, and this is especially true for aqueous liquid phase processing as is the case of fructose dehydration to 5-hydroxymethyl-furfural (HMF). In this work we report the continuous production of HMF from fructose using mesoporous CMK-3 carbon materials functionalized with methyl-, ethyl- and phenyl- sulfonic acid groups. These materials were characterized using scanning electron microscope (SEM), nitrogen adsorption - desorption isotherms, X-Ray diffraction (XRD), elemental analysis and acid titration. The mesoporous carbon materials have the desired long-range order of mesopores, textural properties and have been grafted with the desired sulfonic functional groups. We will discuss the catalytic performance of the sulfonic acid functionalized mesoporous carbons using kinetic data obtained in batch and flow reactors. Preliminary results indicate that the phenylsulfonic catalysts are the most active and selective towards HMF production with 100% conversion of fructose and 44% yield towards HMF in 30 min at 145 °C. We will also present a study of the catalysts stability under reaction conditions in a packed bed reactor. Finally, possible deactivation modes will be presented.

386444 Unraveling Solvation Effects on 5-Hydroxylmethylfurfural Degradation: Insights from Catalytic and Spectroscopic Studies

Abstract:

The prospective use of lignocellulosic biomass as a feedstock of value-added chemicals and fuels requires the effective and synergistic utilization of catalysts and solvents. The necessity to develop efficient and economic processes for the production of biomass derived platform chemicals, such as 5- hydroxymethylfurfural (HMF) has motivated research for the development of carbohydrate dehydration processes using heterogeneous[1, 2] and/or homogeneous[3, 4] catalysts. The use of organic co-solvents is one of the methods used to enhance HMF yields by reducing the rates of its rehydration and humin-formation reactions. Although many solvent mixtures have been tested in this respect, the reasons of enhanced HMF stability in aqueous aprotic co-solvent mixtures are still not clearly understood.

In the present work, vibrational spectroscopy is coupled with catalytic experiments in an effort to elucidate the origins of HMF stability and to provide "solvent-solute interactions" – activity relationships. Attenuated total reflection FTIR spectra of HMF in a wide composition range of aqueous aprotic (i.e. DMSO, g-valerolactone, THF etc.) and protic (i.e. methanol, ethanol etc.) solvent mixtures were measured in order to unravel the effect of the co-solvent on the structural and vibrational properties of dissolved HMF. Deconvolution of pertinent wavenumber regions was carried out in an attempt to discriminate HMF-solvent interactions at different mixed solvent compositions. Different intermolecular interaction have been ascribed that are responsible for the non-conventional change of the HMF C=O and O-H vibrational modes under different solvent composition[5]. Our spectroscopic analysis has revealed the first experimental evidence of preferential interaction of HMF by aprotic solvents that that also depends on the co-solvent nature and amount.

HMF rehydration experiments were carried out in several mixed-solvent mixtures under different conditions (i.e. temperature, and acid concentration), and the effect of co-solvent type and concentration on the HMF stability is quantified. In addition to the possible effects of co-solvent on the intrinsic reaction rate constant, the rehydration and humin formation reaction rates can also be affected by the water activity, and proton concentration. The realative contribution of each of these factors on HMF stability/ reactivity is determined and will be analyzed in an effort to provide solvation/activity relationships based on our spectroscopic analysis. Our findings will also be discussed with respect to the possible HMF rehydration and humin formation reaction mechanisms.

References

[1] V.V. Ordomsky, J. van der Schaaf, J.C. Schouten, T.A. Nijhuis, 5 (2012) 1812-1819.

[2] Y. Zhang, J.J. Wang, J.W. Ren, X.H. Liu, X.C. Li, Y.J. Xia, G.Z. Lu, Y.Q. Wang, Catal Sci Technol, 2 (2012) 2485-2491.

[3] Y. Roman-Leshkov, J.A. Dumesic, Top Catal, 52 (2009) 297-303.

[4] J.N. Chheda, Y. Roman-Leshkov, J.A. Dumesic, Green Chemistry, 9 (2007) 342-350.

[5] G. Tsilomelekis, T.R. Josephson, V. Nikolakis, S. Caratzoulas, Chemsuschem, 7 (2014) 117-126.

378449 Kinetics and Insights into Glucose Dehydration Catalyzed By Homogeneous Lewis Acidic Metal Salts in Water

Abstract:

Dehydration of glucose catalyzed by various Lewis acidic metal salts was studied with water as reaction media. The glucose conversion kinetic profile was used to reflect the Lewis acid character of different metal ions. It was found that the pH value of the aqueous solution played an important role in controlling the Lewis activities. For the lanthanide chlorides, their Lewis acidity was comparable under the pH values studied (from 2.5 to 5.5). However, the Lewis acidity strength of other metal salts, such as aluminum chloride, showed dependence on the pH value of the solution. Activation energies with various Lewis acids were also calculated with both glucose and fructose to obtain more insight about the strength of the catalyst-substrate interaction as well as the dehydration reaction. The kinetic isotope effect with labeled glucose molecules was also studied to explore a more mechanistic understanding of the dehydration, which likely involves the 2-H atom of the glucose molecule in the transition state.

371604 Solvent-Induced Frequency Shifts of 5-Hydroxymethylfurfural and Their Role in Its Stability

Abstract:

5-Hydroxymethylfurfural (HMF) is a promising platform-chemical for sustainable fuels and plastics. A major issue in the production of HMF is the side reactions that lead to significant loss of sugars and high cost. Use of polar aprotic solvents improves HMF yields from sugar dehydration by preventing HMF side reactions, but the influence of solvent on the structure of HMF is not fundamentally understood. In an effort to understand solvent effects on HMF stability and yield, solvent-induced frequency shifts (SIFS) of the carbonyl stretching vibration ν(C=O) of 5-hydroxymethylfurfural were measured in seven protic, polar aprotic, and non-polar solvents. The Gutmann Acceptor Number (AN), an empirical measure of Lewis acidity, was found to correlate with the measured frequency shifts. The SIFS were then investigated using *ab initio* electronic structure calculations, treating the solvent implicitly and with an explicit solvent ligand interacting with the carbonyl. The conductor polarizable continuum model (CPCM) solvation model predicted that ν(C=O) shifted as a function of the dielectric constant, in agreement with the analytical predictions of the Kirkwood-Bauer-Magat (KBM) theory for a dipole in a dielectric continuum, but in disagreement with experimental ν(C=O). Experimental ν(C=O) were best predicted using gas-phase complexes of HMF and explicit solvent ligand. NBO analysis and Bader's Atoms in Molecules theory were used to investigate the electronic structure and hydrogen bonding in these complexes. SIFS arose from H-bonding interactions, as observed in delocalization of carbonyl lone-pair electrons to H-bonding solvent σ*(X-H) orbitals, and in increased charge density and decrease in local potential energy at the H-bond (3,-1) critical point in the charge density. Consequently, by predicting the experimental SIFS and examining the electronic structure, we find the first theoretical evidence for treating Gutmann's solvent AN as a measure of solvent Lewis acidity. Application of these findings in understanding the stability of HMF in various solvents will also be discussed.

Simulation and Optimization of production of Cyclohexane from benzene using ASPEN Plus

Utkarsh Maheshwari[1], Sakshi Batra[2], Pratik Pande[2]

[1]Faculty, Department of Chemical engineering, BITS Pilani, Pilani, Rajasthan 333031, utkarsh@pilani.bits-pilani.ac.in
[2]Student, Department of Chemical engineering, BITS Pilani, Pilani, Rajasthan 33303, sakshibatra07@gmail.com, pratikppade@gmail.com

Abstract

Cyclohexane is used as a non-polar solvent for the chemical industry, and likewise as a raw material for the industrial production of adipic acid and caprolactum, both of which are intermediates used in the production of nylon. Nowadays cyclohexane is most important for the production of caprolactum and other nylon intermediate.

In the present work, the production of cyclohexane from benzene and hydrogen is simulated using the ASPEN Plus platform. The most import process in the caprolactum production is the removal of the heat evolved from the reaction. In the present work, the process is optimized using trial and error method by changing some independent parameters and methods for the process. Here the issue of transition due to the fluctuation of the thermodynamic model is considered. RK-Soave and NRTL method is examined for more conversion in the procedure. By using NRTL method the molar flow-rate of cyclohexane from distillate is coming to be 43.16, by changing the method, operating status and using trial and error method we obtain that for RK-Soave method, it is giving highest conversion and molar flow-rate of 43.43 which is supposed to be a substantial divergence.

Keywords: Cyclohexane, Benzene, Aspen Plus

380439 Simultaneous Product Loss and Energy Minimization for the Cryogenic Separation System of an Ethylene Plant

Abstract:

The integrated cryogenic separation system of an ethylene plant consists of a multi-stage cold box subsystem and a de-methanizer column subsystem. It separates H2 and CH4 from the charge gas from cracking furnaces through integrated liquefaction and distillation operations. Since the majority of charge gas needs to be liquefied for separation, tremendous energy is consumed in this system; meanwhile, it is also observed that the main product of ethylene can easily enter the fuel gas stream during the cryogenic separation process, which causes significant product loss. Thus, to optimize the operation of the integrated cryogenic separation system is very important to the profitability of an ethylene plant.

In this paper, a general methodology is developed for simultaneous product loss and energy minimization for the cryogenic separation system of an ethylene plant. Firstly, the simulation model for the integrated cryogenic separation system is constructed and validated. Secondly, the sensitivity analysis is conducted based on the simulation model to identify the most influential process variables for the ethylene product loss and energy consumption. After that, quantitative relations between those influential process variables and various manipulated variables, such as the column tray efficiency, heat exchanger efficiency, and compressor power, are modeled with the help of artificial neural network techniques. Finally, an optimization model based on these surrogate models are developed and solved by an intelligent optimization algorithm. The developed methodology can quickly provide multi-objective optimal solutions that well balance the ethylene product loss and energy consumption of the studied cryogenic separation system.

The falling point of lubricant

Authors Name- *Arafat Husain*
Dept. name of organization-Chemical engineering
Name of organization –Aligarh Muslim University
Name of country-India
Email-arakant78@gmail.com

Abstract

The lubricants are one of the most used resource in today's world. Lot of the superpowers are dependent on the lubricant resource for their country to function. To see that the lubricants are not adulterated we need to develop some efficient ways and to see which fluid has been added to the lubricant. So to observe the these malpractices in the lubricant we need to develop a method. We take a elastic ball and throw it at probability circle submerged in the lubricant at a fixed force and see the distance of pitching and the point of fall .Then we calculate the ratio of distance of falling to the distance of pitching and if the measured ratio is greater than one the fluid is less viscous and if the ratio is lesser than the lubricant is viscous. We will check the falling point of pure lubricant at fixed force and every pure lubricant would have a fixed falling point. After that we would adulterate the lubricant and note the falling point and if the falling point is less than the standard value then adulterate is solid and if the adulterate is liquid the falling point will be more than the standard value .Hence the comparison with the standard falling point will give the efficiency of the lubricant

1. Introduction

A **lubricant** is a substance introduced to reduce friction between moving surfaces. It may also have the function of transporting foreign particles. The property of reducing friction is known as lubricity. (Slipperiness)

A good lubricant possesses the following characteristics:

- High boiling point
- Low freezing point
- High viscosity index
- Thermal stability
- Hydraulic Stability
- Corrosion prevention

High resistance to oxidation
One of the single largest applications for lubricants, in the form of motor oil, is protecting the internal combustion engines in motor vehicles and powered equipment.

Typically lubricants contain 90% base oil (most often petroleum fractions, called mineral oils) and less than 10% additives. Vegetable oils or synthetic liquids such as the one known mostly hydrogenated polyolefin ,esters, silicones, fluorocarbons and many others are sometimes used as base oils. Additives deliver reduced friction and wear, increased viscosity, improved viscosity index, resistance to corrosion and oxidation, aging or contamination, etc.

Lubricants such as 2-cycle oil are added to fuels like gasoline which has low lubricity . Sulfur impurities in fuels also provide some lubrication properties, which has to be taken in account when switching to a low-sulfur diesel; biodiesel is a popular diesel fuel additive providing additional lubricity.

Non-liquid lubricants include grease, powders (dry graphite, PTFE, Molybdenum disulfide, tungsten disulfide, etc.), PTFE tape used in plumbing, air cushion and others. Dry lubricants such as graphite, molybdenum disulfide and tungsten disulfide also offer lubrication at temperatures (up to 350 °C) higher than liquid and oil-based lubricants are able to operate. Limited interest has been shown in low friction properties of compacted oxide glaze layers formed at several hundred degrees Celsius in metallic sliding systems, however, practical use is still many years away due to their physically unstable nature.

The marketer offers a lubricant at a lower cost than rivals either in the same grade or a similar one that will fill the purpose for lesser price. (Stationary installations with short drain intervals.) Alternative may be offering a more expensive lubricant and promise return in lower wear, specific fuel consumption or longer drain intervals. (Expensive machinery, un-affordable downtimes.)

The importance of lubricant lies in its Quality for a given quantity so the importance to develop easy methods of oil analysis are becoming important.It is important to know oil analys Oil analysis involves sampling and analyzing oil for various properties and materials to monitor wear and contamination in an engine, transmission or hydraulic system. Sampling and analyzing on a regular basis establishes a baseline of normal wear and can help indicate when abnormal wear or contamination is occurring.

Oil analysis works like this. Oil that has been inside any moving mechanical apparatus for a period of time reflects the exact condition of that assembly. Oil is in contact with engine or mechanical components as wear metallic trace particles enter the oil. These particles are so small they remain in suspension. Many products of the combustion process also will become trapped in the circulating oil. The oil becomes a working history of the machine.

Particles caused by normal wear and operation will mix with the oil. Any externally caused contamination also enters the oil. By identifying and measuring these impurities, you get an indication of the rate of wear and of any excessive contamination. An oil analysis also will suggest methods to reduce accelerated wear and contamination.

The typical oil analysis tests for the presence of a number of different materials to determine sources of wear, find dirt and other contamination. Oil analysis can detect:

- Fuel dilution of lubrication oil
- Dirt contamination in the oil
- Antifreeze in the oil
- Excessive bearing wear
- Misapplication of lubricants

Some wear is normal, but abnormal levels of a particular material can give an early warning of impending problems and possibly prevent a major breakdown.

Adulteration in fuels and lubricants-

Adulteration is the practice of adding foreign material to the original substance in order to increase its quantity so as to make profit from it.

This processes of adulteration are done in a number of forms mixing of genuine products bought from licensed players with low quality solvents and put up for sale as authentic products of approved players beneath the same brand; mixing of novel products bought from approved players with degradable solvents and put on the market as real products with no branding; cheaper brands poured into exclusive branded containers and thereby misleading the retailers and customers; refining the used oils and mixing them with solvents; and re-packaging of certified players products and selling them under re-packers name.

The traitor's modus operandi is quite unique. What they do is they acquire worn oil from different service stations, then empty the old oil into pits and leave it until the sediments are dropped to the bottom . After that, sediment-free layer is collected and is mixed with the real oil purchased from the authorized dealers thus resulting in the adulteration of oil. Adulteration of fuels mainly involves adding kerosene or diesel to petrol. World statistics on adulteration indicate that 70 per cent of drivers and mechanics have experienced this problem in petrol engines and rarely on diesel and compressed natural gas (CNG) engines.

Most of lubricants you find today are adulterated. As is the recent case happened in Cuttack, police have recovered a comprehensive automobile lubricant adulteration unit operating out of a borrowed house in Pithapur. According to the sources, police had seized some 500liters of adulterated lubricant oil, massive quantity of stickers, logos and sealed caps of top notch brands like Castrol, Elf and Servo. Because of high prices of Lubricant oil, adulteration of this petroleum product is rampant in many states.

Oil Analysis-

Records show that some equipment can safely run two or three times longer than recommended intervals.

The oil analysis may show that you are changing the oil more often than necessary — or not often enough.

By eliminating too frequent oil changes, you reduce the cost for oil and servicing and also reduce the amount of used oil to deal with. This is an important pollution prevention method — reducing the source.

Oil sample analysis saves you repair and maintenance dollars, has the potential to reduce used oil and increases resale value of equipment.

These are average numbers used but depending on your type of equipment may be higher or lower. Most reports have charts listed on the back to explain the severity of that component in ppm.

The Falling point-

The distance travelled by the ball after pitching on the probability circle.

Probability circle-

The principal sphere where the ball should be pitched in order to be valid for the calculation of falling point.

Falling point Ratio-

(Falling point distance/distance covered before pitching on probability circle)

The principal is based upon the principal that the skidding distance of elastic ball is a function of density and viscosity cancelling out each other with least amount of mass transfer and gravitational force.

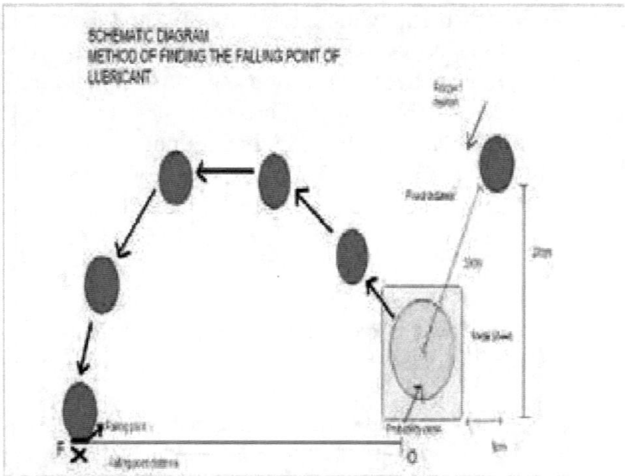

Figure 1.a shows the method to find the falling point

Conditions Required-

- Sliding friction- 0.2-0.4
- Standard temperature- 25°Celsius
- Total volume – 1liter= (contaminant lubricant + original lubricant)
- Metal surface with small curvature
- Lubricant to be used for testing should be anti knocking, sufficient octane number and good viscosity index.
- Probability circle ratio to that of the ball to be not more than 10.
- Force exerted = One Newton.
- Angle of impact—60 degree

Method to find falling point-

The experiment was conducted to determine the falling point of lubricant

Oil analysis was conducted in terms of Falling point ratio of –

(Falling point distance/distance covered before pitching on probability circle)

Apparatus required were –a elastic ball ,a system to produce a constant force, gasoline(VI-11O,octane no 98), water, Vaseline and charcoal .

The force was applied from a constant point and the ball was pitched at the probability circle and the falling point of the ball was noted for different cases.

Readings-

Distance covered before pitching on the probability circle- 30 cm

Force exerted- 1Newton

Sliding friction-0.30

Falling point distances for different lubricants (FPR) in pure state with negligible contamination-

- **Gasoline-160.0 cm**
- **Kerosene-146.0cm**
- **Water- 173.0cm**
- **Vaseline-45.0cm**
- **Charcoal-13.0cm**

Observations-

Falling point ratio (FPR)-

(Falling point distance/distance covered before pitching on probability circle)

Contamination in gasoline-

- Case 1- Gasoline in pure state (1litre)

FPR- 160cm/30cm =5.33

- Case 2- Gasoline + Kerosene (950ml+50ml)

FPR- 151cm/30cm=5.033

- Case 3- Gasoline + Vaseline (950ml+50ml)

FPR-102cm/30cm=3.40

- Case 4- Gasoline + Charcoal(950ml+50ml)

FPR=72/30=2.4

Results-

- Due to contamination the falling point ratio of gasoline varies from- 5.33 to 2.40 and every time different contaminants were used for testing.
- So one can notice the change or deviation of gasoline Falling point ratio as the sign of adulteration.
- From the trends in Falling point ratio(FPR) one can also find out the kind of lubricant added to the gasoline and to how much extent with the help of oil analysis.
- The range of falling point distance(FPD) after contamination was found to be between –
- FPDcontaminent ---- (FPDcontaminent + FPDgasoline)/2

Chemical properties effected-

1) **COMBUSTION-**

The combustibility was found to decrease with the decrease in Falling point ratios.

FPR ratios

5.33-5.23 (good combustion)

5.23-4.94 (partial combustion)

4.94-2.40 (bad combustion)

2.40—so on (no combustion range)

2) Wear on the metal sheet decrease with decrease in falling point ratios.

3) Knocking of the engine is delayed with the decrease in falling point ratios.

4) Emission of harmful gases is also controlled with minimization of deviation in the falling point.

5) The use of correct oil additives can be used to fix the deviation in the falling point and the ratios.

6) Corrosion characteristics increase with decrease in falling

✖ Merits-

1) Potential to be one of the cheapest test for oil analysis for lubricant contamination.
2) Could be used to investigate the future fuels.
3) Easily differentiate between phases of lubricants so they can be used for different purposes.
4) Reduce the problem of corrosion, oxidation and late knocking of the engine in the machines.
5) Reduction of harmful gases from the engine.

✖ Demerits-

1. There can be errors while measuring the force applied ,mass transfer of the ball and water and the angle of impact
2. Can't be used to check the contamination of more than three lubricants.
3. Tribological errors can lead to some deviation in falling point and the ratios.
4. Only contaminants on the larger basis has been identified and micro contaminants like copper, sulphur, magnesium in ppm etc cannot be identified through this analysis.

Relationship between falling point of water and its contamination -

Adding water does increase the falling point ratio but the margin for error is very low.

The major difference in densities leads to layer separation and hence combustibility decrease with increase in the volume of water and so does the falling point and the falling point ratios increase with increase in volume of water. Water can be present in hydraulic fluids and other lubricants as either dissolved, emulsified or free water. The point at which the fluid cannot hold any more dissolved water is called the saturation level. If more dissolved water is present than the fluid can hold, the excess water (or free water) can be present either as a separate bulk water phase or as an emulsion.

Water can originate from a variety of sources. Examples of environmental ingression are rain leakage into external reservoirs, seepage through reservoir covers, access panels, breathers or worn seals, and condensation from air in reservoirs and other system areas. Water can also enter the fluid system from the process side, from leaky heat exchangers or coolers, or direct ingression of process water, such as cooling water, wash down water or steam. Water ingression can be minimized with clever system design and good housekeeping, but it is difficult (and costly) to eliminate all sources of water entirely.

The presence of water in hydraulic fluids and lubricants can have wide-ranging effects on system components. Surface corrosion, probably the most obvious effect, is directly linked to the presence of free bulk water. Accelerated metal surface fatigue, such as in bearings, can be promoted even if all the water present in the fluid is dissolved. This was studied by Cantley in 1977[1], who investigated the effect of dissolved water on the fatigue life of tapered roller bearings. Cantley developed an equation linking the relative bearing life and the water content of the lubricant used in the tests, an SAE 20 fluid containing rust- and oxidation-inhibiting additives. He showed that the bearing life could be extended by a factor of five, if the bearing lubricant contained only 25 ppm dissolved water compared to 400 ppm, close to the saturation level at the test temperature of 150°F. Figure 2 shows an adaptation of Cantley's findings and illustrates the strong correlation between water content and relative bearing life.

Results

But all the falling points for different states of water is not the same.

The results are found to be different for emulsified water, dissolved water and free water.

Water is a significant contaminant in hydraulic and lubrication fluid systems, resulting in degradation of system components and fluid.

Substantial cost savings can be gained from operating with dry fluids. Recommended saturation level: 50 percent or less for typical mineral oil-based fluids.

Water monitoring techniques must be accurate, repeatable and in real time so that increases in water content can be promptly rectified. On-line monitoring of water via a capacitive water sensor offers an optimal and cost-effective solution to monitoring but the falling point also provide help to study the detection of water impurity in the lubricants but not in the amount of water present as a contaminant in the fuel.

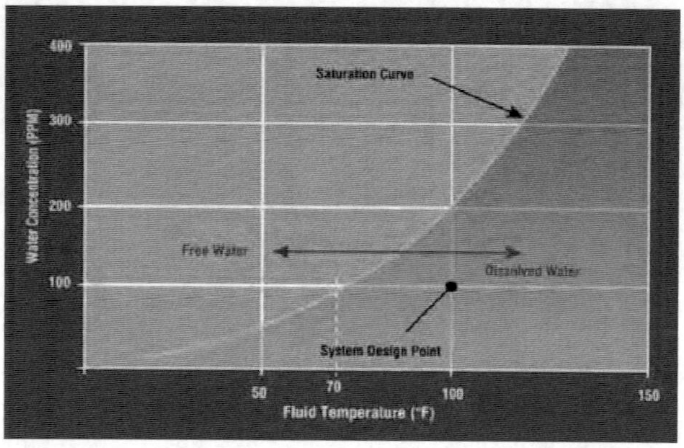

Figure 1b. Saturation Curve for a Typical Turbine Lube Oil

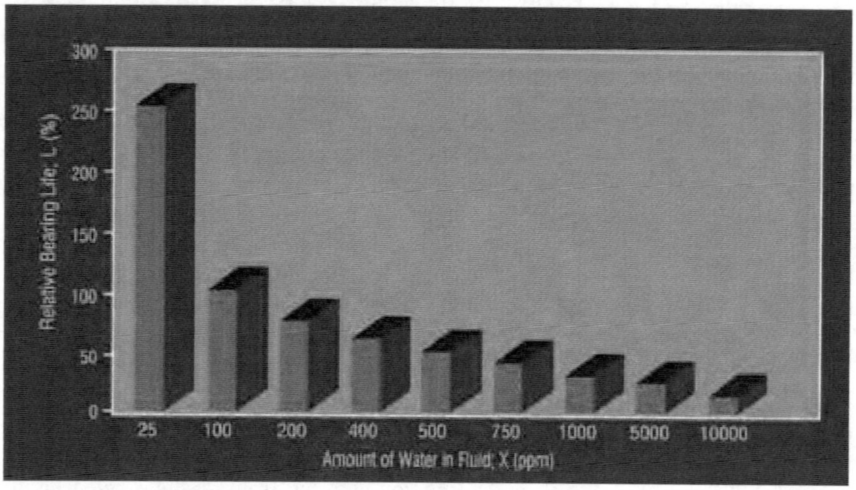

Figure 1c. Relative Bearing Fatigue Life

Scope in the future-

- From these observations one can say that falling point helps in not only identifying the lubricants but also classifying them. Adding impurities to the lubricants helps to observe that when impurities added to lubricants there is a observable changes in falling point of lubricant and the ratios.
- The relationship between water contamination and the falling point and the falling point ratios can be very useful in the coming future.
- An easy kit will help even a common man to check the impurity added to the lubricant injected in his truck

References-

- American Society for Testing Materials, Institute of Petroleum, Portland Place London. Reduction of observed density to density at 15 C and volume at 15 C, Copyright Publication, 1953 Indian Oil Corporation, Industry quality control
- R.E. Cantley. "The Effect of Water in Lubricating Oil on Bearing Fatigue Life." ASLE *Transactions, American Society of Lubrication Engineers,* Volume 20, No. 3, p. 244-248, 1977; from a presentation at the 31st Annual ASLE Meeting, Philadelphia, Penn.
- J. Fitch and S. Jaggernauth. "Moisture, the Second Most Destructive Lubricant Contaminant and its Effect on Bearing Life." *P/PM Technology*, p. 50-53, 1994.
- M. Weinschelbaum. "A Study of the Invisible but Measurable Particulate Contaminants in Hydraulic Systems." *Proceedings* of the National Conference on Fluid Power, Volume XXIII, p. 265-277, 1969.
- K . Farooq and R. Fowler. "Comparison of Water Measurement Results in Polyol Ester-based Lubricant Fluids by the Coulometric Karl Fischer Method and a Thin Film Polymer Capacitive Water Sensor." JOAP International Condition Monitoring Conference, April 3-6, 2000.
- Arafat Husain ."The falling point of lubricant with the help of dynamic ratios."AICHE annual meet Atlanta,November-16,2014

Evaluation of Flare gas Recovery (FGR) Process, Case Study: India, Norway and Middle East

In refineries, gas processing plans, oil wells and other oil and gas related facilities flammable waste gases are released from processing units during operation and transmitted to flare stack for safe disposal. The main purpose of flare is to maintain the safety of facilities, staffs and environment. However, burning this waste gas in flare stack causes huge energy losses and also environmental problems such as local air pollution. Therefore, flare gas reduction has considerable benefit to environment and preserve energy resources.

Flare Gas Recovery (FGR) is the process of recovering the waste gases that are normally flared, so they can be used as fuel gas elsewhere in the facility. In this work, a step by step approach and calculations are presented to compute and evaluate the FGR process for three different flare gas lines. The method comprises data gathering of flare lines' compositions and properties, design of process using simulation software and economic analyses for each case study. The flare gas line for three oil and gas facilities located in Norway, India and Middle East are used as case studies.

The FGR processes were simulated in HYSYS simulator based on a single stage compression and cooling for three flare gas lines. After economic analyses of each case, it was demonstrated that FGR process is not economically feasible in India and Norway flare lines due to the low concentration of hydrocarbons and current prices of fuels. However, FGR for Middle East flare gas is profitable project due to the high hydrocarbon load. The payback period for the project was estimated at 9 months with the current price of natural gas. The financial gain of this process is around 5 million USD per year. Considering the return on investment (ROI) for Middle East flare gas, the FGR process would be economical.

COMPARATIVE STUDY OF ENHANCED OIL RECOVERY EFFICIENCY BY CO2 INJECTION AND CO2 HUFF-N-PUFF IN STIMULATED SHALE OIL RESERVOIRS

Tao Wan, Yang Yu, James J. Sheng, Texas Tech University, Lubbock, TX

Abstract

Multi-stage fracturing applied in horizontal wells is considered as the most effective method for producing shale gas/oil reservoirs. However, the primary oil recovery from natural depletion is only few percent. The low recovery rates and the abundance of shale resource plays provide huge potential for enhanced oil recovery. Gas or oil production performance from nano-Darcy permeable shale rocks is characterized by steep decline rates. The primary shale oil recovery from hydraulically fractured horizontal wells is around 5%. Without enhancement of oil production from secondary recovery by solvent or gas injection, the huge cost of hydraulic fracturing treatment makes it economically unviable to produce shale oil reservoirs. It is imperative that we implement an enhanced oil recovery method in going beyond conventional natural depletion means to achieve profitable production. In this paper, we will compare the improved oil recovery efficiency in shale oil reservoirs by designed cyclic gas injection in a fractured horizontal well and gas flooding in two fractured wells. The biggest challenge for gas flooding in fractured shale oil reservoirs is that the injected solvent may break through to production well via the fracture network or fissures. In contrast, cyclic gas injection is a good candidate for fractured reservoirs that is not subject to earlier breakthrough.

The oil recovery factor obtained from these two injection approaches were compared under both miscible and immiscible displacements. Our composition simulation results showed that under miscible flooding condition cyclic gas injection is more effective than gas flooding given at the same injection pore volume and identical fracture spacing. One reason for low recovery in gas flooding is the loss of injected fluids and decrease of displacement efficiency due to earlier breakthrough to the producer via natural or hydraulic fractures. Oil response from cyclic gas injection well performed steadily even after many cycles in shale oil reservoirs, but oil rate from gas flooding falls off rapidly. Considering the existence of natural fractures, the oil recovery by cyclic CO2 injection increased from primary 5% to ultimate 18% after 10 cycles in two hydraulically fractured horizontal well. On the other hand, the enhanced oil recovery by CO2 flooding is only 11%. The disadvantage of cyclic gas injection is that it consumes more time than gas flooding because injection and production processes are separate. If the injection time is too long, oil will be pushed far away from the fractures that makes it more difficult to produce back. The objective of this paper is to evaluate the IOR performance of cyclic gas injection and gas flooding approach in shale oil reservoirs.

Introduction

Cyclic CO_2 injection has been successfully applied to the recovery of light oil with encouraging results. Both previous laboratory coreflood and field tests have shown the potential of cyclic CO_2 injection to improve oil recovery. Monger and Coma (1988) conducted an experimental investigation of CO_2 cyclic injection and indicated that overall process performance was benefitted to recover more oil. A second cycle of CO_2 injection recovered additional oil, but a decline in process performance was noted. Asghari and Torabi (2007) presented an experimental study of CO_2 huff-and-puff in fractured conventional rocks (100md-1000md). The effectiveness and efficiency of CO_2 huff-and-puff performance to increase oil recovery is observed at both miscible and immiscible conditions. Wang et al. (2013) reported experimental study results of CO_2 huff-and-puff process operated in a 973 mm long composite core with an average permeability of 2.3 mD. Their results suggested that the enhanced oil recovery factor could be as high as 34.6% by CO_2 huff-and-puff operating in a low permeability and low pressure with a closed boundary rocks. They (Karim et al. 1992; Thomas et al. 1992; Wang et al. 2013) observed that the first operation cycle contributes nearly or above half of the total oil production of the operation and additional oil produced from subsequent cycles is significantly decreased when compared to previous cycles.

Cyclic CO_2 injection for heavy oil recovery has been field tested and five case histories indicate that oil production is enhanced (Khatib et al.1981; Saner and Patton, 1986; Bardon et al. 1986; Olenick et al.1992). Feasibility of using CO_2 in Halfmoon field was confirmed and incremental oil can be recovered by immiscible CO_2 injection, using either cyclic or drive injection mode. With the increasing depletion of conventional resources, the application of cyclic CO_2 injection has been transplanted to improve oil recovery in ultra-tight shale oil formations. In shale gas or oil reservoirs, the presence of fissures or induced hydraulic fractures provides highly conductive paths for injected gas to diffuse or penetrate into the nano-permeable matrix. The experimental results that demonstrated the potential of cyclic N_2 and CO_2 injection to improve tight shale oil recovery has been reported by Gamadi et al. 2013 and 2014. Both miscible and immiscible operation schemes were performed to show the effectiveness of cyclic gas injection recovering oil from shale matrix. Recent studies (Chen et al. 2013; Wan et al. 2013; Yu et al.2014) showed that cyclic gas injection could be a viable method to improve the oil recovery in shale oil reservoirs. Studies showed that cyclic gas injection combined with modern technologies such as horizontal well drilling and hydraulic fracturing has achieved promising results in low permeability formation.

The objective of this paper is to compare the flooding performance by CO_2 injection in two fractured horizontal wells and cyclic CO_2 injection mode.

Model input data

Kurtoglu (2013) presented a dual porosity simulation model that was run with oil rate control to history match the pressure and gas production profile of a middle Bakken horizontal well with 1.2 year of primary production in Reunion Bay. **Figure 1** displays reasonable model prediction against the history data (oil rate, gas rate). Later on, Yu et al. (2014) used this model to investigate the effect of cyclic CO_2 injection on improved oil recovery. We used their model as input data to compare the efficiency of cyclic CO_2 injection in a single fractured horizontal well and CO_2 flooding in two fractured horizontal well in shale oil reservoirs. Since this model has been calibrated, we will not repeat the model validation procedures in this paper.

Table 1- Model input parameters

Region	Parameters	Value	Unit
Reservoir	Area	1277	Acres
	Well Length	8800	ft
	Width	5280	ft
	Length	10537	ft
	Formation Thickness	287	ft
SRV	Area	146	Acres
	Width	724	ft
	Length	8800	ft
	Height	50	ft

Table 2-Matrix and Fracture properties

	Non-SRV	SRV
Region	Middle Bakken	Middle Bakken
Thickness, ft	50	50
Matrix Permeability, md	3.0E-04	3.0E-04
Matrix Porosity	0.056	0.056
Fracture Porosity	0.0022	0.0056
Fracture Permeability, md	2.16E-03	3.13E-02
Fracture Spacing, ft	2.27	0.77

Table 1 presents the dimensions of the reservoir and SRV. The simulation model includes two regions: the stimulated reservoir volume and un-stimulated reservoir volume. A hydraulic fractured horizontal well in the Middle Bakken shale with a well length of 8800-ft is selected for simulation. The base model consists of a horizontal well with 15 stages of hydraulic fracturing completion. The hydraulic fracture spacing apart from each other is 592-ft. The hydraulic fracture half-length is 160-ft. The hydraulic fracture width is 0.5-ft. Table 2 summarizes the input matrix and fracture properties in the Non-SRV and SRV regions in the Middle Bakken shale (Kurtoglu, 2013). Matrix permeability was obtained from core analysis and fracture half-length was estimated from the fractured vertical well pressure buildup test (Kurtoglu, 2013). It is important to notice that the stimulated reservoir volume is only a small fraction of the reservoir volume. The primary oil recovery primarily depends on the stimulated reservoir volume. The SRV is 8800 ft X 724 ft X 50 ft and centered in the reservoir model. The dual permeability model was used to simulate the naturally and hydraulically fractured shale reservoirs. The shale matrix permeability is 0.0003 mD. The natural fracture effective permeability in the SRV is 0.0313 mD. The natural fracture permeability in the un-stimulated reservoir region is 0.00216 mD that is much lower than the stimulated region. The hydraulic fracture is discretized by using a 21 X 21 LGR cells. We used a fine gridding surrounding the hydraulic fractures and near the wellbore. Inside the SRV we use a dual permeability model with LGR gridding to simulate the hydraulic fractures.

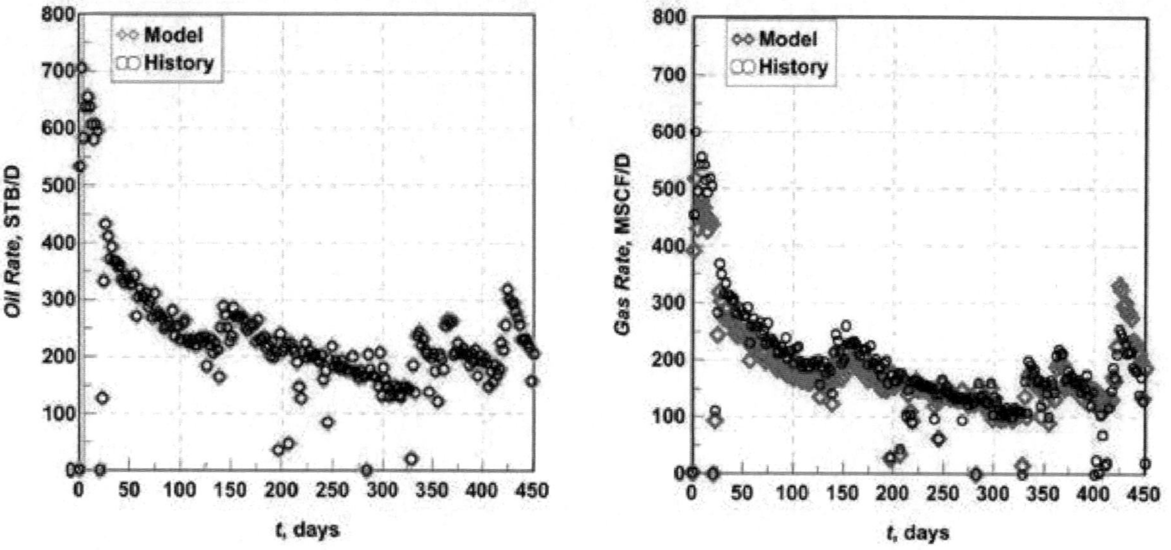

(a) Oil rate (b) Gas rate

Figure 1- History match results for a 15-stage horizontal well in Bakken shale oil, dual-porosity model (By Kurtoglu)

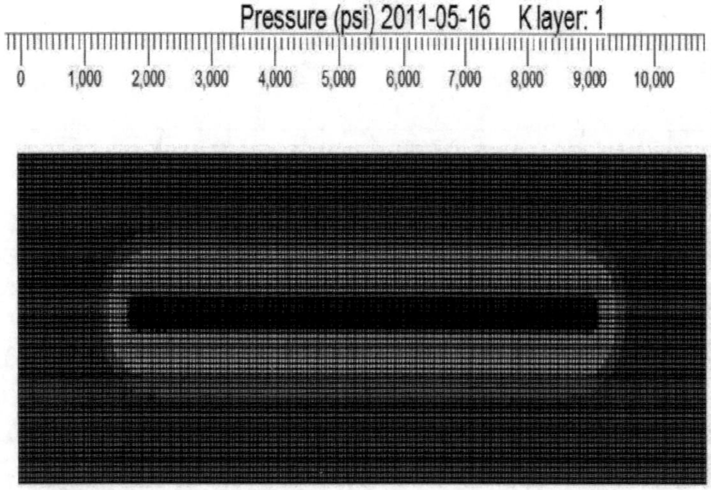

Figure 2a-Middle Bakken reservoir simulation model

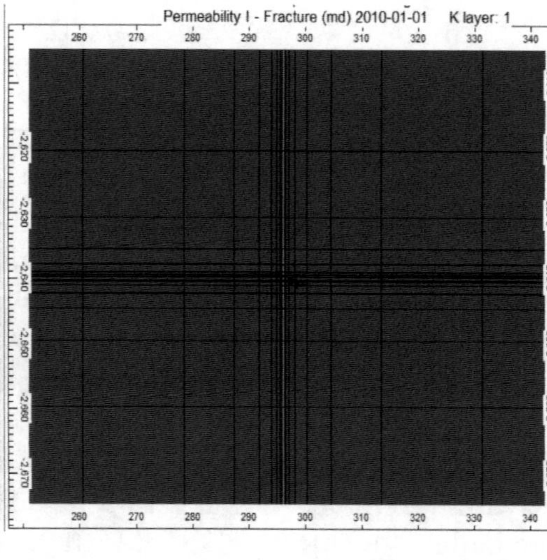

Figure 2b- A single stage model

Figure 1 shows the oil rate and gas rate output by the simulation model against the history field production data by Kurtoglu (2013). The dual permeability by considering the contributions of oil recovery from micro and macro fractures produced a good agreement results with the field production data. **Figure-2a** demonstrates the entire Middle Bakken reservoir simulation model. Due to the computationally expensive nature of using such a large number of grids to run the compositional simulation, we simplified the model by selecting a single stage fractured SRV (as shown in **Figure 2b**). The long-term EOR recovery forecast by using a single stage fractured SRV may exhibit a slight difference with the entire SRV simulation, but the recovery trend prediction is on the same track.

Discussion of Results

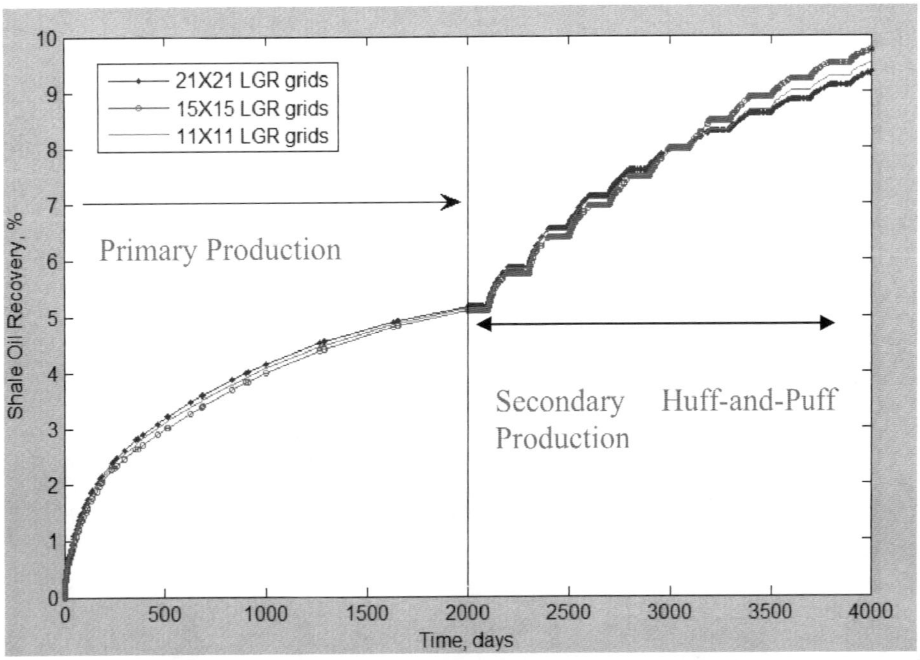

Figure 3- Effect of grid block size on shale oil R.F

In this study, we examined the impact of grid refinement on oil recovery by performing a series of numerical sensitivity calculations. **Figure 3** illustrates that using 11x11x1 2D gridblocks is producing similar results with more refined 21x21x1 gridblocks. Grid refinement near the transition zone between highly conductive fracture and shale matrix is needed to produce steady numerical solutions. The base model we used is a 21x21x1 LGR gridding to simulate the fluids flow in hydraulic fractures.

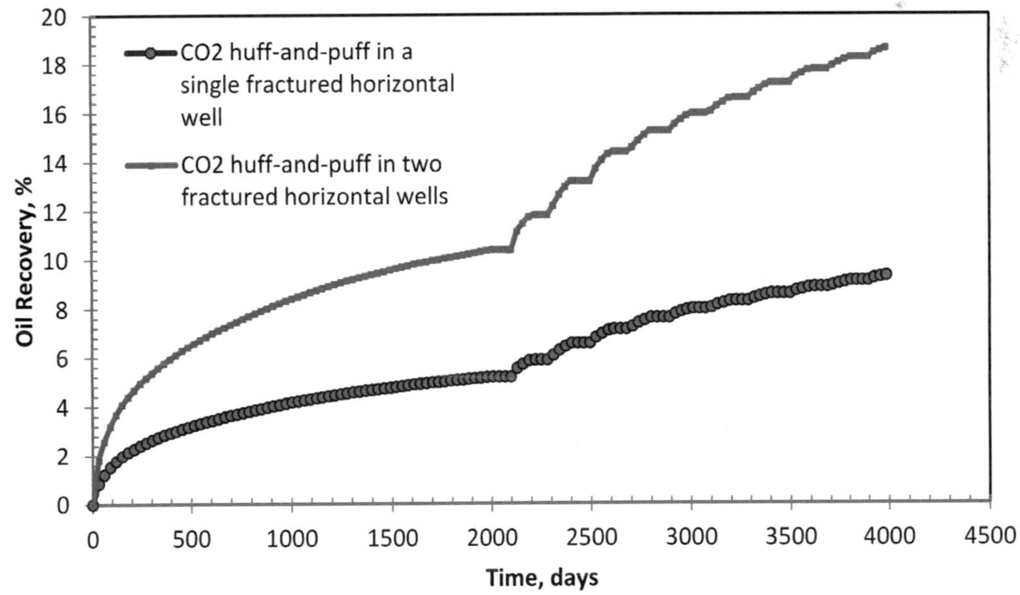

Figure 4-Comparison of CO2 huff-and-puff efficiency in a single and two fractured wells

The reservoir initially operates in naturally depletion for 2000 days. The production well is subject to the minimum bottom-hole pressure control with 500 psi. Then, we start 10 cycles of CO2 huff-and-puff secondary recovery processes. Each cycle consists of 100 days gas/solvent injection and 100 days production. The entire injection well is constraint to inject at a maximum 1MMSCF/D and maximum injection pressure is 5000 psi. **Figure 3** shows that the primary oil recovery is 5% and the secondary recovery by CO2 huff-and-puff in a single hydraulically fractured well is 4.5%. The low primary recovery attributes to the low percentage of SRV. Well productivity in shale oil and gas reservoirs primarily depends upon the size of fracture network and the stimulated reservoir volume (SRV) which provides highly conductive conduits to communicate the matrix with the wellbore. The SRV complexity is critical to the well production performance and it also provides an avenue for injected fluids to displace the oils.

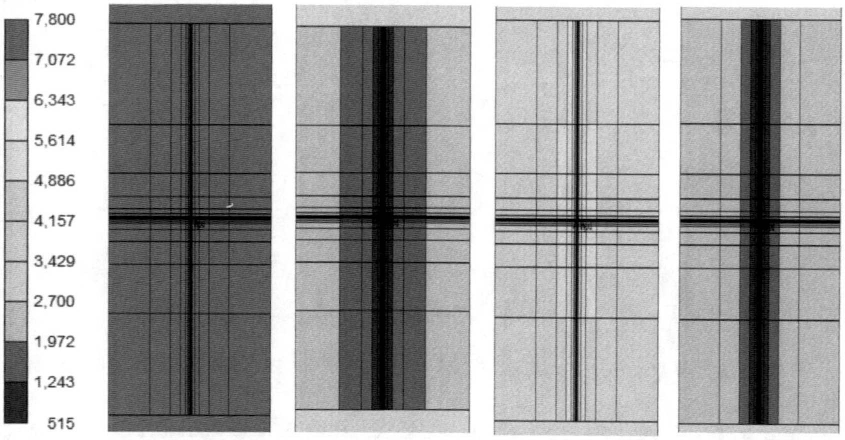

Figure 5-Reservoir pressure variations during cyclic CO2 injection processes (A single Well)

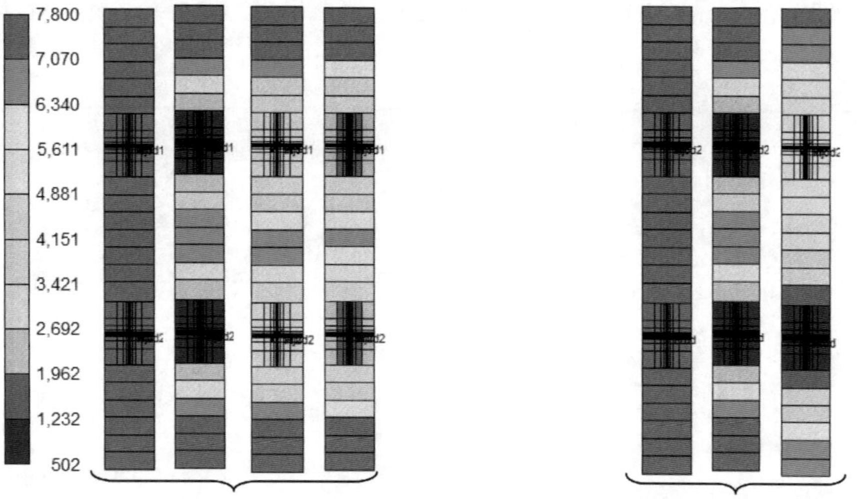

Figure 6-Reservoir pressure variations in 2 fractured wells:
a.Cyclic CO2 injection processes (Left)   b. Flooding process (Right)

Figure 4 presents the comparison results of CO2 huff-and-puff efficiency in a single and two fractured horizontal wells. As shown in the graph, the oil recovery by CO2 huff-and-puff in two fractured horizontal wells is approximately twice as high as the production from a single well. Figure 4 emphasizes the role of stimulated reservoir volume in shale reservoirs which can significantly enhance the oil production recovery.

Figure 5 shows reservoir pressure variations during CO2 huff-and-puff processes in a single fractured horizontal well at initial condition (time 0), after 2000 days of primary production, after the first 100 days of injection and after the first 100 days of production. **Figure 6a** shows reservoir pressure variations during CO_2 huff-puff processes and CO2 flooding in two fractured horizontal well at initial condition (time 0), after 2000 days of primary production and after secondary recovery. It is seen from Figure 6 that the reservoir pressure variations are confined within the stimulated reservoir volume. Without the presence of highly conductive natural or hydraulic fractures, it is difficult for the injected CO2 to penetrate the ultra-low permeability shale matrix to contact matrix oil. The pressure dissipation or depletion occurs surrounding the stimulated reservoir volume that is not propagated to the shale matrix. The pressure in the shale matrix maintains fairly high during the production process. **Figure 6b** shows that the CO2 flooding in a two fracture horizontal wells pattern is not able to build-up the pressure in the producing well. The hydraulic fractures in the injection well are not communicated with the producing well. The injected CO2 contributes to increasing the pressure in the SRV in the injection well. Owing to a lack of fluid flow communication paths between the injection SRV and production SRV, the higher pressure in the injection well cannot propagate to the producing well. **Figure 7** shows the comparison results of CO2 huff-and-puff and flooding efficiency in two fractured wells. As discussed before, the pressure profile for CO2 flooding case concentrating the injection well maintained high, which is difficult to transport to the producing well. The un-stimulated reservoir volume acts as a barrier that undermines the transmissibility of pressure from high pressure region. The impedance of pressure transmission results in a low oil recovery. **Figure 7** shows that the shale oil recovery gained by CO2 huff-and-puff is much higher than CO2 flooding. The enhanced oil recovery by CO2 flooding is immaterial provided that the pressure communication between the injection well and producing well is impaired by the shale matrix. Gas molecular diffusion effect is not addressed in this paper.

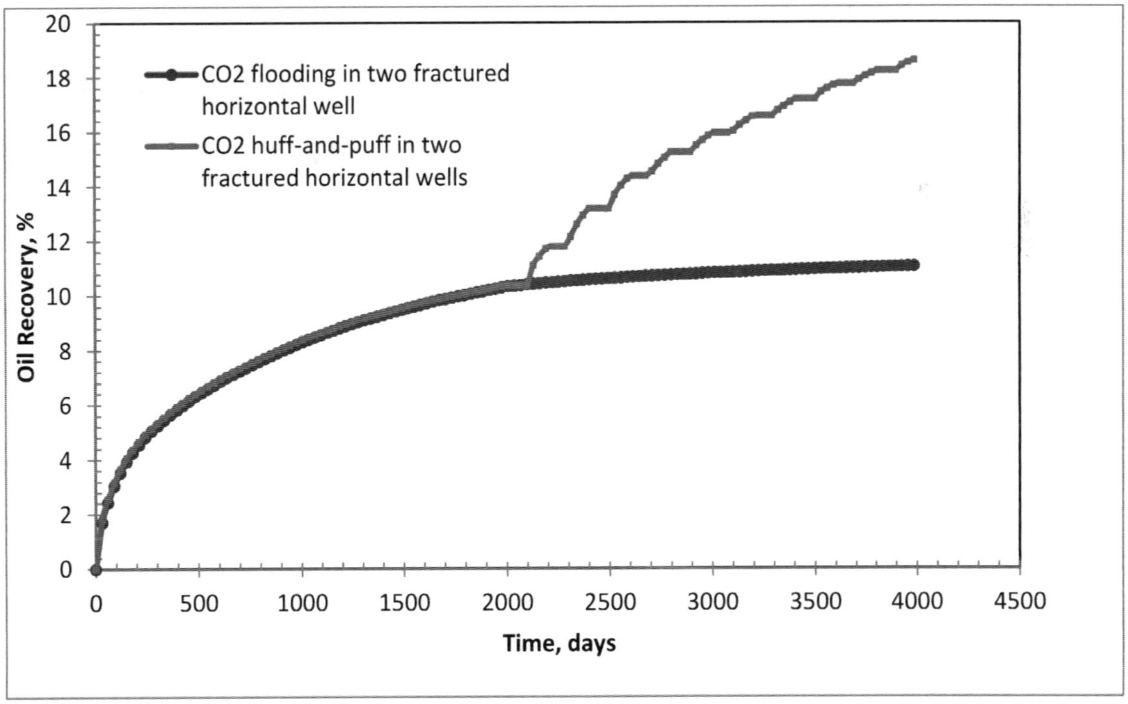

Figure 7-Comparison of CO2 huff-and-puff and flooding efficiency in two fractured wells

Summary

The objective of this paper is to evaluate the response of cyclic CO2 injection and CO2 flooding efficiency in naturally fractured and hydraulically fractured reservoirs. The reservoir simulation model was calibrated against the field production data before performing the EOR recovery simulation. Well productivity in shale oil and gas reservoirs primarily depends upon the size of fracture network and the stimulated reservoir volume (SRV) which provides highly conductive conduits to communicate the matrix with the wellbore. The enhanced oil recovery by CO2 flooding is not significant provided that the pressure communication between the injection well and producing well is impaired by the shale matrix. The pressure dissipation or depletion occurs surrounding the stimulated reservoir volume that is not propagated to the shale matrix. The pressure in the shale matrix maintains fairly high during CO2 flooding process. It is suggested that implementing hydraulic fracturing treatment to create more SRV and communicate the injection well and producing well to some extent (not resulting in early breakthrough). The shale oil recovery gained by CO2 huff-and-puff is more significant than CO2 flooding in two hydraulically fractured horizontal wells at the same operation condition.

References

1. Asghari, K., Torabi, F. (2007), "Laboratory Experimental Results of Huff-and-Puff CO2 Flooding in a Fractured Core System," SPE 110577 presented at SPE Annual Technical Conference and Exhibition, 11-14 November, Anaheim, California.
2. Monger, T.G. and Coma, J.M. (1988), "A Laboratory and Field Evaluation of the CO2 Huff 'n' Puff Process for Light-Oil Recovery," SPERE 1168-76.
3. Khatib, A.K., Earlougher, R.C., and Kantar, K. (1981), "CO2 Injection As an Immiscible Application for Enhanced Recovery in Heavy Oil Reservoirs," paper SPE 9928 presented at the 1981 SPE California regional Meeting, Bakersfield, March 25-26.
4. Karim, F., Berzins, T.V., Schenewerk, P.A., et al. (1992), "Light Oil Recovery from Cyclic CO2 Injection: Influence of Drive Gas, CO2 Injection Rate, and Reservoir Dip," Paper SPE 24336-MS presented at the SPE Rocky Mountain Regional Meeting, Casper, Wyoming, 18-21 May. http://dx.doi.org/10.2118/24336-MS.
5. Thomas, J., Berzins, T.V., and Monger-McClure, T.G. et al. (1990), "Light Oil Recovery from Cyclic CO2 Injection: Influence of Gravity Segregation and Remaining Oil," Paper SPE 20531-MS presented at SPE Annual Technical Conference and Exhibition, New Orleans, Louisiana 23-26 September. http://dx.doi.org/10.2118/20531-MS.
6. Saner, W.B. and Patton, J.T. (1986), "Recovery of Heavy Oil: Wilmington Field Test," JPT (July 1986) 769-76.
7. Bardon, C.P., Karaoguz, D., and Tholance, M. (1986), "Well Stimulation by C02 in the Heavy Oil Field of Camurlu in Turkey," paper SPEIDOE 14943 presented at the 1986 SPEIDOE Symposium on Enhanced Oil Recovery, Tulsa, April 20-23.
8. Olenick, S., Schroeder, F.A., Haines, H.K., Monger-McClure, T.G. (1992), "Cyclic CO2 Injection for Heavy-Oil Recovery in Halfmoon Field: Laboratory Evaluation and Pilot Performance," SPE 24645 presented at SPE Annual Technical Conference and Exhibition, 4-7 October, Washington, D.C.
9. Wang, Z., Ma, J., Gao, R., Zeng, F., Huang, C., Tontiwachwuthikul, P., Liang, Z. (2013), "Optimizing Cyclic CO2 Injection for Low- permeability Oil Reservoirs through Experimental

Study," SPE 167193 presented at SPE Unconventional Resources Conference Canada, 5-7 November, Calgary, Alberta, Canada.

10. Gamadi, T.D., Sheng, J.J., Soliman, M.Y. (2013), "An Experimental Study of Cyclic Gas Injection to Improve Shale Oil Recovery," SPE 166334 presented at the SPE Annual Technical Conference and Exhibition held in New Orleans, Louisiana, USA, 30 September–2 October 2013.
11. Gamadi, T.D., Sheng, J.J., Soliman, M.Y. Menouar, H., Watson, M.C., Emadibaladehi, H. (2014), "An Experimental Study of Cyclic CO2 Injection to Improve Shale Oil Recovery," SPE 169142 presented at the SPE improved oil recovery symposium held in Tulsa, Oklahoma, USA, 12-16 April 2014.
12. Chen, C., Balhoff, B., Mohanty, K. K., (2013), "Effect of Reservoir Heterogeneity on Improved Shale Oil Recovery by CO2 Huff-n-Puff," SPE 164553 presented at 2013 SPE Unconventional Resources Conference, Woodlands, Texas, USA,10-12 Apr. Doi: 10.2118/164553-MS.
13. Wan, T., Sheng, J.J., Soliman, M.Y. (2013), "Evaluate EOR Potential in Fractured Shale Oil Reservoirs by Cyclic Gas Injection," Paper 168880 presented at the Unconventional Resources Technology Conference held in Denver, Colorado, USA, 12-14 August.
14. Yu, W., Lashgari, H., Sepehrnoori, K. (2014), "Simulation Study of CO2 Huff-n-Puff Process in Bakken Tight Oil Reservoirs," SPE 169575 paper was prepared for presentation at the SPE Western North American and Rocky Mountain Joint Regional Meeting held in Denver, Colorado, USA, 16–18 April 2014.
15. Kurtoglu, B. (2013), "Integrated reservoir characterization and modeling in support of enhanced oil recovery for Bakken," Ph.D. dissertation at petroleum engineering at the Colorado School of Mines.

373310 Refinery Challenges When Processing Tight Oil

Abstract:

Tight oils from shale are more readily available in North America with incentives on cost. However, these typically lighter crude oils don't come without some processing challenges. Some of the challenges include high variability in API gravity, H_2S entrained in the crude, amine-based H_2S scavengers added in the pipeline or railcars prior to transporting, paraffin waxes, high filterable solids content, crude incompatibilities from blending, energy balancing, and balancing the intermediate cut point yields for downstream utilization.

This presentation will provide an overview on challenges with processing tight oils, including examples on how refiners are tackling these difficulties.

375640 Impact of Recycling Light Olefins to the FCC

Abstract:

Recycling light olefin product streams back to a fluid catalytic cracking (FCC) unit has potential applications in US refineries due to the growing level of Shale Oils in the crude slate as well as in maximum propylene units such as many built in Asia. In refineries that process high levels of Shale Oil, light naphtha streams may be a potential target for recycle to obtain more valuable C4 olefins or possibly heavier gasoline range compounds. The light naphtha products from the FCC such as C5 compounds have relatively low value due to the abundance of these hydrocarbons in the crude and RVP requirements for gasoline. In the case of max propylene units, C4 olefins have been proposed for recycle to further maximize the production of more desirable propylene through either oligomerization and re-cracking or shifting the light olefin equilibrium distribution. The goal of this work is to determine how product yields shift when light olefins are recycled back to the FCC under different operating conditions. To this end, propylene, 1-butene, and 1-pentene were each co-fed with a Mid-Continent vacuum gasoil into a circulating FCC pilot unit to study the impact on the process and the resulting product distribution. The experiments were performed using a commercial high matrix FCC catalyst alone and with varying amounts of ZSM-5 additive to simulate a max propylene operation. In addition, the reactor outlet temperature was changed during the course of the study to evaluate the effect of operating conditions. Incremental olefin yields as a function of olefin type, reactor temperature and ZSM-5 content are reported. These results help to understand the potential impact on the FCC process and resulting products from different olefin recycling schemes.

Process Modeling of Shale Oil Extraction in Rotary Reactor Using ASPEN Simulation

Hassan Golpour, Joseph D. Smith, PhD., Anand Alembath

Missouri University of Science and Technology, Rolla, Missouri, USA

ABSTRACT

Declining worldwide crude oil production and increasing energy needs has focused attention on developing existing unconventional fossil fuels including oil shale. This hydrocarbon resource represents a major energy reserve and can increase U.S. energy security and support sustained economic growth. America's richest oil shale deposits are found in the Green River Formation of western Colorado, eastern Utah and southwestern Wyoming. Various extraction processes have been developed but none yet has been commercialized to produce synthetic crude from this oil shale deposit. The current work describes process simulation of an *ex-situ* process using a horizontally configured pyrolytic rotating reactor using a novel heating process involving external and internal heating.

Efforts to improve process yield for commercial operation relies on first developing a complete Aspen based process model of a proposed shale refining plant, identifying the key process parameters for the reactor and then optimizing the overall process. Simulation results are compared to earlier experimental yield date collected from a pilot scale rotary kiln operated by Combustion Resources. This work identified the critical impact bed temperature had on crude production with bed temperature <400°F producing <10% conversion and bed temperatures of 450-500°F maximizing yield while minimizing carbon dioxide formation from carbonate minerals in the shale. The residence time for shale in the reactor was also shown to be a critical parameter with times. The shale feed rate and kiln rotation speed are two parameters which affect the residence time. The work shows that at constant rpm, conversion decreases by 20% when feed rate is increased from 3.5 to 4.5 tpd (tons per day) and at a constant feed rate, conversion increases by 10% when kiln rotation speed increased from 6 to 12rpm. ASPEN process parameters were identified to match this data and used as part of a complete process model that includes a gas burner to supply the heat for shale pyrolysis. A cyclone separator is used to separate carry over solids from the product vapors with the solids recycled back to the reactor. The process model also includes a recovery and separation section to process the product oil to transportation fuel and recover the light gases which is directed back to the gas burner. This work was completed to optimize the production of syncrude from oil shale which also enhanced process environmental and economic sustainability.

387845 Enhanced CO2 Interactions with Shale for Sustainable Extraction of Tight Oil and Gas

Abstract:

According to the International Energy Agency (IEA), fossil fuels continue to supply more than 80% of the world's energy, with natural gas and tight oil extraction as the fastest growing fossil fuel resources. Therefore, there is considerable interest in investigating subsurface gas-liquid-solid interactions, particularly for more efficient and sustainable recovery of oil and gas from shale reserves. One of the current technologies to extract oil and gas is hydraulic fracturing which involves injecting water, chemicals, and proppants at high pressure to form new fractures and to keep them open which involves the use of large quantities of water. Hydrofracking is often accompanied by the mobilization of heavy and radioactive metals such as arsenic, radon, and radium, which is environmentally hazardous. Therefore, the aim of this study is to use CO_2 as an alternative processing fluid. However, considerable uncertainties exist around the reaction rates of shale with CO_2. Shale comprises carbonates, clays, and quartz at varying compositions. The interactions of CO_2 with shales containing variable compositions of carbonates, clays, and quartz are expected to result in significant differences in the rates of dissolution and carbonation, and changes in the pore spaces, depending on the pH, temperature, and partial pressure of CO_2, which have not yet been well characterized in existing literature. Therefore, the aim of this study is determine the effect of variable pH (1-12), temperature (T_{max} = 90 °C), and partial pressure of CO_2 ($P_{CO2,\ max}$ = 150 atm) on *in-situ* chemical and morphological changes on clay-rich and carbonate-rich shales, and determine the corresponding reaction mechanisms that result in the alteration of the pore spaces of clay-rich and carbonate-rich shales.

EVALUATE THE EOR POTENTIAL OF CO_2 DISPLACEMENT IN SHALE RESERVOIRS USING MODIFIED ZIPPER FRACTURED HORIZONTAL WELLS

T. Wan, X. Meng, J. J. Sheng, Texas Tech University, Lubbock, TX, U.S.

Abstract

Modified zipper frac technique is developed in a manner different from zipper frac in which the fractures are stimulated in a staggered pattern. The benefit of the modified zipper frac is that it will improve the contact area with the reservoir and increase the effective stimulated volume. Studies showed that enhancing fracture complexities in shale gas resources is critical to improving stimulation treatment and well production performance. CO_2 injection EOR process under the miscible flooding condition can significantly reduce oil viscosity. Oil viscosity reduction combined with the increased contact area by hydraulic fractures could be the dominant recovery mechanism. Problems associated with gas injection in conventional well patterns such as early breakthrough and channeling through high permeability zones will not likely happen in nano-permeable shale oil or gas reservoirs.

In this paper, we propose gas injection to enhance gas condensate recovery in a horizontal well pair, a gas injection well and a production well, which is stimulated in a modified zipper frac pattern. The approach integrates the advantages of hydraulic zipper fracturing, horizontal wells and miscible gas flooding. Miscible gas flooding has shown the IOR potential in shale gas and oil reservoirs in this simulation and experiment work.We develop a compositional model to simulate complex interactions between the injected gas and reservoir fluids that were not modeled precisely by black-oil simulation. Our simulation results of the Eagle Ford Shale indicate that the secondary recovery increased to 86% for 4000 days of secondary production from the 39% of 5 years primary recovery in a 100-ft fracture spacing staggered zipper fracture pattern. The investigation of CO_2 injection in a modified zipper fractured horizontal well pair provides an insight into the EOR performance in nano-shale reservoirs.

Introduction

In this paper, we use modified zipper frac technique to stimulate two adjacent horizontal wells to maximize the exposure of new reservoir rock. Although the laterals expose more of the nano Darcy-permeability shale rock, the desired end result in hydraulic fracture stimulation is to maximize coverage around each lateral.The benefit of modified zipper frac technique from a geomechanical consideration is to increase stimulated reservoir volume and complexity development in successive fracturing stages because the net pressure created by the stimulation stage on the adjacent first well help divert the fracture direction (Belhadi, J. et al 2011).Numerous numerical simulations (Mayerhofer et al. 2006; Warpinski et al. 2009) show that well productivity in shale oil and gas reservoirs primarily depends upon the size of fracture network and the stimulated reservoir volume (SRV), which provides highly conductive conduits to communicate the matrix with the wellbore. The natural fracture complexity is critical to the well production performance and it also provides an avenue for injected fluids to displace the oils (Cipolla et al. 2008; Wan et al 2013; Cipolla et al. 2011 and Moinfar et al. 2013).

Two horizontal wells are drilled through the reservoir and casing is set and cemented in place. The two horizontal wells are closely spaced, one above the other, forming a gas injector and producer pair like steam-assisted gravity drainage (SAGD) system.Hydraulic fracturing is introduced to improve the flow capacity of the reservoir and well productivity performance in shale reservoirs. The horizontal well fracture stimulation requires pumping fluids into a wellbore at high rate and pressure that is too high for the formation to accept without breaking. The perforating design plays a critical role in the stimulation treatment. The effectiveness of the perforating process in this two horizontal well pair cased-hole completion system depends on the perforating location. Before selecting components for a perforating job, the first task is to understand how to avoid the hydraulic fracturing in the injection well directly communicating with producer. This requires perforating a staggered pattern of effective entrance hole through the pipe and cement. Once the perforation and staging design were completed for one lateral, the perforation scheme for successive laterals was based on a staggered design with respect to adjacent wells (Belhadi, J. et al 2011). There is no perforating in the injector at the location where there are perforations in the producer. The primary objective is to prevent the created two wings of the injector from communicating with the producer.In this case, there is no connection between the propagated hydraulic fractures in the injector and the production horizontal well. The injected solvent would not breakthrough to the producer by the created hydraulic fractures of the injector. The injected CO_2 from the hydraulic fractures of the injector effectively pushes the oil towards the hydraulic fractures of the production well, components in the oil transfer to the vapor phase, and components in the injection gas dissolve in the oil phase as chemical equilibrium is established.

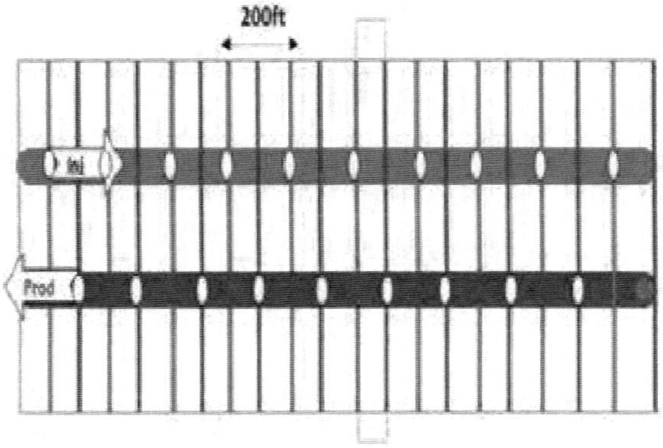

Fig 1A.The horizontal well pair perforated and stimulated in a staggered pattern

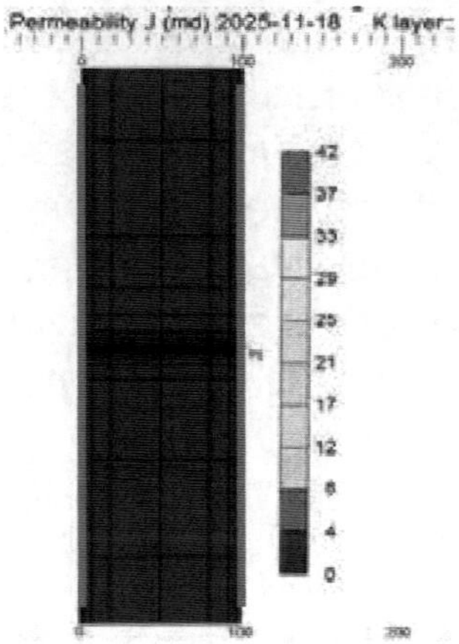

Fig 1B. Unit fracture SRV

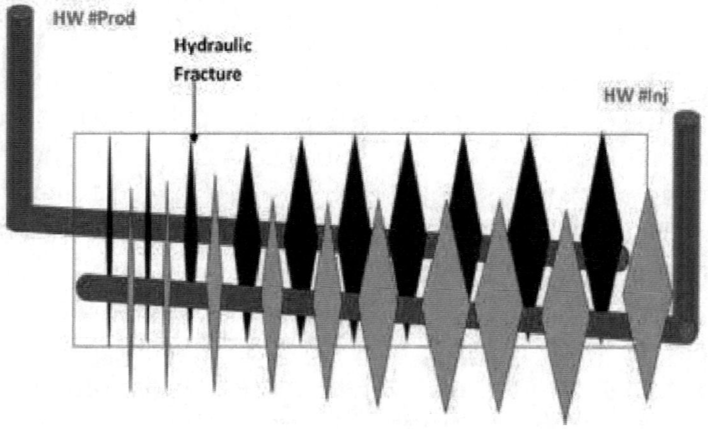

Fig 2. The horizontal well pair stimulated in a staggered pattern

Model Description

The reservoir fluid composition description and phase behavior is from fifth SPE published data (Pope et al; Wu et al; Wang et al) which is aimed to model the behavior of a gas condensate fluid using six hydrocarbon components. Table 1 presents the pseudocomponent description and input for Peng-Robinson equation of state calculations. The initial reservoir oil compositions are also given in Table 1 which representa gas condensate, as shown in Fig. 3.The reservoir properties we used in this model arebased on the published data in Eagle Ford shale as was previously used (Hsu and Nelson, 2002; Chaudhary et al., 2008; Bazanet al., 2010; Wan et al., 2013). The dimension of the shale reservoir is 2000-ft long×1000-ft wide×200-ft thick, as shown in Fig 4. The producer is located at the beneath of reservoir domain and the injection well is placed at the upper layer. The producing horizontal well is stimulated with 10 transverse fractures each placed 200-ft apart as well as the injection well.The fracture spacing between the injector and producer is 100-ft.The initial reservoir pressure for this field is 6,425 psi. The producer is subject to minimum bottom-hole pressure constraint (BHP) of 500 psi and is produced for 1800 days (5 years) as the natural depletion. Our operational flowing bottom-hole pressure 500 psi is far below the dew-point pressure which results in the condensate dropout.

Table 1- Peng-Robinson EOS Fluid Description of Eagle Ford Condensate lumping

Components	Initial Comp.	Pc (atm)	Tc (k)	Accentric Fac.	MW	Shift
CO_2	0.1618	72.8	304.2	0.225	44.01	0.094
C_1	0.7098	45.4	190.6	0.008	16.043	0.099
C_{2-3}	0.079	45.7	330.5	0.119	36.0	0.16964
C_{4-6}	0.026	34.9	453.8	0.226	72.0	0.2986
C_9	0.02	26.0	606.0	0.359	128.0	0.4904
C_{22+}	0.0034	14.9	869.7	0.788	310.0	1.0589

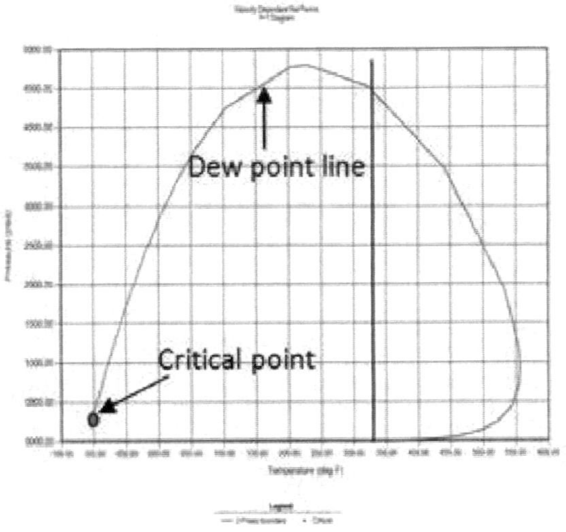

Fig 3. Phase diagram of gas condensate behavoir

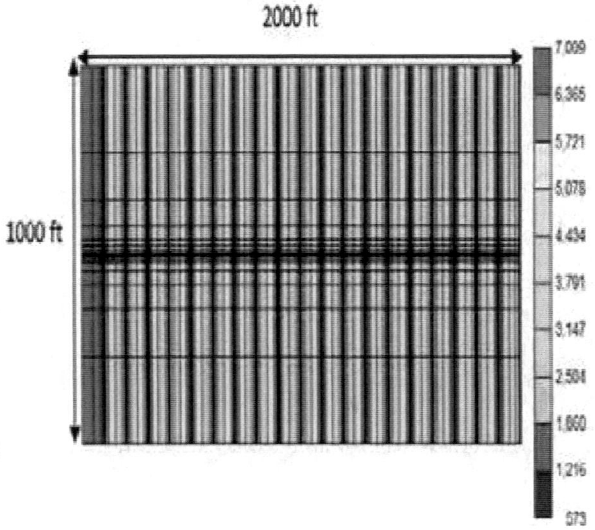

Fig 4. Reservoir pressure changes during gas injection

Table 2-Reservoir properties for the model input

Initial Reservoir Pressure	6425 psia
Reservoir Temperature	335F°
Saturation Pressure	4456 psia
Rock Compressibility	5.0E-06
Porosity	6 %
Permeability of shale	100 nano-Darcy
Water Density	62.4 lb/cuft
Hydraulic fracture conductivity	83.3 md-ft

CO_2 recovery mechanisms include CO_2 dissolution in the oil leads to an increase in its volume, oil viscosity reduction, vaporizationof intermediate to heavy hydrocarbons and development of multicontact miscibility (Whitson et al). The swelling test was simulated by varying proportions of injection gas mixed with original reservoir oil. The swelling test provides useful phase and volumetric data to investigate how a reservoir fluid reacts with gas injection. Fig 5 shows that the reservoir fluids remain single gas phase preceding to gas injection because the initial reservoir pressure at 6425 psi is higher than saturation pressure. With increasing injection percentage of CO_2, the saturation pressure decreases which forms first contact miscible with the equilibrium reservoirfluid gas mixture. The swelling factor increases with increasing CO_2 solubility. Fig 6 shows the simulated relative volumes in a CCE experiment at 335 F for the gas condensate mixture. Fig 7 shows the simulated liquid dropout curve in a CVD experiment at 335 F for the gas condensate mixture. Liquid dropout curve, expressed as $V_{ro}=V_o/V_S$. relative oil volume is defined as the volume of oil, V_o at a given pressure divided by the original saturation volume. The relative volume provides a measurement of the average reservoir oil saturation that will develop during depletion of a gas condensate reservoir (Whitson et al). The reservoir oil saturation can be calculated from V_{ro} with$S_o=(1-S_w)V_{ro}$.

Liquid dropout starts at the dew point pressure 4456 psi and continue to increase until the pressure reducing to 1500 psi when a maximum of condensate liquid reached.

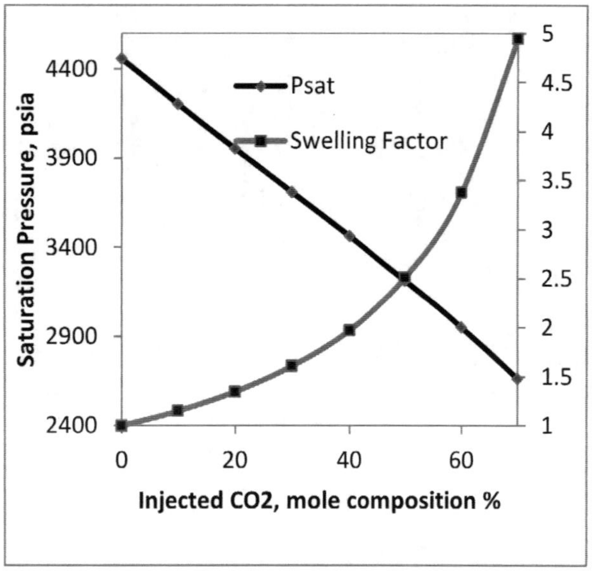

Fig 5. Effect of solution gas on swelling of reservoir fluid by CO_2

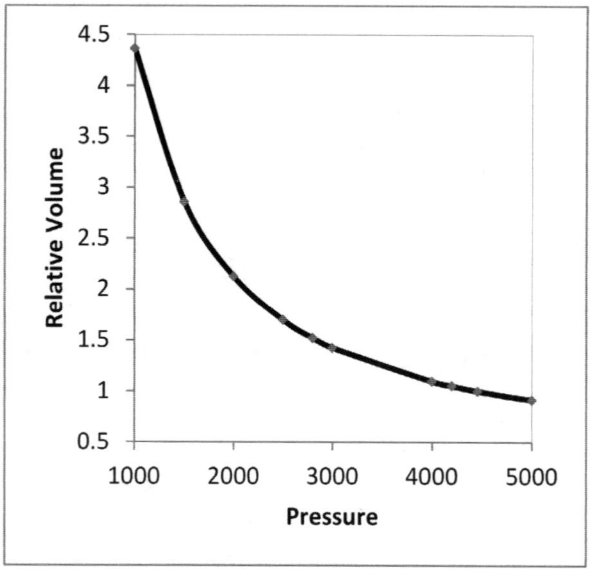

Fig 6. Relative volume curve

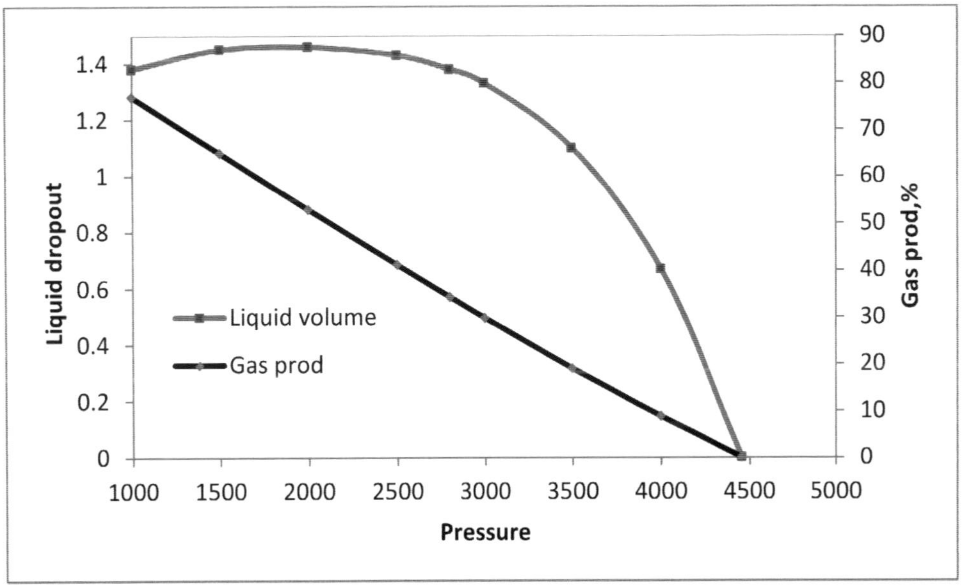

Fig 7. The liquid dropout curve for constant-composition expansion experiment at 335F on the gas condensate mixture.

In this study, we examined the impact of grid refinement through numerical sensitivity calculations. Fig 8 and 9 show the reservoir model using 10 x 21grids produces similar results to far more refined 22 x 21grids. Fig 8 illustrates that using 9 grids for hydraulic fracture representation (as shown in Fig.1B) is getting close to the result as 21gridblocks that is enough to eliminate the error caused by numerical dispersion.From these results we conclude that a 10 x 21 grid discretization around the fractures is able to properly model the rapidly varying pressure in and near the fractures.In Fig 8 and 9, the gas recovery 'tail' flatten out is due to the software resultsoutputissue which does notpreciselyreflect the simulation cases.On the basis of symmetry, Fig. 1B simulates only a unit fracture SRV consisted of two 'mirror 'half fractures stimulated reservoir volume. Fig. 10 shows the production performance from the unit facture controlled SRV produces the similar results with the simulated entire fractured horizontal well controlled SRV.Initially the reservoir operates in naturally depletion for 1800 days. Then, we start CO_2injection for 4000 days. The injection well is constraint to inject at a maximum 800MSCF/D and maximum injection pressure is 7000 psi. Fig. 11 shows the results for the different flowing bottom-hole pressure on the gas recovery factor. The BHP used in the base case simulation is 500 psi. The dew point pressure is 4456. The largest pressure drops occur near producing wells and hydraulic fractures. Condensate liquid saturation will build up near a well because of drawdown below the dewpoint pressure.The results in Fig. 11 did not take into account the effect of condensate blockage, namely gas productivity index reduction due to build up of condensate in the near wellbore region.The benefit of large pressure drawdown in the producing well to the gas inflow performance overwhelms the gas recovery loss due to condensate dropout. There is no visible distinction for the BHP of 500 psi and 1000 psi that self-evidently explains the gas recovery loss is equivalent to the benefit from pressure drawdown.

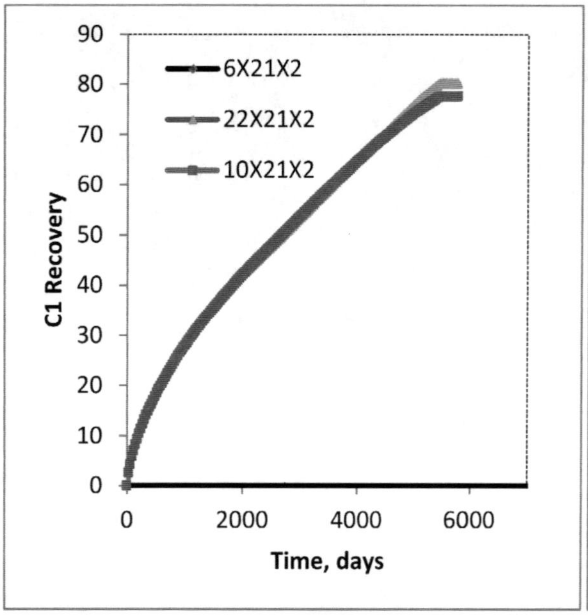

Fig 8. Effect of numerical dispersion on C1 R.F. vs. time

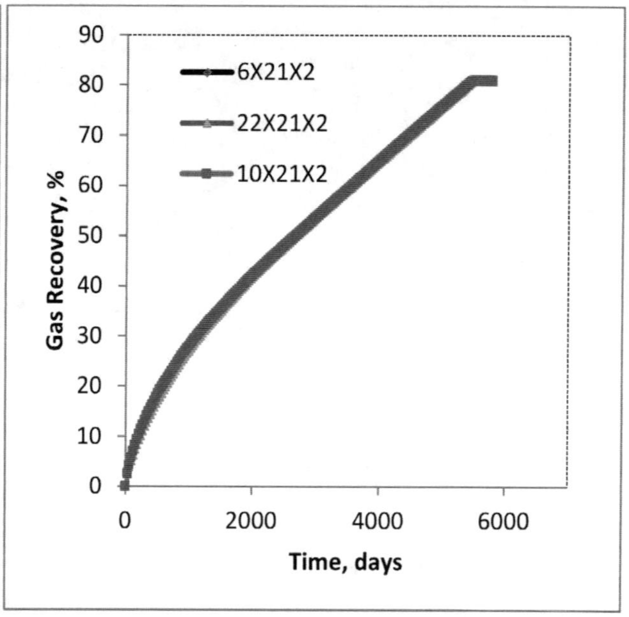

Fig 9. Effect of numerical dispersion on gas R.F. vs. time

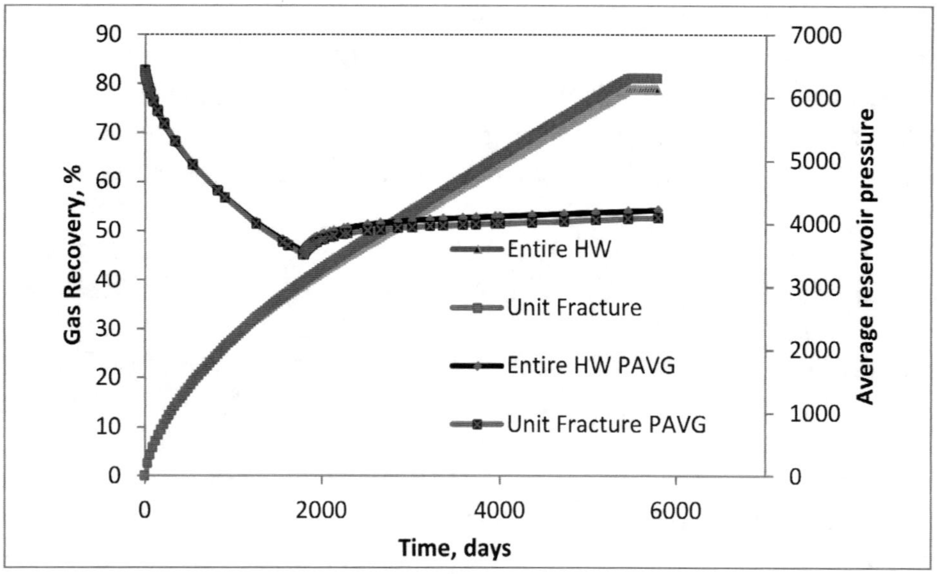

Fig 10. Unit fracture controlled stimulated reservoir volume vs. entire horizontal well SRV results

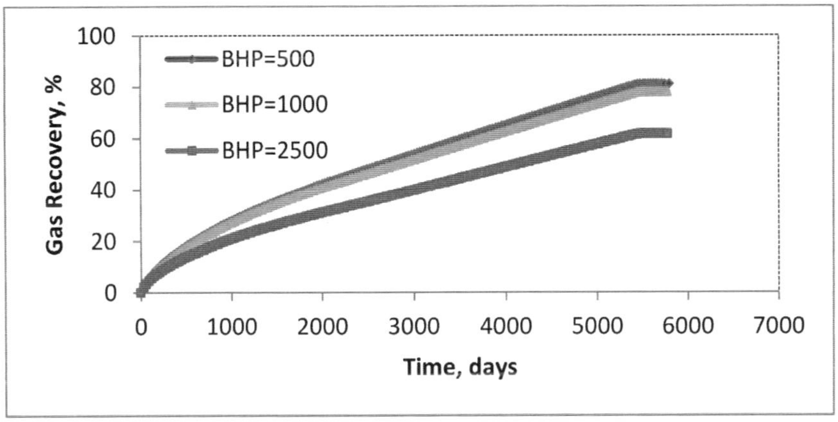

Fig 11. BHP impact on gas recovery performance

The non-Darcy flow effect should be considered for gas injection in fractured reservoirs. The Forchheimer equation can be written as (Dake, 1978)

$$\frac{dp}{dr} = \frac{\mu}{k}v + \beta\rho v^2, \qquad \beta = \frac{2.73 \times 10^{10}}{k^{1.0145}}$$

$$\frac{dp}{dr} = \frac{\mu}{k}(1 + Fo)v \qquad Fo = \frac{\beta\rho v k}{\mu}$$

F_o is referred to as the Forchheimer Number. The correction to the Forchheimer Number is given by $K_{corr} = (K_f/K_{feff})^{(2-N1g)}$. The non Darcy flow Forchheimer correction factor was used in the hydraulic factures in this model to correct a 0.001 ft wide fracture to a represent the non Darcy flow in the 2-ft pseudo-fracture. The hydraulic fracture conductivity was assumed to be 83.3 md-ft. The effective fracture permeability for the 2-ft width of fracture is 41.65mD to conserve the fracture conductivity. The value used in the fractures equals to 903.8= (83300/41.65)^(2-1.1045).

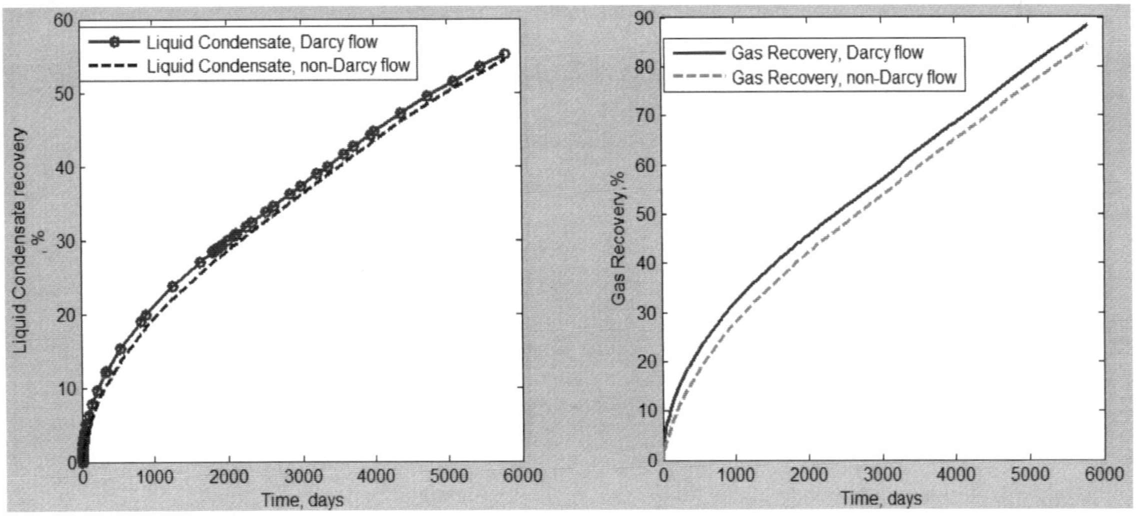

Fig. 12 Gas recovery factor-Darcy flow vs. gas recovery-non-Darcy flow

Fig.12 compares the gas and condensate recovery results of Darcy flow model with the non-Darcy flow model. The non-Darcy flow corrected model in 2-ft wide fracture conduits exhibit a little bit lower recovery than Darcy flow model. Rubin (2010) showed that the non Darcy correction could be used to accurately model a pseudo fracture 1000 times wider than the actual fracture size in fractured shale gas reservoirs for horizontal wells. Fig. 13 shows the impact of fracture spacing on the cumulative gas recovery factor. As shown in the graph, reducing fracture spacing from 200-ft to 100-ft would result in a more than two fold increase in cumulative gas recovery factor. The well production performance is the total sum of all fracture network segments which is strongly dependent on the fracture spacing. Increasing stimulation stages can be used to increase fracture-network size and SRV.Mayerhofer et al. (2010) illustrateed that well production in shale reservoirs is related directly to reservoir volume stimulated during the fracture treatments which is in context with the fracture spacing.

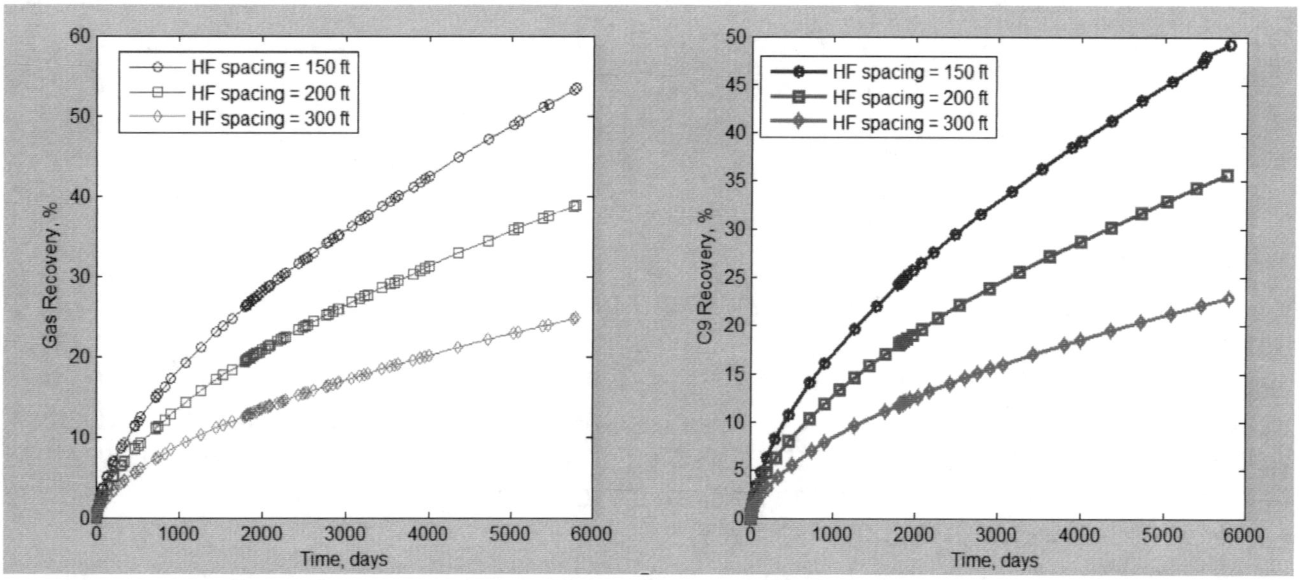

Fig.13 Impact of hydraulic fracture spacing between injection well and production well on gas recovery and C9 recovery

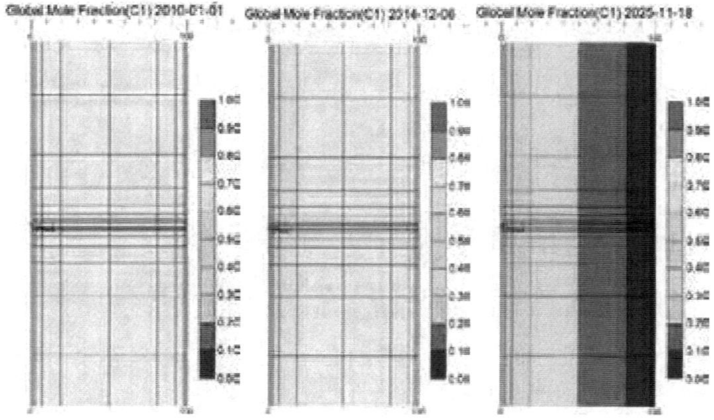

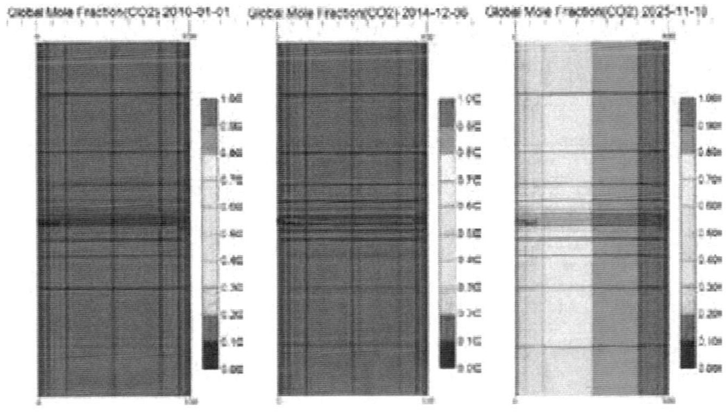

Fig. 14 Global mole fraction of CO_2 and C1 changes during CO_2 flooding

Fig.14 shows global mole fraction of CO_2 and C1 changes at time 0, after 1800 days of primary depletion and after 4000 days of CO_2 flooding. The initial reservoir gas viscosity at 335 Fcalculated by Jossi-Stiel-Thodos (JST) correlation (Fong and Nghiem, 1980) is 0.0373 cp. The CO_2 viscosity at 335 F at 6425 psi is 0.052 cp. When injected CO_2contacts reservoir fluids, the injected CO_2 effectively pushes the C1 and other components towards the hydraulic fractures of production well without suffering from large viscosity contract between injected CO_2 and the reservoir fluids that makes it is not possible for CO_2 to channel or break through to the producing well bypassing most of the oil.After 4000 days of continuous CO_2injection from the top of the reservoir, most of the C1 and other components have been uniformly swept to the hydraulic fracture of the producing well.

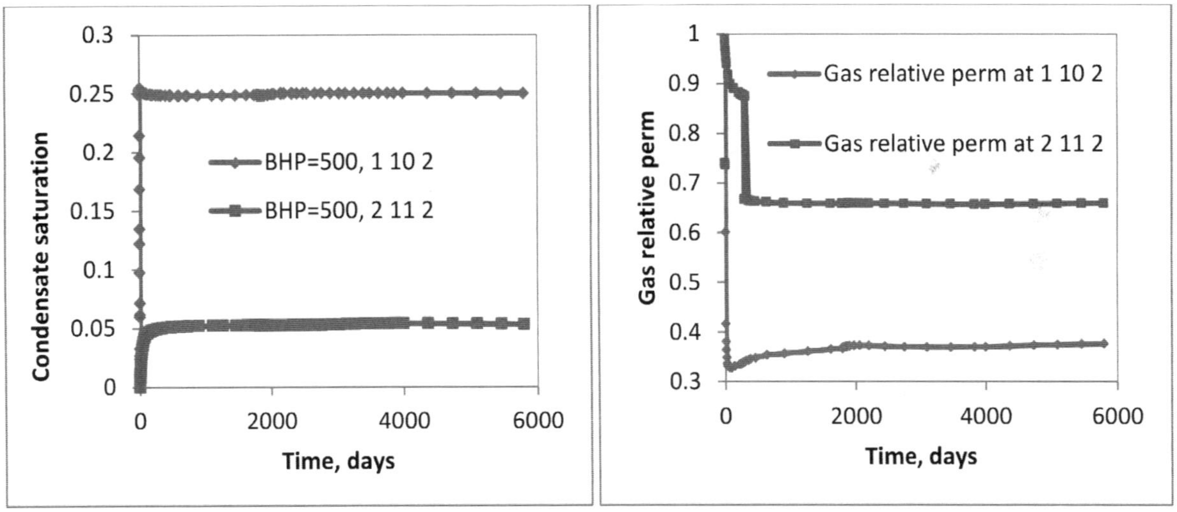

Fig. 15 Condensate saturation and gas relative permeability on or vicinity of the fracture

It is well known that well deliverability would be compromised by accumulation of condensate near the wellbore. However, the presence of hydraulic fracture can also contribute to permeability and the well productivity. Pope et al (1998) and Fan et al (2005) developed a radially symmetric and single well model to demonstrate that condensate blockage could result in a rapid falloff in productivity. Narayanaswamy et al. (1999) concluded that when both non-Darcy and capillary number effects were considered, the capillary number effect can in some case overshadow the two-phase non-Darcy effects which results in smaller descrease in PI than the case with capillary number effect only.In Pope's model, they intentionally set the initial reservoir pressure below the dewpoint that allows the condensate saturation goes up immediately. Fig. 15 shows that as pressure declines rapidly along the fracture block (1,10,2) condensate saturation builds up near (2, 11,2) and on the fracture, which results in the gas relative permeability decreasing. In our model, the grids near the wellbore and hydraulic fractures are not fined enough to capture the condensate dropout and rapid pressure variations. It is out of this paper's scope to investigate the well productivity index loss due to gas condensate blockage.

Conclusions

The objective of this work is to propose a new approach to implant the modified zipper frac technique to stimulate a horizontal well pair. This requires perforating a staggered pattern of effective entrance hole through the pipe and cement. Once the perforation and staging design were completed for one lateral, the perforation scheme for successive laterals was based on a staggered design with respect to adjacent wells. Gas injection EOR process was performed in this modified zipper fractured horizontal wells. Unit fracture controlled volume simulation was able to represent the entire horizontal well recovery on the basis of symmetry. The phase behavior of CO_2mixed with reservoir fluids were investigated and proved to be efficient for EOR. From our compositional modeling of the gas injection in a fractured horizontal well pair, it is demonstrated that injected CO_2 from the injection well fracture uniformly pushes the reservoir fluids to the producing well and its fracture, which increases the cumulative gas recovery more than twice. The well production performance is the total sum of all fracture network segments which strongly dependents on the fracture spacing. Although the well productivity index (PI) loss due to condensate dropout induced by the large drawdown in the fracture was not considered, the condensate saturation builds up in and near the fracture and gas relative permeability reduction results were presented. The main goal of this paper is to examine the EOR potential by gas injection in modified zipper fractured horizontal well pair, which tends to produce promising results.

References:

1. Wu, Wei-Jr, P. Wang, M. Delshad, C. Wang, G.A. Pope and M. Sharma (1998), "Modeling Non-Equilibrium Mass Transfer Effects for a Gas Condensate Field," paper SPE 39746 presented at the Asia Pacific Conference held in Kuala Lumpur, Malaysia.
2. Wang, P., G.A, Pope and K. Sepehrnoori (Nov. 1997), "Development of Equations of State for Gas Condensates for Compositional Petroleum Reservoir Simulation," Submitted to Industrial & Engineering Chemistry Research.
3. Pope, G.A., Wu, W., Narayanaswamy, G., Delshad, M., Sharma, M., Wang, P. (September 1998), "Modeling Relative Permeability Effects in Gas-Condensate Reservoirs," Paper SPE 49266 presented at SPE Annual Technical Conference and Exhibition, New Orleans, Louisiana. Doi: 10.2118/49266-MS.
4. Whitson, Curtis H. & Brule, Michael R. (2000), "Phase Behavior," SPE Monograph Series Vol. 20. Richardson, Texas.
5. Belhadi, J., Ramakrishnan, H., Yuyan, R. (2011), "Approach Optimizes Frac Treatments," American Oil and Gas Report.
6. Mayerhofer, M.J., Lolon, E.P., Youngblood, J.E., and Heinze, J.R. (2006), "Integration of Microseismic-Fracture-Mapping Results With Numerical Fracture Network Production Modeling in the Barnett Shale," Paper SPE 102103 presented at the SPE Annual Technical Conference and Exhibition, San Antonio, Texas, USA. Doi: 10.2118/102103-MS.
7. Cipolla, C.L., Warpinski, N.R., Mayerhofer, M.J., Lolon, E.P., and Vincent, M.C. (2008), "The Relationship Between Fracture Complexity, Reservoir Properties, and Fracture Treatment Design," Paper SPE 115769 presented at the SPE Annual Technical Conference and Exhibition, Denver, doi: 10.2118/115769-MS.
8. Warpinski, N.R., Mayerhofer, M.J., Vincent, M.C., Cipolla, C.L., and Lolon, E.P. (2009), "Stimulating Unconventional Reservoirs: Maximizing Network Growth While Optimizing Fracture Conductivity," JCPT, Vol.48, No 10. Pp 39-51. Doi: 10.2118/114173-PA.
9. Wan, T., Sheng, J.J., Soliman, M.Y. (2013), "Evaluate EOR Potential in Fractured Shale Oil Reservoirs by Cyclic Gas Injection," Paper 168880 presented at the Unconventional Resources Technology Conference held in Denver, Colorado, USA.
10. Cipolla, C., Weng, X., Mack, M., Ganguly, U., Gu, H., Kreese, O., and Cohen, C. (2011), "Integrating micro-seismic mapping and complex fracture modeling to characterize fracture complexity," SPE 140185 presented at the SPE Hydraulic Fracturing Technology Conference, The Woodlands, Texas. Doi: 10.2118/140185-MS.
11. Moinfar, A., Varavei, A., Sepehrnoori, K., Johns, R.T. (2013), "Development of a Coupled Dual Continuum and Discrete Fracture Model for the Simulation of Unconventional Reservoirs," Paper SPE 163647 presented at the SPE Reservoir Simulation Symposium, The Woodlands, TX, USA. Doi: 10.2118/163647-MS.
12. Fan, L., Harris, B.W., Jamal, A.,Kamath, J., Mott, R., Pope, G.A., Shandrygin,A., Whitson,C.H.(2005), "Understanding Gas-Condensate Reservoirs," Oilfield Review, Vol.17.
13. Hsu, S.-C., and Nelson, P.P. (2002), "Characterization of Eagle Ford Shale. Engineering Geology," Volume 67, PP. 169-183.

14. Chaudhary,A.S., Ehlig-Economides,C., Wattenbarger, R. (2011), "Shale Oil Production Performance from a Stimulated Reservoir Volume," Paper SPE 147596 presented at Annual Technical Conference and Exhibition, Denver, Colorado, USA. Doi: 10.2118/147596-MS.
15. Bazan, L.W., Larkin, S.D., Lattibeaudiere, M.G., Palisch, T.T. (2010), "Improving Production in the Eagle Ford Shale with Fracture Modeling, Increased Fracture Conductivity, and Optimized Stage and Cluster Spacing Along the Horizontal Wellbore," SPE 138425 presented at Tight Gas Completions Conference, San Antonio, Texas, USA. Doi: 10.2118/138425-MS.
16. Dake, L.P. (1978), "Fundamentals of Reservoir Engineering," Elesvier Scientific Publishing Company.
17. Rubin, B. (2010), "Accurate Simulation of Non Darcy Flow in Stimulated Fractured Shale Reservoirs," Paper SPE 132093 presented at Western Regional Meeting, Anaheim, California, USA. Doi: 10.2118/132093-MS.
18. Straight, D.(1989), "Antelope Field: Preliminary Horizontal Drilling Evaluation, Bakken Formation," Simtech Consulting Services, Inc., Golden, CO.
19. Mayerhofer, M.J., Lolon, E.P., Warpinski, N.R., Cipolla, C.L., Walser,D., Rightmire, C.M., et al. (2010), "What Is Stimulated Reservoir Volume?" SPEJ, Vol.25, No1. Pp 89-98. Doi: 10.2118/119890-PA.
20. Fong, D.K.S., and Nghiem, L.X. (1980), "A Viscosity Model for Reservoir Fluids," Computer Modelling Group Research Report R7.02.
21. Narayanaswamy, G.,Pope, G.A., Sharma, M.M. (1999), "Predicting Gas Condensate Well Productivity Using Capillary Number and Non-Darcy Effects," SPE paper 51910 presented at SPE Reservoir Simulation Symposium, Houston, Texas.

388411 Nanoparticle Stabilized High Internal Phase CO2-in-Water Foams: Effects of Surfactants and Polymers

Abstract:

Chemical formulations of nanoparticles and surfactants are developed to stabilize high internal phase carbon dioxide-in-water (C/W) foams for low water hydraulic fracturing fluids. Silica nanoparticles and low concentrations of surfactants that weakly interact with nanoparticles were used to stabilize the C/W foams. The surfactants are adsorbed onto the nanoparticles to influence the contact angle of the nanoparticles to stabilize the dispersed CO_2 phase droplets against coalescence. Also, the free surfactants reduce the interfacial tension between the aqueous phase and CO_2 phase to facilitate the break-up of the CO_2 phase into bubbles. The ability to achieve stable high internal phase C/W foams with properly designed nanoparticles could offer an alternative to conventional surfactants to raise the foam stability to be adequate for hydraulic fracturing operations. Further, water-soluble polymers were demonstrated to greatly improve the viscosities of the C/W foams. The ability to generate viscous fluids with low amounts of water for fracturing of shale reservoirs is of great interest for both reducing global water utilization and mitigation of post-processing of contaminated fracturing fluids.

394219 Reverse Electrodialysis: Sustainable Energy from Hydraulic Fracturing Water Recycle

Abstract:

Due to the growing limitations in dependence on fossil fuels, attention has been turned towards searching for alternative means for harvesting energy. One emerging source is energy produced from natural salinity-gradients between freshwater and seawater. Two different processes used to accomplish this are pressure-retarded osmosis and reverse electrodialysis. Significant amounts of research are currently under way regarding pressure-retarded osmosis, however very few studies have been conducted on reverse electrodialysis due to heavy resistance in the diluate chamber of a reverse electrodialysis cell. The focus of this research is to reduce this resistance by using wafers and ion-exchange beads in order to maximize net power gain. Surrogate brackish/saltwater feed streams were used to measure overall power density.

By reducing resistance to levels previously deemed impossible, we are exploring the fundamental limits of separation technology. This research proposes an innovative resolution to high resistance levels through the use of ion exchange wafers to promote ionic movement. Here we show that such wafers have a great impact on the reverse electrodialysis process, and succeeded in achieving a sizeable gain in power density.

Traditional applications include harnessing energy from areas where freshwater bodies meet saltwater, as in an estuary. The future of reverse electrodialysis, however, lies in fracking. Produced water from a fracking well site in Oklahoma was used for power density calculations under realistic conditions. Preliminary results show that our system is capable of producing power densities of up to $10 W/m^2$, which are five times higher than the current state-of-the-art. Environmental concerns will be all but eliminated if fracking can evolve to be a sort of cyclic process, where brackish fracking water is converted to energy to, in turn, power the overall operation. This poster presentation will model the power increase of electrode ionization and comment on future potential for improvement, especially in the realm of hydraulic fracturing applications.

394215 Modeling of the Aerobic Cometabolic Transformation of Chlorinated Ethenes By the Mycobacterium Elw-1

Abstract:

ELW1 is a recently isolated aerobic bacterium, in the mycobacterium family, that uses isobutene as a growth substrate. ELW1 was also discovered to have the ability to cometabolically transform substances with chemical structures similar to isobutene, such as chlorinated ethenes, into chloride ion and water. Chlorinated ethenes are notorious carcinogens, and groundwater pollutants, whose remediation has been a worldwide problem for over 40 years. Bioremediation of chlorinated ethenes by the augmentation of contaminated soils with microorganisms such as ELW1 is a proposed method of reducing contaminant concentrations in groundwater. To determine the viability of using ELW1 for groundwater remediation, an initial kinetic analysis of transformation rates must be performed. The focus of this research was to determine the rates of contaminant transformation when this organism was exposed to high concentrations of chlorinated ethenes such as trichloroethylene (TCE), *cis*-dichloroethene (c-DCE), and vinyl chloride (VC).

The ELW1 culture was grown and sustained in both batch culture and steady-state chemostat environments with isobutene as a growth substrate on a minimal growth media. Cells that were used for experiments were concentrated by centrifugation, washed with excess media and placed in a new media solution containing high concentrations of the compound to be tested. Initial rates of transformation were determined using gas chromatography to analyze 100μL samples of headspace gas over a period of one to three days to determine the rate of compound loss. The rates at which ELW1 can utilize isobutene were compared with rates of chlorinated ethene transformation, which overall quantifies the practicality of using this organism bioremediation purposes in the field.

Kinetic parameters of these cometabolic processes were determined using an analytical equation based on a modified Monod equation in order to account for decrease in cell activity over time. Results of this investigation indicate that the ELW-1 culture would be effective for the bioremediation of c-DCE and VC, but would be less effective for treated TCE.

394214 The Enzymatic Hydrolysis of Alfalfa Stalks for Use As a Biofuel Resource

Abstract:

Most biofuel resources, especially plant sources, have two key issues: farming space and slow growth. After overexpressing the sucrose phosphate synthase gene (SPS) in Medicago sativa (alfalfa) plants, large differences in phenotype and growth rate were seen in the plant. While the leaves of the alfalfa plants can be used for animal forage, the stems of the alfalfa can be fractionated into sugars that can then be fermented into bioethanols. This technique could solve the growth problem faced by bioethanol producers.

The purpose of the project was to fractionate and analyze the extractability of fermentable sugars from a control and several transgenic species that show overexpression in the SPS gene. Using a strong acid pre-treatment and enzymatic hydrolysis to break down the cellulose into glucose, the alfalfa stalks were fractionated then analyzed using ultraviolet spectroscopy with anthrone as an indicator for glucose content. Two of the three tested transgenic species were shown to have more glucose content per gram than the control species (2-4% more glucose/gram). A further development of the overexpression technique can lead to faster growth, increased concentration of fermentable carbohydrates, and larger crop yields.

UNIVERSITY OF SOUTH CAROLINA
College of Engineering and Computing

SAGE
Power through Engineering

Synthesis and Characterization of FeAlCr via High Throughput Techniques

Randy Fang
Jonathan Kenneth Bunn and Dr. Jason Hattrick-Simpers
Department of Chemical Engineering
University of South Carolina
AICHE National Conference 2014
11/17/2014

UNIVERSITY OF SOUTH CAROLINA
Honors College

1

Fukushima Daiichi Nuclear Disaster

https://coto2.files.wordpress.com/2011/03/2-fuel-pellet-assembly.jpg

http://diggingintheclay.files.wordpress.com/2011/03/5525887859_e1934af238_o_d1.jpg

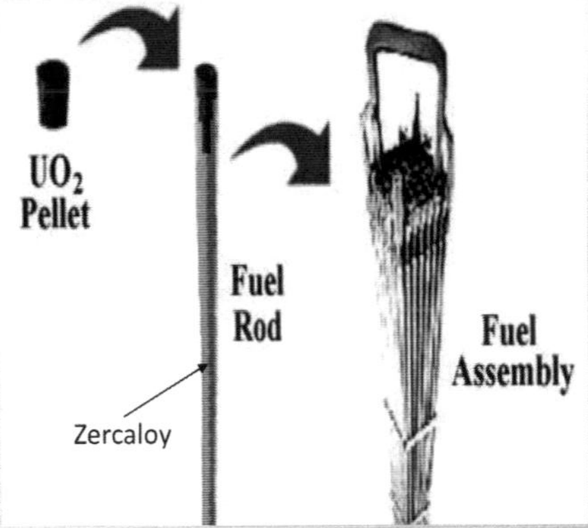

2

FeAlCr Alloy

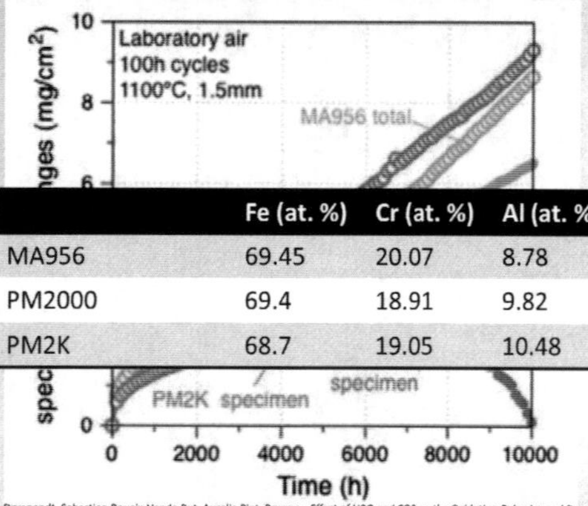

	Fe (at. %)	Cr (at. %)	Al (at. %)
MA956	69.45	20.07	8.78
PM2000	69.4	18.91	9.82
PM2K	68.7	19.05	10.48

Dryepondt, Sebastien Rouaix-Vande Put, Aurelie Pint, Bruce a., *Effect of H2O and CO2 on the Oxidation Behavior and Durability at High Temperature of ODS-FeCrAl*, Oxidation of Metals, Vol 79,2003

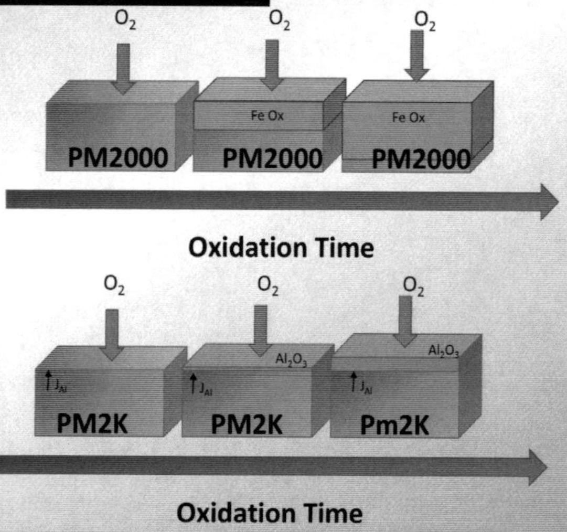

3

Introduction to High Throughput Techniques

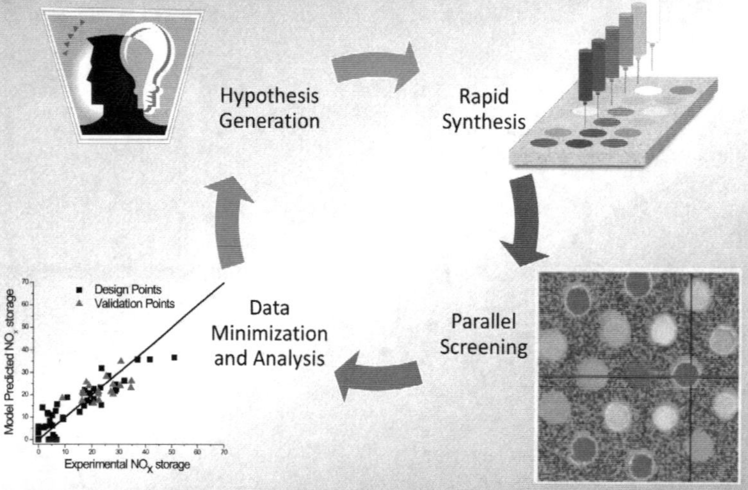

4

Research Objective

- **Implementing established high-throughput methods for FeAlCr high temperature oxidation resistance screening**
- Non-destructive characterization
- Gain comprehensive understanding of:
  - Time dynamics of oxide formation at high temperature
  - Role of metallurgical phase in promoting growth and continued growth of passivating oxide
- Identification of promising compositions for passivating oxide growth

High Throughput Approach to Explore the Ternary Phase Diagram

Qu Materials Science and Engineering A Vol 562 (2013)

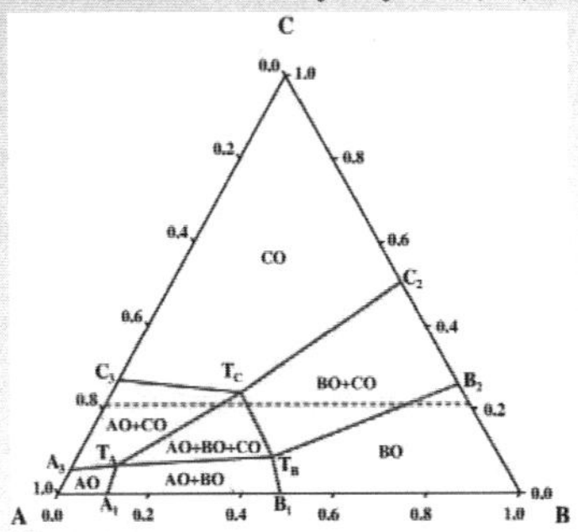

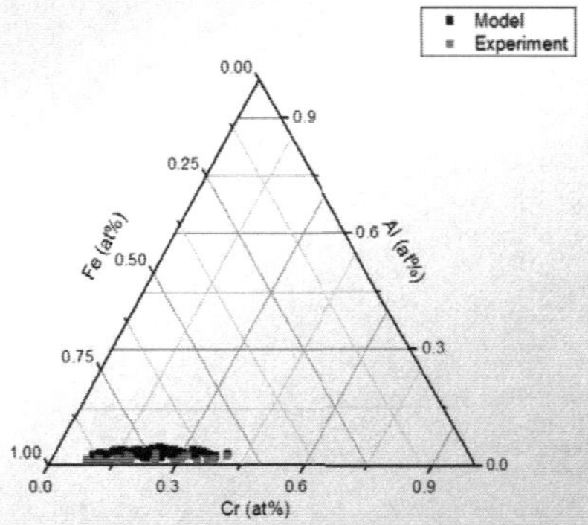

J. K. Bunn, C. J. Metting, and J. Hattrick-Simpers, "A Semi-Empirical Model for Tilted-Gun Planar Magnetron Sputtering Accounting for Chimney Shadowing ", JOM, invited paper, Accepted January 2014 Special Issue

Sputtering Geometry and Sample Locations

Ternary Deposition

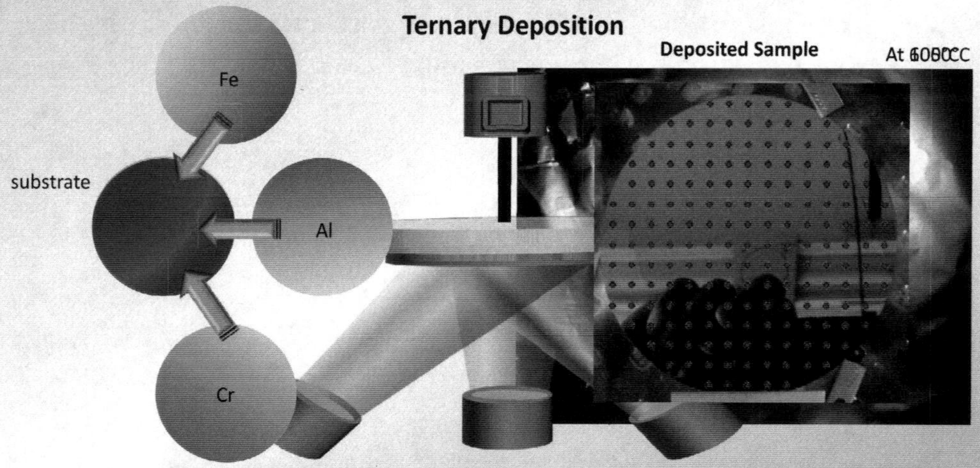

Magnetron Sputtering

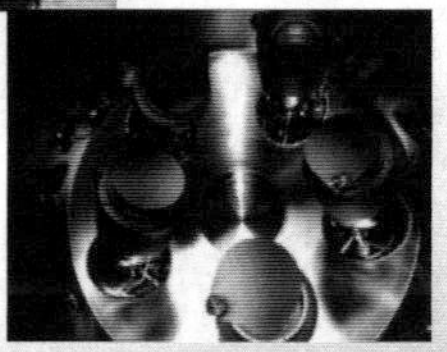

- **5 Guns (only used 3)**
- **DC/RF power sources**
- **Ultra high vacuum**
- **Deposition temperature at room temperature**

Experimental Design for High Temperature

Sample fabrication
-12 FeAlCr samples prepared on Al_2O_3 substrate and Ti with adhesion layer
-1 micron nominal thickness

Heat treatment
- Tube furnace treated
- Treated at 1050°C
- Air quenched

After deposition

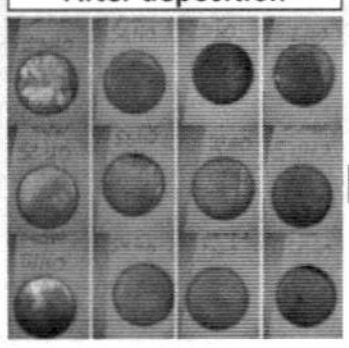

Cycled 7 times for cumulative heat treatments at 1050°C for:

-5 min -10 min
-20 min -40 min
-60 min -120 min
-240 min -360 min

After heat treatments

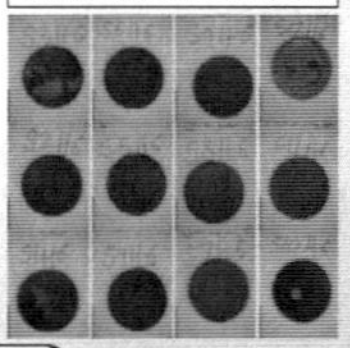

Characterization
-SEM/EDS
-XRD

Characterization Techniques
- **Raman**: Fe_2O_3, Fe_3O_4, and Al_2O_3
- **XRD**: FeCrAl, Surface Oxides, and the substrate
-**Optical Microscopy**

Oxide Formation at 5 minutes, 1050°C

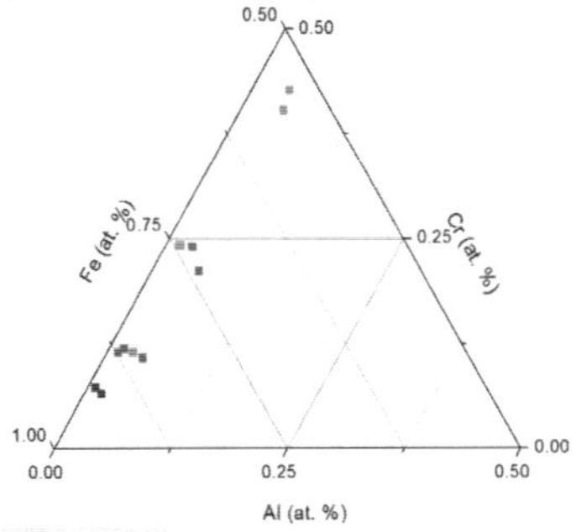

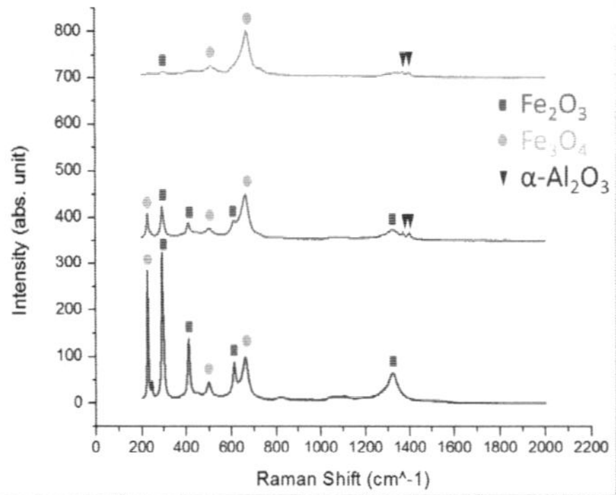

Blue: mostly Fe_2O_3, some Fe_3O_4
Red: mostly Fe_3O_4, some Fe_2O_3, some α-Al_2O_3
Green: very little Fe_2O_3, some α-Al_2O_3, some Fe_3O_4

- No α-Al_2O_3 growth was observed when the composition is Fe heavy

Oxide Formation at 360 minutes, 1050°C

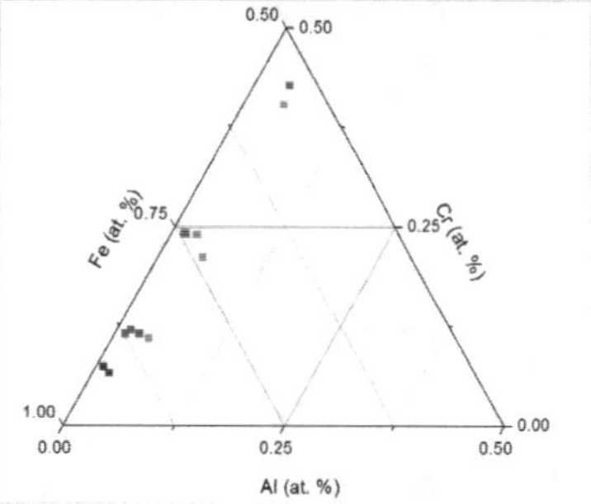

Blue: some Fe_2O_3 and Fe_3O_4, very little α-Al_2O_3
Red: some α-Al_2O_3, some Fe_3O_4
Green: majority α-Al_2O_3, some Fe_3O_4

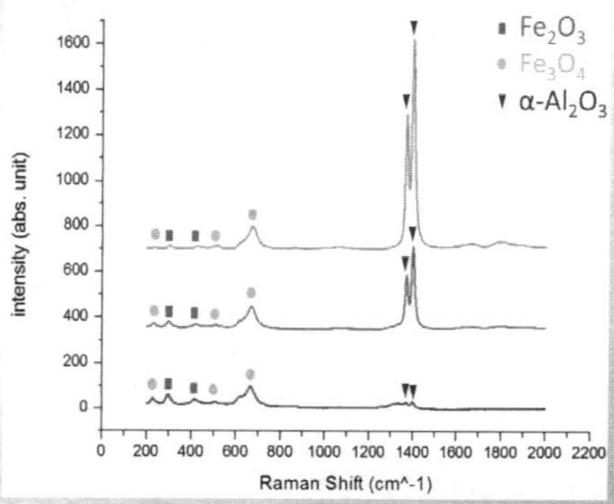

- Large amount of α-Al_2O_3 growth was observed at the Cr heavy region, indicating that Cr promotes the growth of the alumina oxides

Experimental Design at Low Temperature

Sample fabrication
-FeAlCr samples prepared on Si substrate with Ti diffusion barrier
-1 micron nominal thickness

Cycled 6 times for cumulative heat treatments 1 hour each (with the exception of 600°C):

-room temperature	-200°C
-300°C	-400°C
-500°C	-600°C (for 2hr)

Characterization
-*in-situ* GIXRD

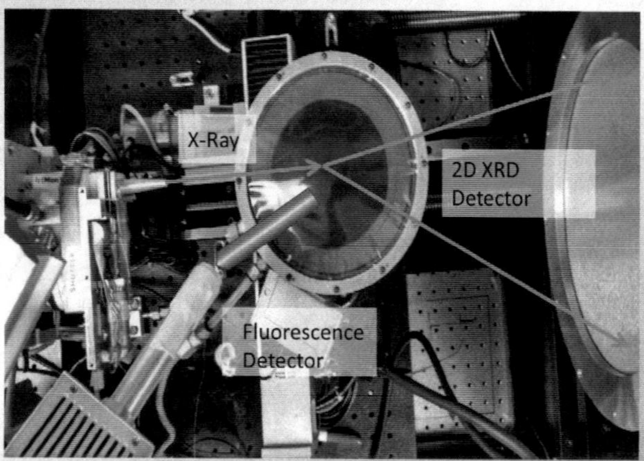

Oxide Growth Study through *in-situ* XRD

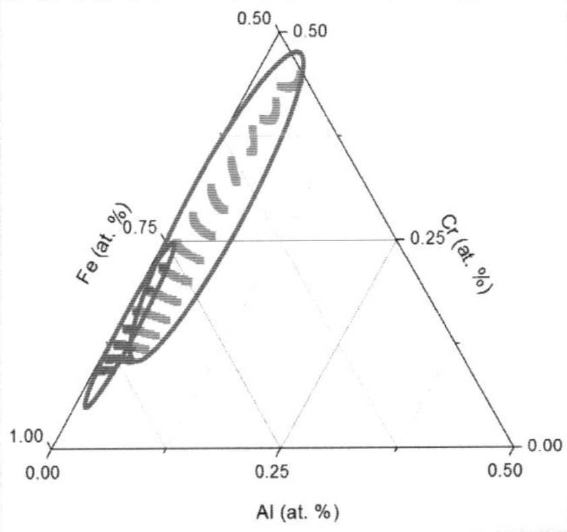

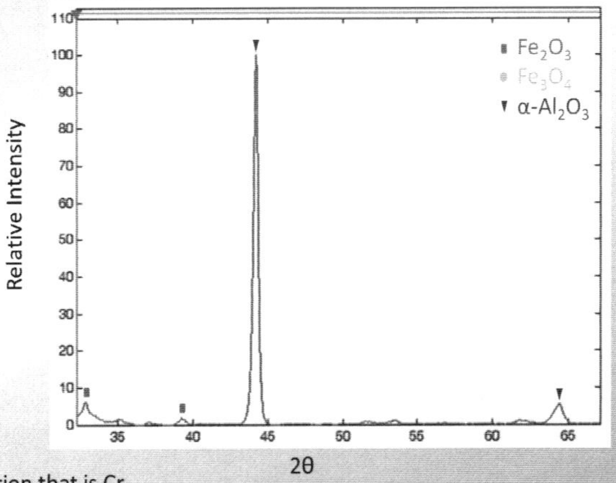

- Iron oxides are suppressed when the composition that is Cr rich (small amount are observed)

13

Comparison of Trends Between Low and High Temp Oxidation

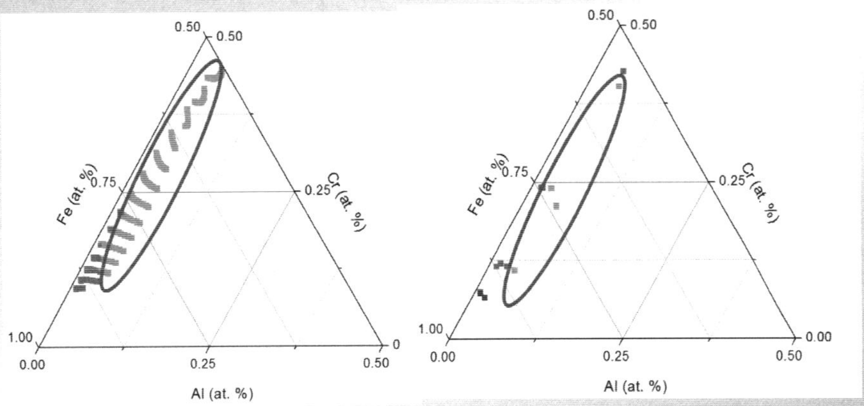

- XRD and Raman matched up where the more Al and Cr rich region are a lot easier to form α-Al_2O_3

14

Summary

- Probed the temporal effect of temperature on the metallurgic and oxide phase transitions
- A method for the **initial high-throughput and non-destructive screening** of FeCrAl thin film was developed
- **Cr promotes** α-Al_2O_3 growth on the surface of the samples to form a protective layer preventing further development of other oxides
  - This indicates that Cr rich FeCrAl are more favored to operate under high temperature conditions
- Oxidation studies indicate that severe oxide growth was not observed until 400°C
- The methodology can be applied to more complicated systems for high temperature oxidation resistant materials.
- Detailed studies of **oxidation and phase transitions of single compositions at different temperatures** will refine the results

15

Acknowledgement

- **Dr. Jason Hattrick-Simpers**
- **Jonathan Kenneth Bunn**
- **SmartState Center for Strategic Approaches to the Generation of Electricity**
- **USC Honors College (Science Undergraduate Research Fellowship Program)**
- **University of South Carolina**
  - **College of Engineering and Computing Science**

This project was funded in part by the South Carolina Honors College

16

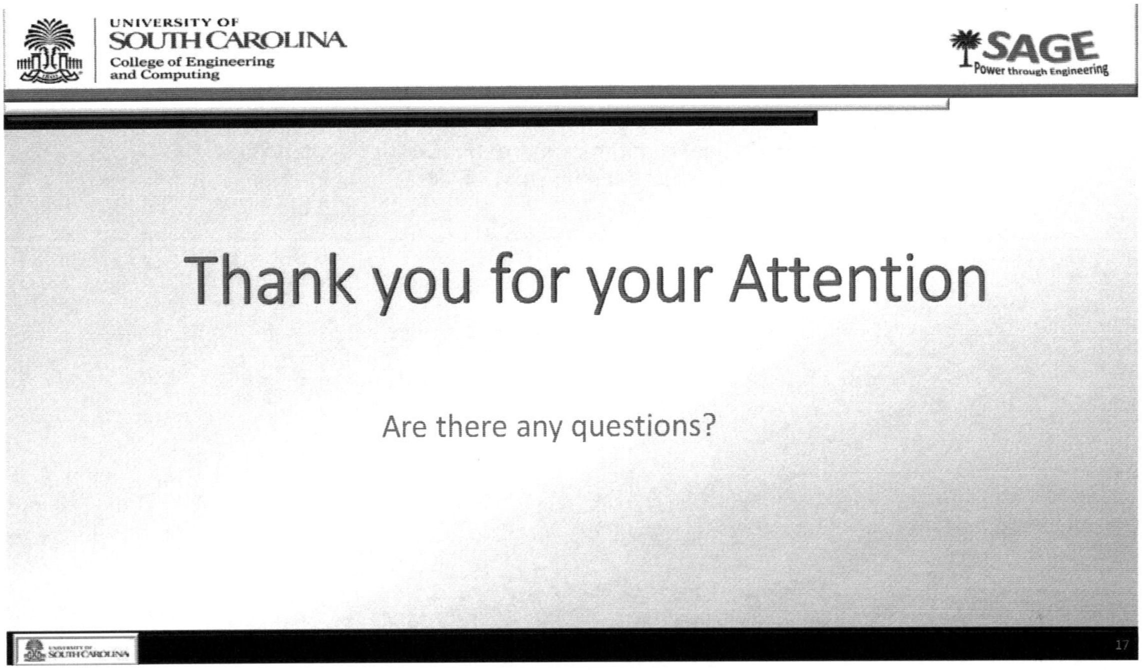
UNIVERSITY OF
SOUTH CAROLINA
College of Engineering
and Computing
SAGE
Power through Engineering
Thank you for your Attention
Are there any questions?
UNIVERSITY OF
SOUTH CAROLINA
17

Crossing Borders and Communities for Healthy Water

Joshua Gomez, Meng Zhou, Dr. Shuguang Deng

Chemical and Materials Engineering Department, New Mexico State University, Las Cruces, New Mexico, 88003

In Partnership with:

NM Water Resources Research Institute

NMSU Chemical Engineering Department

The New Mexico Alliance for Minority Participation

New Mexico Environment Department

Border Partners (Non-profit)

EPA 2020

National Science Foundation

New Mexico State University NM STATE

Puerto Palomas, Chihuahua, MX

- Located just south of Columbus, NM.
- Population of 4,688.
- Arsenic contamination (0.030-0.1 mg/L).
  - Horrible Carcinogen.
- Fluoride contamination (5-40 mg/L).
  - Bone Fluorosis and neurological damage.
- Concentration depends entirely on well water.
  - Naturally occurring concentrations of arsenic and fluoride exceed the permissible limits of Mexico and the U.S.

New Mexico State University NM STATE

New Mexico Transboundary Aquifers

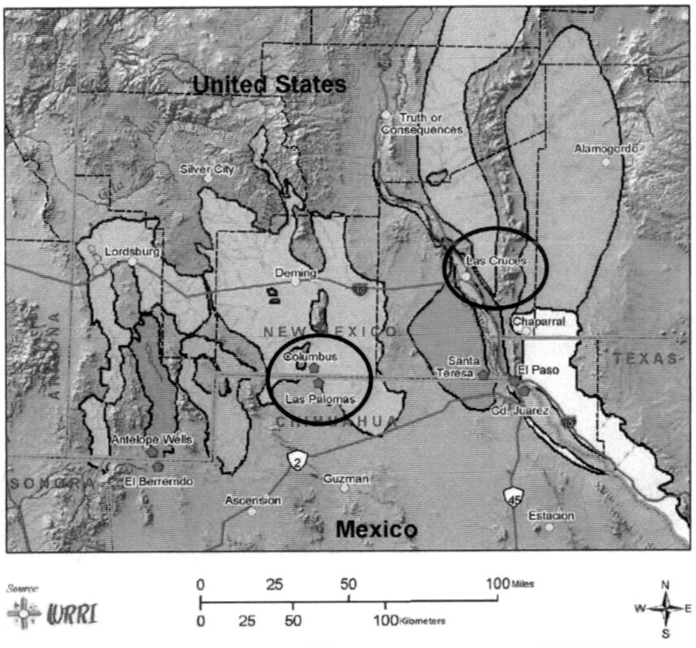

New Mexico State University

NM STATE

Proposed Solution

- Homemade water filter
  - Low cost <$15.00
  - Easy to assemble
  - Removes arsenic
  - Removes fluoride
  - Utilizing activated carbon (AC)
  - Utilizing activated alumina(AA)
    - Iron coated
  - Improves the taste

New Mexico State University

NM STATE

Filter Designs

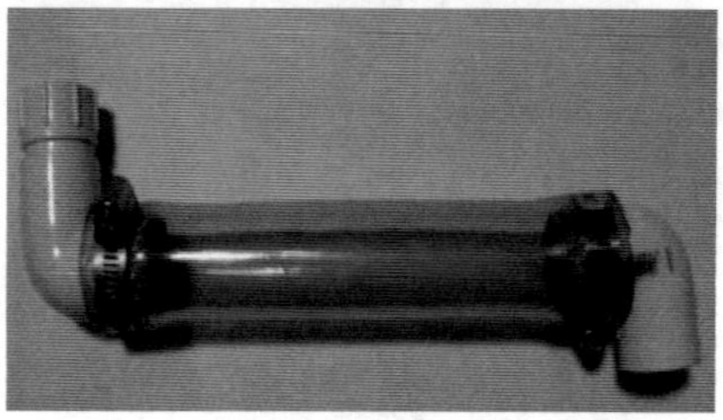

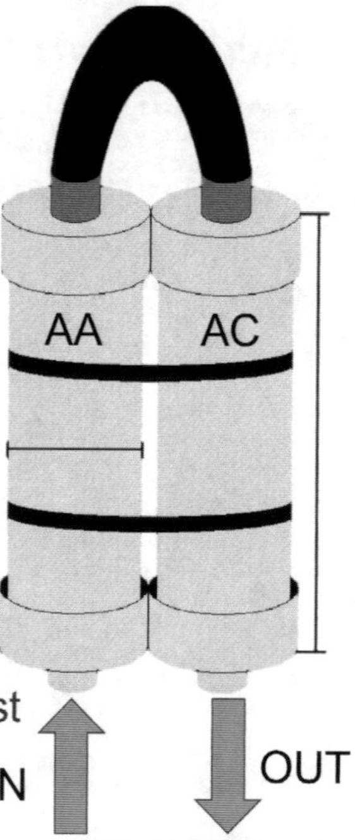

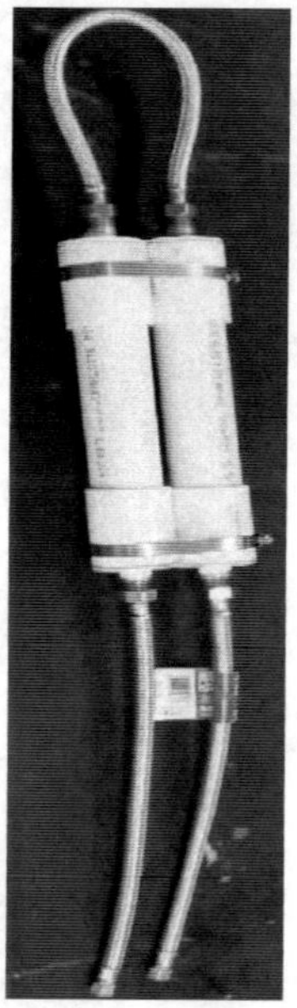

- 1 filter of the final design costs $5.00 to make.
- The Activated Alumina costs $1/kg.
- The Activated carbon used cost $35/lb.

New Mexico State University

Our Approach

NMSU demonstrated to Palomas science teachers how to assemble a simple water filter.

Science teachers Juan Carlos Huerta and Joel Carreon pose with the filter they assembled for installation at the Palomas primary school.

New Mexico State University

What's Working

Hall Environmental Laboratory, Albuquerque

CONTAMINANT	RESULTS	STANDARDS					
		U.S.	Exceeded	%	Mexico	Exceeded	%
Fluoride							
Unfiltered	4.90 mg/L	4.00 mg/L	Y	22.5	1.50 mg/L	Y	226.7
Filtered	0.240 mg/L	4.00 mg/L	N	-	1.50 mg/L	N	-
Arsenic							
Unfiltered	0.030 mg/L	0.01 mg/L	Y	200.0	0.02 mg/L	Y	20.0
Filtered	ND	0.01 mg/L	N	-	0.02 mg/L	N	-

ND = Not Detected

New Mexico State University

Optimization of Algae Derived Activated Carbon (AC)

- Low cost alternative to commercial activated carbon.
- More effective adsorbent than commercial carbon.
- Chlorella algae is used as a carbon source.
- Potassium Hydroxide and Iron (II) Chloride
  - Activating agent.
- Carbonization times of 500° C, 600° C, and 700° C.
- Shaker experiments to determine most effective adsorbent for Arsenic(III).

New Mexico State University

Carbon Experimental data

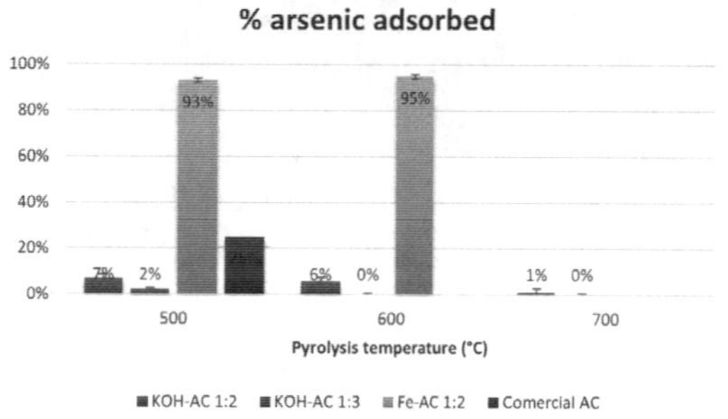

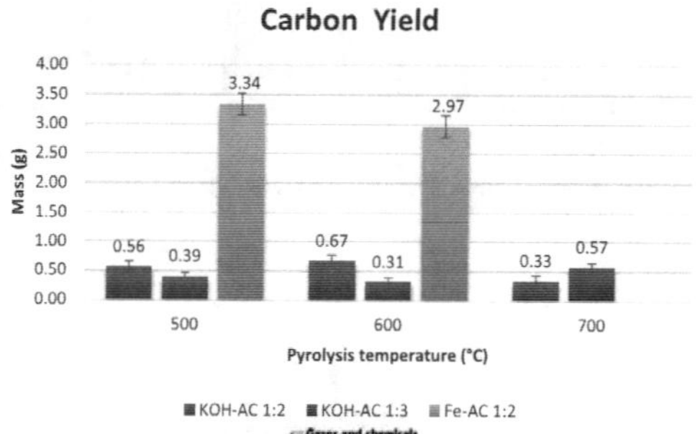

adsorption capacity (mg/g)			
	500°C	600°C	700°C
KOH-AC 1:2	0.082	0.067	0.010
KOH-AC 1:3	0.024	0.000	0.000
Fe-AC 1:2	1.143	1.168	-
Comercial AC	0.275	-	-

$$capacity = \frac{(InitialConcentration - FinalConcentration)Volume}{\text{Mass of adsorbent}}$$

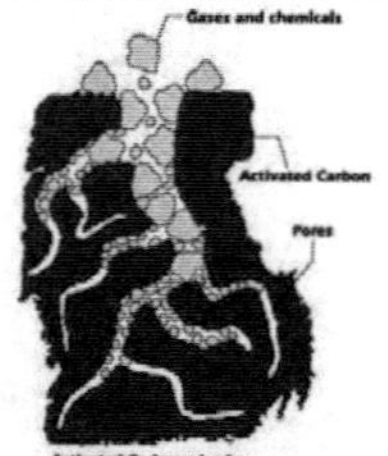

Fasci, M.
http://www.carbonanswers.com/carbonadsorptionabsorption.html

New Mexico State University

X-Ray Difraction Pattern

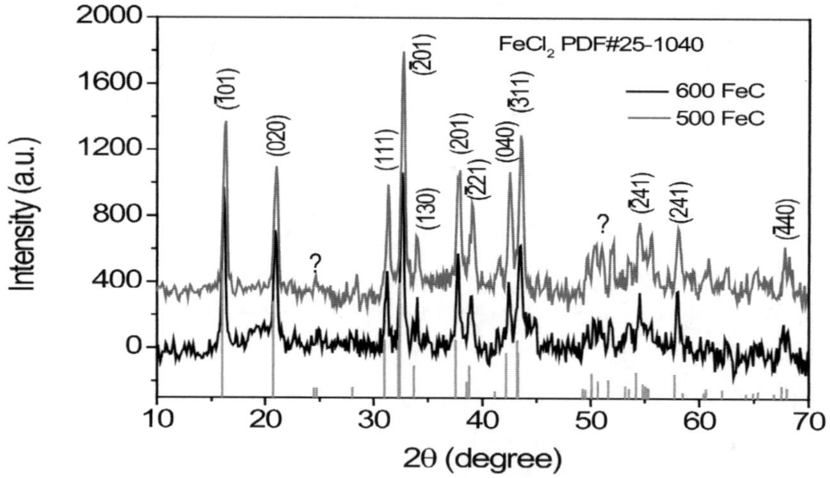

New Mexico State University

Scanning Electron Microscope Imaging

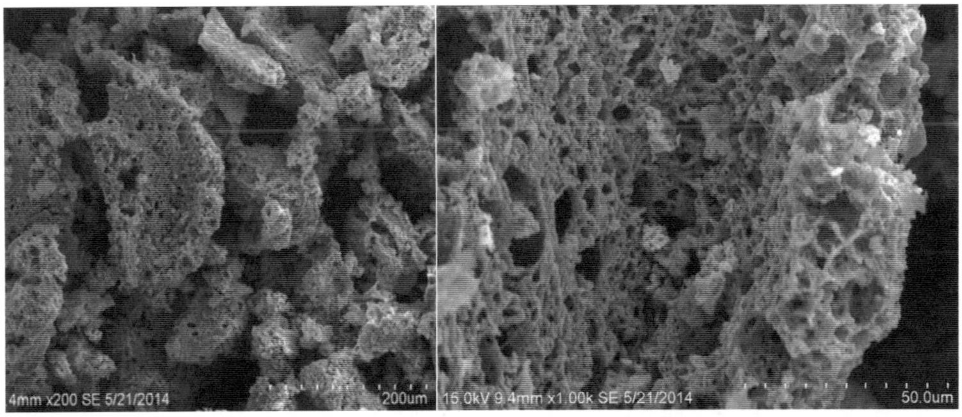

New Mexico State University

Unwashed FeCl(II) Activated Carbon Textural Properties

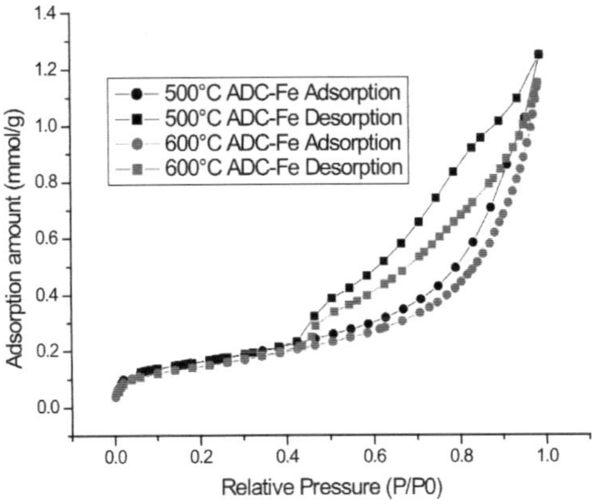

N_2 adsorption and desorption isothermal @ 77K for unwashed samples.

	500° C FeCl2/ADC	600° C FeCl2/ADC
BET surface area (m^2/g)	12.9893	11.6336
Average pore size (nm)	8.6408	6.5769
Pore volume (cm^3/g)	0.0394	0.0476

New Mexico State University

Washed FeCl(II) Activated Carbon Textural Properties

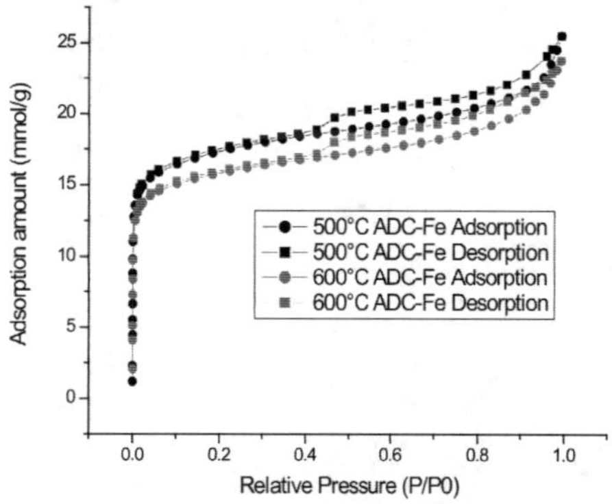

	500 ° C ADC (after washing)	600 ° C ADC (after washing
BET surface area (m^2/g)	1326.0368	1208.1831
Average pore size (nm)	7.8704	7.9340
Pore volume (cm^3/g)	0.3362	0.3305

N2 adsorption and desorption isothermal @ 77K for washed samples.

New Mexico State University

Shaker Experiments

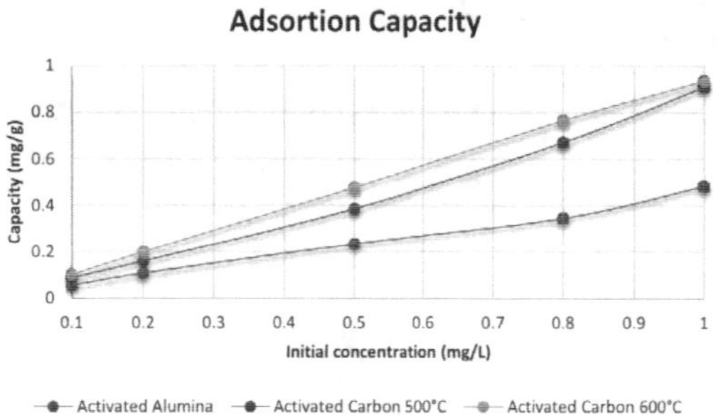

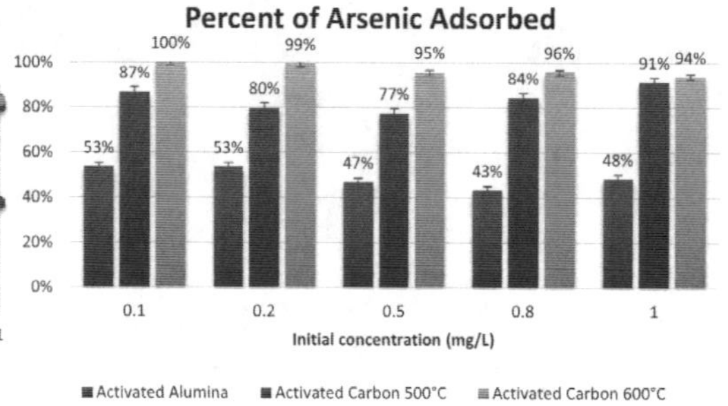

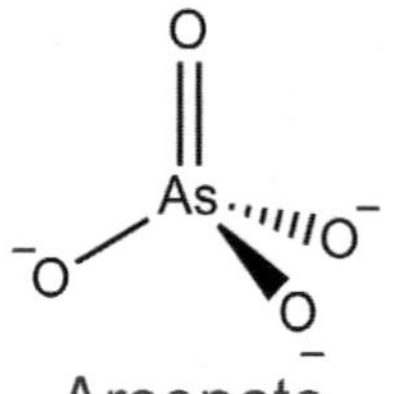

Arsenate

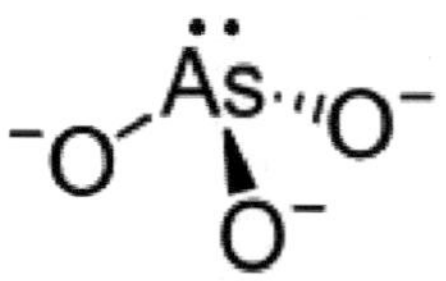

Arsenite

New Mexico State University

Conclusions

- Filters are still being made in Palomas by the teachers today and will continue to be made.
- The results of the shaker experiments and textural property data shows that the carbon activated with Iron (II) Chloride is a viable replacement.

New Mexico State University NM STATE

Next Steps

- Further Development of the filter.
- Reach out to other communities.
- Advance research with optimized carbon.

New Mexico State University NM STATE

Acknowledgments

- Dr. Shuguang Deng, Chemical & Materials Engineering dept.
- Meng Zhou, Chemical & Materials Engineering dept.
- Dr. David A. Rockstraw, Chemical & Materials Engineering dept.
- Erin Ward, Water Resources Research Institute.
- Mark Chidester, Chemical & Materials Engineering dept.
- Peter Edmunds, Border Partners.

NEW MEXICO AMP

NSF HRD # 1305011

Deng's Renewable Energy and Advanced Materials

New Mexico State University

Questions?

Email: jgomez38@nmsu.edu

New Mexico State University

394186 Ferroelectric BTO on Si (001) for High-Efficiency Solar Cell Heterostructures

Abstract:

Ferroelectric oxides such as barium titanate (BTO) have recently become a topic of interest in relation to thin-film solar cell production due to the material's stability at a wide range of temperatures, its spontaneous polarization, and the potentially low-cost of production. Integration of BTO on silicon could enable the creation of highly efficient solar cells, as BTO's ferroelectric properties encourage the detachment of photo excited electrons. We have previously demonstrated successful integration of BTO on silicon carbide (SiC) substrates, using a magnesium oxide (MgO) heteroepitaxial layer to enable single-crystal orientation of the BTO film deposited on SiC.[1] BTO has also been successfully grown on silicon substrates, but with the use of strontium titanate (STO) and MgO buffer layers to resist formation of SiO_2.[2] Integration of BTO on Si with a single MgO interface layer would decrease interfacial losses in a final solar cell device. Based on previous work with SiC, we believe a single interface can be possible by controlling the surface interactions during film growth.

In this research, MgO is grown on Si (001) substrates in an ultrahigh vacuum (UHV) environment using a number of growth methods including molecular beam epitaxy (MBE) and sputtering. In comparing differences in surface structure using reflection high-energy electron diffraction (RHEED) and surface chemistry using X-ray photoelectron spectroscopy (XPS), we can understand surface interactions between Si, Mg, and O. By understanding the relationship between atomic level interactions and ultimate film characteristics, we can engineer the most effective process for BTO integration on Si. The substrates are cleaned using wet chemicals to create a hydrogen-terminated surface that resists contamination from laboratory exposure. XPS and RHEED are used to verify the lack of contamination on the surface and proper crystallographic orientation. Contaminant levels of 7.61 % oxygen (0.15 O/Si ratio) and 18.0% carbon (0.31 C/Si ratio) are the most successful results that have thus far been attained. XPS and RHEED will serve to confirm proper growth of MgO (001) on Si and determine thickness of the MgO film. BTO is then grown on the correctly-oriented MgO film using MBE, as has already been demonstrated on MgO/SiC substrates.[1] This presentation will discuss cleaning and growth conditions for successful production of high quality BTO on Si using the single MgO interface layer.

393873 Natural Gas, the Bridge Fuel

Abstract:

Medium and heavy duty trucks and buses produce almost 18 percent of all carbon emissions from transportation. Alternative energy sources need to be used on a larger scale to reduce these emissions, as well as to reduce dependence on foreign oil. Natural gas is the future of fuel in large vehicles. This paper tackles different issues surrounding the usage of natural gas as the main fuel for natural such as public awareness, public policy, existing infrastructure, consumer choice and controversies surrounding hydraulic fracturing. Current pilot programs, in which public city buses are using natural gas as the fuel of their choice, will be analyzed to determine the best way to implement these programs in other areas of the United States as well.

397811 The Federal Role in Fostering an Innovative U.S. Energy Ecosystem

Abstract:

Although renewable sources of energy – solar, wind, hydro, and geothermal – have been around for over one hundred years, they only generate 12.5 percent of electricity in the United States today. Renewable technology has advanced in cost, reliability, performance, and efficiency, but not enough to allow renewables to be a large contributor to electricity generation and widely used in the U.S. Renewables need continued innovation and progression to become more efficient, more reliable, and more cost competitive with fossil fuels, as well as expanding the possibilities of their use. Increased R&D funding for renewable sources is the key to advancing the technology and giving renewables the tools it needs to provide a larger portion of electricity in the U.S.

With renewables generating a larger portion of electricity, the U.S. can reach environmental sustainability, energy security, and a new economic opportunity. In reference to reaching environmental sustainability, renewables do not emit carbon and greenhouse gases, which is important for a cleaner future and mitigating climate change. Because the "fuel" for renewables includes the sun and wind, electricity generation from these sources does not depend on obtaining fossil fuels inside or outside of our country. They also extend the lifetime of these resources to be used for base load power, providing energy security to the U.S. Finally, if the U.S. is able to quickly and effectively develop competitive renewable technologies and deploy them into the electricity sector, it can provide competitive renewable technologies to the global community, boosting the economy and maintaining world leadership.

Although renewables have the capability to provide many benefits for the U.S., they face obstacles in receiving the funding they need to be widely used. These obstacles include climate change skeptics, the fossil fuel industry, the polarized political environment, and having to cut funding for other areas such as education and defense in order to allocate more money to R&D. The demand for renewables has also been relatively weak because electricity generated by renewables is no different than electricity generated by fossil fuels when the consumer uses it. The main way for renewables to be competitive with fossil fuels is through the cost and since wind is the only renewable that is cost competitive with fossil fuels, a strong customer demand has not been present to spur large private sector investment in renewable technologies.

In order for renewables to accomplish necessary advancements and contribute a larger portion of electricity to the U.S., financing to R&D for the technologies needs to be increased through government funding, capital generated through a carbon tax, public-private partnerships, and incentives for investment in the technology and its R&D. A commitment to innovating renewables in order to reach environmental sustainability, energy security, and a new economic opportunity is necessary from governmental officials to make the tough decisions advocating for widespread use of renewables in the U.S.

397856 A Biowall for Improving Indoor Air Quality

Abstract:

The Biowall is a living plant wall that filters air in the HVAC system of a home or small commercial building. This new technology improves indoor air quality be removing Volatile Organic Compounds and has the potential for energy savings by reducing the need for heating and cooling large amounts of outside air in a building. The presentation provides an overview of biofiltration for buildings and discusses the new research being conducted in this emerging field.

370097 Layered Double-Hydroxide Catalysts for Lignin Depolymerization

Abstract:

A critical step in the conversion of lignin to renewable value-added aromatic chemicals is selective depolymerization of the biopolymer to its constituent monomers. For optimal process economics, this depolymerization should occur over heterogeneous catalysts in the absence of high-pressure hydrogen, but few catalysts exist that perform well and meet these criteria. One promising class of catalyst is layered double-hydroxide materials, such as hydrotalcite. Our group has recently shown that nickel-loaded hydrotalcite selectively cleaves the *β*-O-4 bond in a lignin model dimer, and reduces the molecular weight of two types of whole lignin. Here we expand on this finding by varying the many tunable parameters of hydrotalcite catalysts, including the type and location of metal and the identity of the charge-compensating anion(s), as well as the reaction solvent. Application of these catalysts to a variety of lignin model dimers and real lignin streams shows that these catalysts do not rely only on basic or metal sites for activity, and that the solvent has an impact on catalyst activity and product spectrum.

Catalytic Depolymerization of Lignin Extracted from Kraft Black Liquor and Lignin Model Compound 4-phenoxyphenol

Bethany D. Carter and David A. Bruce*

Department of Chemical and Biomolecular Engineering, Center for Advanced Engineering Fibers and Films, Clemson University, Clemson, SC, 29634-0909, USA.

* *Corresponding author. E-mail: dbruce@clemson.edu*

Lignin is the only renewable source of aromatic species, consisting of a network of guaiacyl, syringyl, and p-hydroxyphenyl units connected through various types of ether linkages forming an alkyl aryl ether backbone functionalized with methoxyl, benzyl alcohol, and phenolic hydroxyl species. Lignin conversion into lower molecular weight, monomeric type species with valuable functionality, such as phenols and aromatic acids, would shift dependence away from fossil fuels to a renewable source of aromatics.

Lignin was extracted from Kraft black liquor via the continuous Sequential Liquid Lignin Recovery and Purification (SLRP) process whereby the pH of the incoming black liquor (pH ~ 13.5) was lowered to a pH ~ 9.5 by acidification with CO_2 resulting in lignin precipitation. The precipitated, liquid lignin phase is then acidified with sulfuric acid until a final pH ~ 2.5 is achieved, whereby the lignin is fully acidified.

Catalytic conversion of SLRP lignin and lignin model compound 4-phenoxyphenol was investigated using a Pd/SBA-15 heterogeneous solid catalyst. The high surface area, large pore diameter, and hydrogenation sites enable the cleavage of ether linkages in lignin and lignin model compounds as well as stabilization of the resulting product thus preventing repolymerization which would result in low conversion. The nanometer length scale of the catalyst particles also contributes to these advantages. The heterogeneous nature of the catalyst allows for separation and reuse of the catalyst.

Lignin model compound depolymerization experiments were carried out in small batch reactor in the presence of the catalyst and under hydrogen pressure. Characterization of depolymerized lignin fractions was done using gas chromatography-mass spectrometry to analyze lighter molecular weight fractions. Catalyst reusability experiments were also performed to assess potential industrial application.

377869 Conversion of Guaiacol on Noble Metal Catalysts

Abstract:

Catalytic hydrodeoxygenation (HDO) is a rapidly developing technology for oxygen removal from bio-oils obtained from pyrolysis of biomass. In this study, Guaiacol, a phenol derived compound produced by the thermal degradation of lignin, was selected as a model compound to study the HDO process. Guaiacol is among the major components of pyrolysis bio-oils; however it is thermally unstable which leads to catalyst deactivation. Four noble metal catalysts (Pt, Pd, Rh and Ru) supported on activated carbon were tested in a fixed-bed reactor at atmospheric pressure and their performance for the guaiacol HDO process was compared. The screening criteria were as follows: (1) High degree of deoxygenation, (2) Low hydrogen consumption, (3) High carbon recovery in liquid phase, and (4) Long catalyst lifetime. Our results showed that, among the tested catalysts, Pt has higher deoxygenation activity and shows relatively little deactivation. Additionally, the operating temperature for Pt catalyst was optimized and found to be 300 ^{o}C. Finally, results from catalyst characterization demonstrated that polyaromatic deposits, especially the condensed ring compounds, are the most likely cause for catalyst deactivation. [1]

[1]D. Gao, C. Schweitzer. H. Hwang and A. Varma, Conversion of Guaiacol on Noble Metal Catalysts: Reaction Performance and Deactivation Studies, *Ind. Eng. Chem. Res*., DOI: 10.1021/ie500495z

384936 2,5-Dimethylfuran As a Model Compound to Investigate Hydrodeoxygenation of Complex Furanyl Compounds

Abstract:

Ying Lin Louie, Konstantinos Goulas, and Alexis T. Bell

Department of Chemical and Biomolecular Engineering,

University of California-Berkeley, Berkeley, California 94720

Several innovative pathways have been explored to produce diesel-range fuel additives via C-C bond coupling reactions of furanyl platform molecules, 5-hydroxymethylfurfural (HMF) and furfural (FUR), obtained from the dehydration of sugars [1, 2]. In each case, however, a final hydrodeoxygenation (HDO) step is required in order to produce a product that is oxygen-free, has a high cetane number, is compatible with conventional diesel fuel, and has minimal impact on the performance current diesel engines. While a number of studies of the hydrodeoxygenation of biomass-derived fuels have been reported [3], little is understood about the reaction pathways and how catalyst composition affects these pathways. The aim of this work is to develop such knowledge based on studies conducted with model compounds containing structural motifs representative of those found in second-generation biomass-derived diesel.

To understand the HDO chemistry of furanyl compounds and develop suitable HDO catalysts for this task, 2,5-dimethylfuran (DMF) was chosen as a model compound. The sequence of reactions by which DMF undergoes HDO was then investigated in both gas and liquid phase over various bi-functional metal/acid catalysts. Screening studies with carbon-supported Pt, Pd, and Rh revealed that Pt/C exhibits the highest C-O hydrogenolysis activity. Hydrogenation of DMF over Pt/C was found to proceed via parallel pathways of direct ring opening and ring saturation existed. It was observed that under mild reaction conditions, C-O bond hydrogenolysis of DMF to 2-hexanone is preferred over saturation of the furan ring to form 2,5-dimethyltetrahydrofuran (DMTHF). The resulting tetrahydrofuran ring is resistant to ring-opening to form ketones and alcohols. The distribution of products was found to be sensitive to the reaction conditions used. High temperatures and low H2 pressures were found to favor the furan ring-opening pathway, and under optimal conditions, we were able to obtain greater than 95% selectivity toward C-O hydrogenolysis products. Removal of oxygen from 2-hexanone and DMTHF occurred very effectively when Pt was supported on an oxide, such as TiO2, Nb2O5, and NbOPO4, which contain Lewis acid centers. Of the acid catalysts screened, niobium based oxides had stronger acidity and the highest ratio of Lewis acid to Brønsted acid sites [4], measured by infrared spectroscopy of adsorbed pyridine absorption. Pt/NbOPO4, in particular, was found to be exceptionally active for promoting the HDO of DMF to hexane. The results obtained from our studies of DMF have enabled us to optimize the HDO of more complex substrates, such as bis(5-methylfuran-2-yl)methane (BMFM, C11) and tetra(5-methylfuran-2-yl)methane (TMFM, C16). Greater than 81% yield to C11 was achieved during the HDO of BMFM at 250oC and 500psi H2, conditions that are much milder than those previously reported in the literature.

1. West, R.M., et al., Liquid alkanes with targeted molecular weights from biomass-derived carbohydrates. ChemSusChem, 2008. 1(5): p. 417-24.

2. Corma, A., O. de la Torre, and M. Renz, Production of high quality diesel from cellulose and hemicellulose by the Sylvan process: catalysts and process variables. Energy & Environmental Science, 2012. 5(4): p. 6328.

3. Zhong He, X.W., Hydrodeoxygenation of model compounds and catalytic systems for pyrolysis bio-oils upgrading. Catalysis for sustainable energy, 2012: p. 28-52.

4. Ziolek, M., Niobium-containing catalysts—the state of the art. Catalysis Today, 2003. 78(1-4): p. 47-64.

Catalytic Oligomer/Polymerization of Furfuryl Alcohol over Homogeneous and Heterogeneous Catalysts

Xiaojun Chan[1], Taejin Kim[1,2]

[1]Department of Material Sciences and Engineering, Stony Brook University, Stony Brook, NY 11794

[2]Department of Chemical and Molecular Engineering, Stony Brook University, Stony Brook, NY 11794

2014 AIChE Annual Meeting, Nov 17, 2014, Atlanta, GA

Introduction

- ***Lignocellulosic biomass (non-edible biomass resources)***
  - Renewable & Recyclable
  - CO_2 neutral

- ***Furfuryl alcohol ($C_5H_6O_2$, FA)*** is an attractive intermediate chemical for the production of levulinic acid and various useful green polymer products.

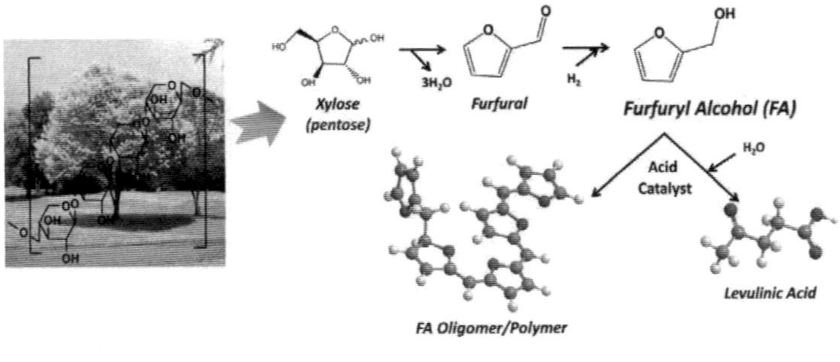

Objects

FA monomer

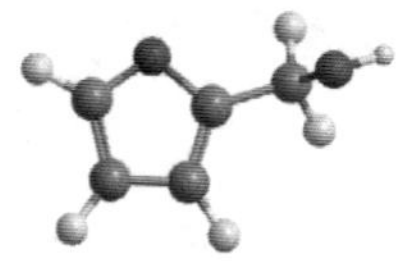

$C_6H_5O_2$

FA Oligomer
(Dimer, Trimer, Tetramer…)

Investigate the FA Oligomerization over Homogeneous and Heterogeneous Catalyst

- ***Intermediate species (Dimer, Trimer)***
- ***Feasibility to use Heterogeneous catalysts***

Experiment

- ***Vibrational Spectroscopy Analysis:***
  FT-IR, Raman(Complimentary methods)
- ***Product Composition Analysis:***
  GC, GC/MS(Quantitive calculation based on Response Factor theory)

FT-IR Spectra of Homogeneous/Heterogeneous Catalystic Furfuryl Alcohol Oligomerization under Room Temperature

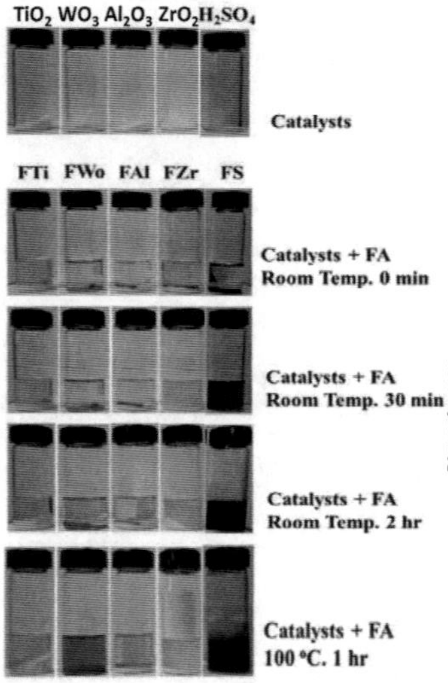

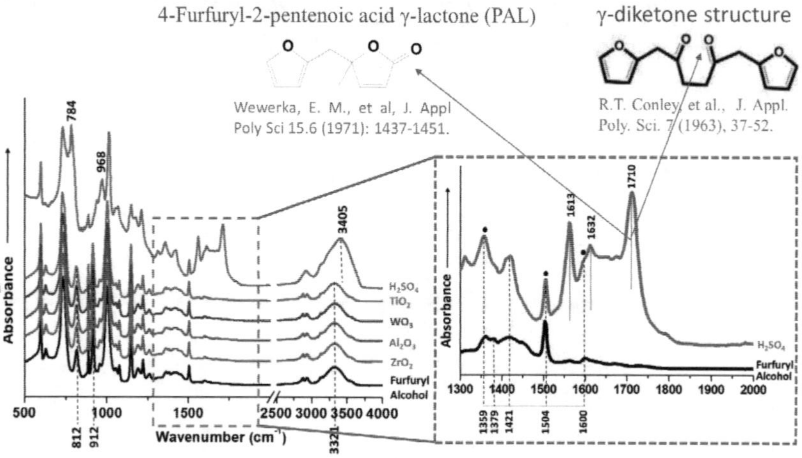

- ➢ Homogeneous catalyst (H_2SO_4) shows tremendous performance under room temp. and at 100°C.
- ➢ Among Heterogeneous catalysts, WO_3 shows sign of reaction at 100°C

WO_3 Catalystic Furfuryl Alcohol Oligomerization under 100°C

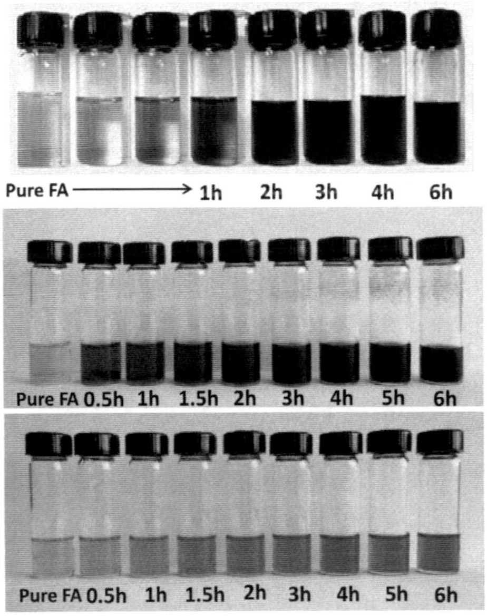

Sulfuric acid + FA
1:100
R.T.

WO_3+ FA
1:10
100°C

FA
100°C

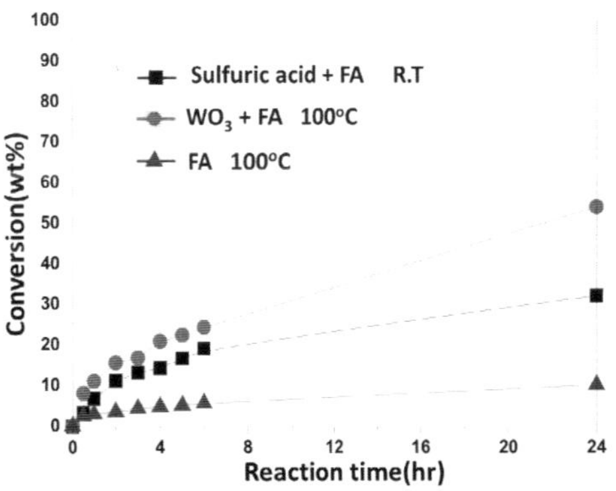

- WO_3 reaches similar conversion at 100°C compared with conversion obtained by Sulfuric acid at room temperature

Kim, T.J., et al, Cata Com, 46 (2014): 66-70.

5

Compare of fresh and used WO_3 for Furfuryl Alcohol Oligomerization

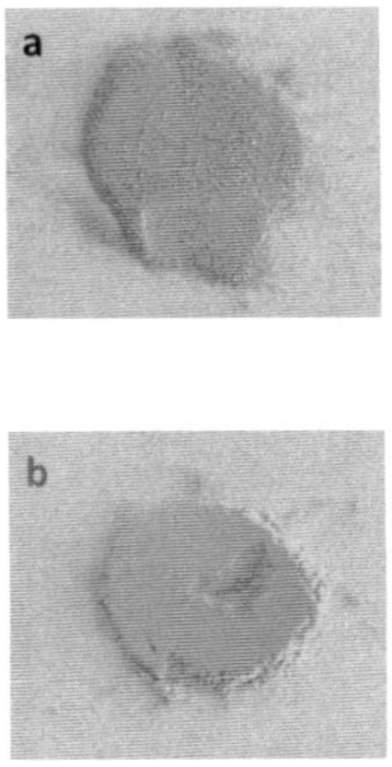

Fresh WO_3

Used WO_3

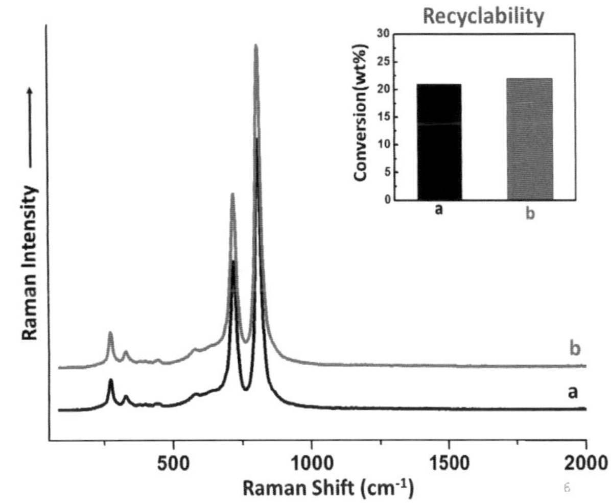

6

Stony Brook University | The State University of New York

Oligomers obtained from Homogeneous(Ho)/Heterogeneous(He) Catalystic Furfuryl Alcohol Oligomerization

	Oligomer	Name	Structure	
6 Dimers	D1	2,2'-Difurylmethane		(Ho, He)
	D2	2-(2-furylmethyl)-5-methylfuran	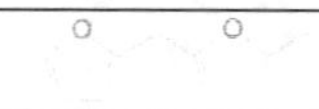	(He)
	D3	2-Furanethanol, α-2-furanyl-	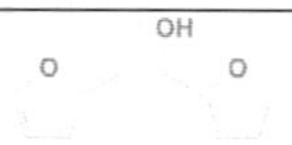	(Ho)
	D4	Difurfuryl ether		(Ho, He)
	D5	4-Furfuryl-2-pentenoic acid γ-lactone		(He)
	D6	5-Furfuryl-furfuryl alcohol	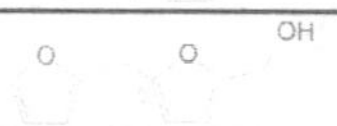	(Ho, He)
2 Trimers	T1	2,5-Difurfurylfuran	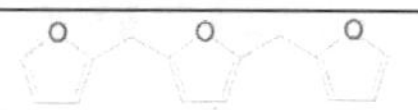	(Ho,He)
	T2	2,2'-(Furylmethylene)bis(5-methylfuran)	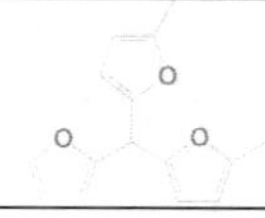	(He)

Concentration and Selectivity of oligomers obtained from WO_3 catalyzed Furfuryl Alcohol Oligomerization under 100°C

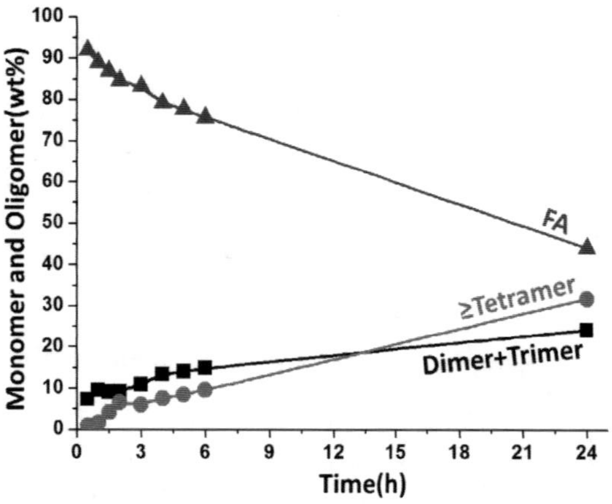

At the beginning, dimer and trimer have higher concentration. At the last several hours, concentration of heavy oligomer exceed dimer and trimer.

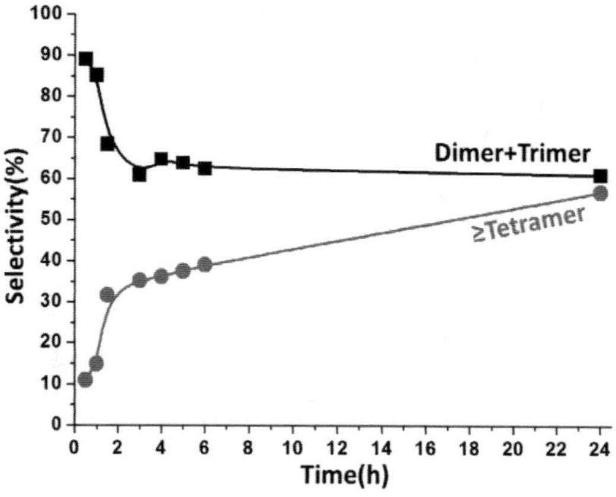

Selectivity to dimer+trimer decrease, while to heavy oligomer increase as a function of time.

FT-IR Spectra of WO_3 catalyzed FA Oligomerization under 100°C

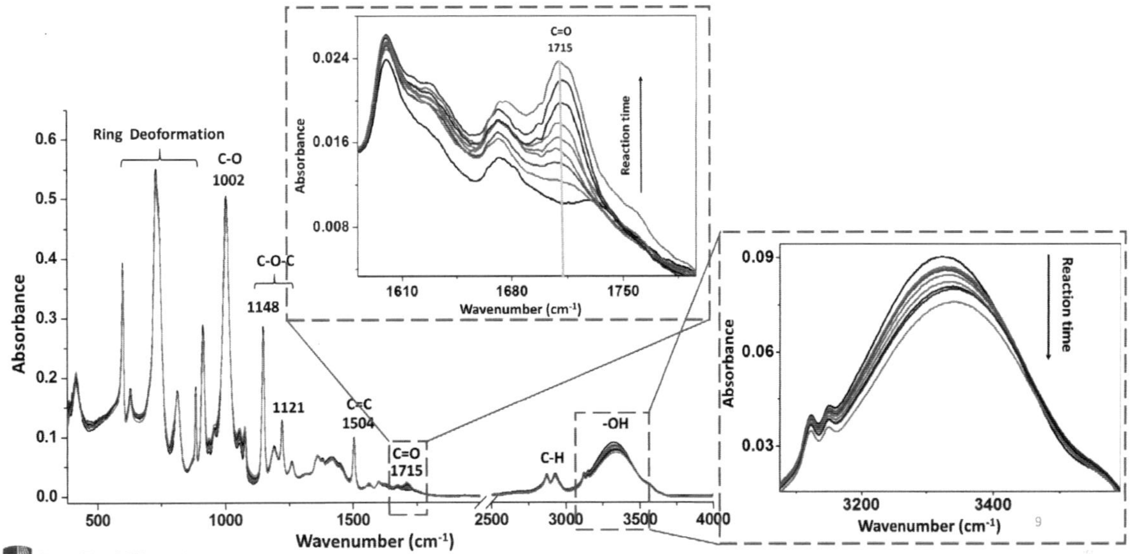

Stony Brook University 9

Raman Spectra of Homogeneous/Heterogeneous Catalystic FA Oligomerization

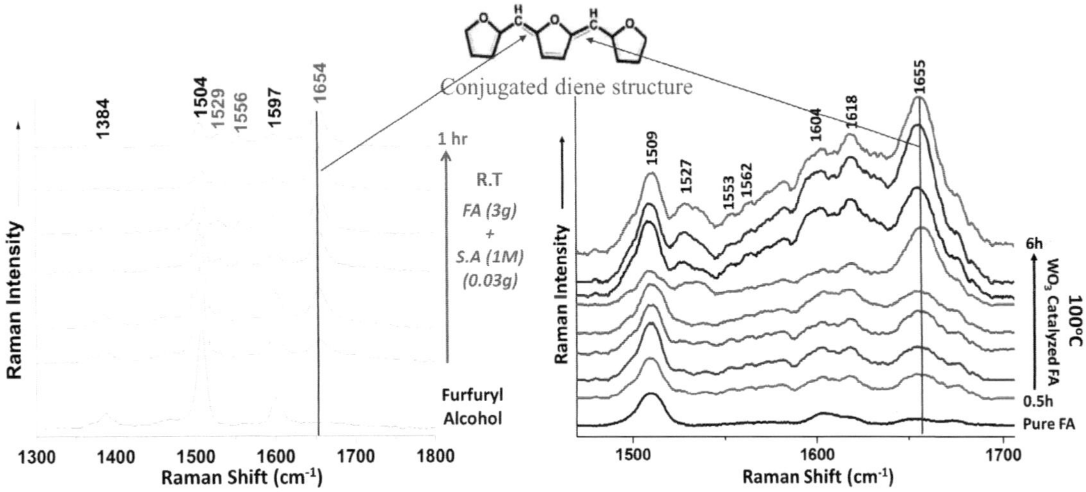

Stony Brook University 10

Macromolecules, Vol. 29, No 11, 1996, 3839-3850, C. Moreau et al., Top. Catal., 2004, 27, 11-30, S. Spange et al., Makromol. Chem., 194, 1537-1544 (1993), C.C. Burket et al., Carbon 44 (2006) 2957-2963

Proposed Initiation Mechanism of Homogeneous/Heterogeneous Catalystic FA Oligomerization

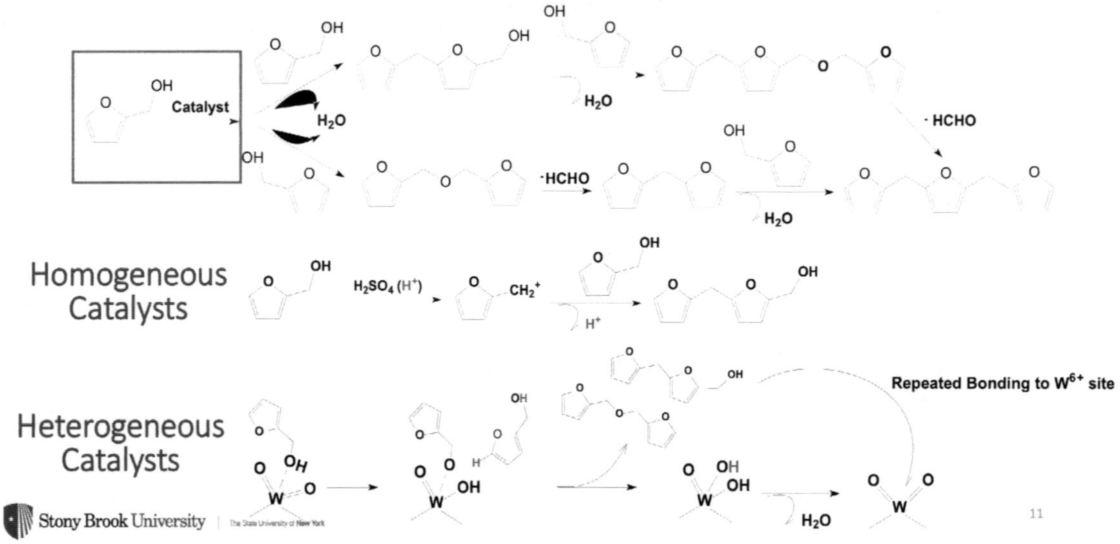

Conclusion

- In this work, we investigated the products and intermediate species for the liquis-phase oligomer/ polymerization of furfuryl alcohol monomer using homogeneous/heterogeneous catalysts
- Six different kinds of dimers, Two different kinds of trimers were observed; most of the maintained functional group, such as methylene, ether and CH_2OH.
- A strong C=O (γ-diketone) and aliphatic C=C band (conjugated diene) observed in the IR and Raman spectra, respectively, provided evidence to understand the oligomer polymerization reaction mechanism.

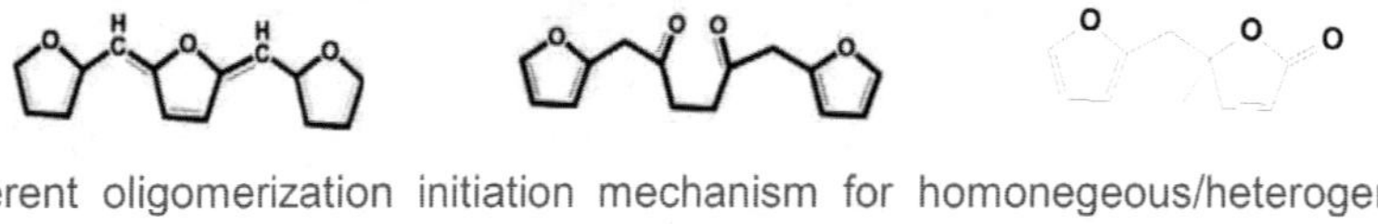

- Two different oligomerization initiation mechanism for homonegeous/heterogeneous were proposed respectively. Further conformation can be fulfilled through In-situ/Operando Spectroscopy.

Thank you!

Acknowledgements

Prof. Israel E. Wachs

Prof. Devinder Mahajan

Dr. Christopher Keturakis

Dr. Wei Nan

Vitoria Barbosa

STONY BROOK UNIVERSITY

ADVANCED ENERGY RESEARCH AND TECHNOLOGY CENTER

Stony Brook University | The State University of New York

13

389211 Selective Oxidation of Sugars and Polyols to Disaccharic Acids on Pt-Based Catalysts in One Pot Process

Abstract:

Catalytic conversion of biomass-based sources to high value chemicals has attracted increasing interest because it provides renewable alternatives to those derived from petrochemical feedstocks. Sugars and polyols, such as glucose, xylose and glycerol are abundantly available from biomass resources. Oxidation of these oxygen functional feedstocks to disaccharic acids such as glucaric acid and tartronic acid shows promising industrial applications, because these acids are valuable pharmaceutical intermediates and food additives. More importantly, disaccharic acids are potential feedstocks for the production of various megaton everyday chemicals such as adipic acid, succinic acid and malonic acid.

Disaccharic acids were conventionally produced from biomass or petroleum crudes via nitroxidation, cyanide and biological routes, which generate large amounts of toxic side products with poor overall productivity and chemoselectivity toward desired products. Catalytic oxidation of biomass in the presence of molecular oxygen on metal catalysts (*e.g.* Pt, Pd) could potentially yield disaccharic acids in one step without producing any toxics. Selective oxidation of sugars and polyols to disaccharic acids is however challenging because of significant side reactions of feedstocks to form smaller mono-acids, CO_2 as well as humic substances under reaction conditions (T = 40~80 ^{o}C in the alkaline medium). In addition, the reaction network of sugars to $C_{1\sim6}$ mono- and di-acids under oxidative conditions is not clearly understood at the current stage. Therefore, design of active and selective oxidation catalysts and fundamental understanding of reaction pathways/mechanism is the key to improve the overall yield of disaccharic acids from biomass. In particular, heterogeneous Pt-based catalysts have been evaluated and characterized for aqueous phase conversion of glucose, xylose and glycerol in the presence of molecular oxygen. Detailed kinetic studies taking into account the effects of Pt particle size, pH, temperature and oxygen pressure on the rate of oxidation, C-C cleavage and dehydration will be presented based on Pt-based catalysts.

368973 Catalytic Oxidation of Cellobiose over Gold-Based Bimetallic Nanoparticles Supported on TiO2

Abstract:

A series of bimetallic catalysts Au-M (M = Cu, Co, Ru, Pd and Ni) were supported on TiO_2 via deposition-precipitation (DP) method using urea as precipitating agent. The resulting catalysts were used in the catalytic oxidation of cellobiose into gluconic acid. The interactions between the metals in the bimetallic catalysts were carefully investigated in order to unravel the effect of the second metal on the chemical and electronic properties of the active sites and their impact on the catalytic performance. Cu and Ru were found to be promising promoters for the oxidation of cellobiose into gluconic acid, promoting oxygen activation in gold-catalyzed oxidation reactions, while Co and Pd were found to enhance the retro-aldol condensation of fructose into glycolic acid. In this case, the catalytic activity is dominated by the second metal added to Au. Hence, Au behaves more like a promoter of the second metal, preventing over oxidation and poisoning by the reaction intermediates and products. Ni specifically promoted the C-C scission, yielding a mixture of lower carbon-containing compounds and glycolic acid as the main products.

The major reaction products obtained were gluconic acid, glycolic acid, ethylene glycol and erythritol. The catalytic activity of the catalysts followed the order: Au-Cu/TiO_2 > Au-Pd/TiO_2 > Au-Ru/TiO_2 > Au-Co/TiO_2 > Au/TiO_2 > Au-Ni/TiO_2. Upon comparison, Au-Ru/TiO_2 was found to be an active and selective catalyst towards the formation of gluconic acid. The nature of the metal interactions in the bimetallic systems greatly influenced the product distribution.

388331 Thermodynamic Modeling of Phase Behavior in Shale Media

Abstract:

Phase behavior in shale remains a mystery due to various complexities and effects. Methane in shale media may have a Langmuir type adsorption isotherm, but CO_2 adsorption may continuously increases with pressure. In conventional permeable media, once pore volume is known, the amount of fluid-in-place can be estimated accurately. This is because the fluid is locally homogeneous and pores are generally more than 100 nm, and surface adsorption will be negligible. In shale media, in addition to knowledge of pore volume, knowledge of pore size distribution, total organic content, and chemistry of the rock is also required. Currently, there is no viable model in the literature that can describe fluid phase behavior in shale rocks.

Fluid molecules in shale media can be found in three different states: 1-free molecules in the pores, 2-adsorbed molecules onto the pore surface, and 3-dissolved molecules in the organic matter. Of the three, the first two mechanisms are discussed in the literatures. In this work, we compute for the first time the amount of dissolved molecules. We present a comprehensive model for phase behavior in shale media that allows the computation of fluid-in-place and knowledge of where the fluid molecules reside.

In order to compute the fluids in shale media, we divide the pores into sizes greater than 10 nm, and sizes less than 10 nm. In pores greater than 10 nm, the interface curvature affects phase behavior and fluid phases are homogeneous. Therefore, they can be described by conventional equations of state. Our calculations show that retrograde condensation increases in nanopores; the upper dewpoint increases, and the lower dewpoint decreases. These calculations are supported by experimental measurements.

In pores less than 10 nm, the fluids become inhomogeneous and the direct use of conventional equations of state cannot be applied even with adjusted critical pressure and temperature. We suggest the use of molecular modeling, such as density functional theory and Monte Carlo simulations. The molecular modeling will allow determination of adsorption isotherm for both pressure increase and decrease (hysteresis). We will also show that the use of Langmuir adsorption based on many assumptions in derivations cannot provide correct results.

Fluid dissolution may provide additional mechanism to fluid-in-place. Our modeling shows that dissolution in kerogen can be up to 20% of total fluid-in-place.

We use available data in shale media, which are mainly limited to excess adsorption of methane and carbon dioxide to compare to our thermodynamic model computations. Fluid adsorption in nanopores of shale is determined by the pore size distributions, but the contribution from fluid dissolution in kerogen is mainly dependent on the organic content. Based on our findings, we suggest two basic measurements in shale media: 1) pore size distribution, and 2) proportion of organic matter.

374517 A Comparison of US Shale Plays with Emerging Non-Marine Shale Plays in Australia

Abstract:

The geological and petrophysical parameters are critical in evaluating the production potential of prospective shale gas plays and their economic viability for commercial development. However, based on the US experience, these characteristics can vary widely. Furthermore, the non-marine shale plays of the Cooper Basin, South Australia are distinctively different from the marine shale plays in the US. Non-marine shale may consist of higher clay content that may be less responsive to hydraulic fracturing due to the increased ductility. Conversely, the availability of sufficient amounts of quartz and siderite may counteract this effect and maintain brittleness. In this study, the mineral, total organic and gas content, thermal maturity, Poisson's ratio and Young's modulus data obtained from the Roseneath and Murteree formations of the Cooper Basin have been compared with US shale formations. The results show that the formations compare well in terms of thickness, gas content, thermal maturity, Young's modulus and Poisson's ratio, but not in clay mineral content, total organic content, permeability and formation temperature. To determine whether such plays are economically viable will require technical, as well as economic, analysis of the hydraulic fracturing process, including the potential for horizontal drilling and multi-well pads. A shale play that has unfavourable prospects may have upfront infrastructure and major capital costs that outweigh the potential receipts from gas production.

390307 Flare Minimization of Boil-Off Gas during LNG Production

Abstract:

Flare Minimization of Boil-off Gas during LNG Production

Yogesh Kurle and Qiang Xu

Dan F. Smith Department of Chemical Engineering

Lamar University, Beaumont, TX 77710, USA

Abstract

With the continuously increasing of clean energy demands, the world-wide LNG (liquefied natural gas) production capacity is being expanded very fast and LNG is actually becoming the world's fastest growing energy sector. United States Energy Information Administration (EIA) states that world natural gas trade, both by pipeline and by shipment in the form of LNG, is poised to increase in the future. New LNG terminals, which are currently under construction, will increase the LNG production by 100 million tons per year (MTPA). Over 250 MTPA of liquefaction capacity has been proposed in North America alone. In 2013, 240 million tons of LNG was delivered worldwide.

LNG industries are also facing problems of LNG evaporation occurring at different places in its supply chain. The evaporated material is called Boil-off Gas (BOG), which is generated during LNG production, storage, loading, transportation, and unloading. Normal BOG generation at exporting terminals alone ranges from 2 to 6% of LNG produced. If this is not recovered and reused, the amount of material lost would be equivalent to at least one large-scale LNG plant capacity. Also due to stringent environmental regulations, flaring of BOG is not a viable option. Since LNG production is increasing, BOG generation problem would become more severe. If this issue is not addresses properly and in time, losses of valuable materials and energy plus air pollutions would be significantly greater than ever.

In this paper, the steady-state simulation tool Aspen Plus v8.2 was used to explore various options for BOG recovery. LNG plant with 1100 m^3/hr production rate and 10,000 m^3/hr LNG loading rate was considered. At LNG plant and exporting terminal, BOG is generated due to five main factors: (1) depressurization of LNG, (2) heat leaks through containers and pipelines, (3) tank breathing, (4) heat added by equipment like pumps, and (5) LNG carrying vessels being hot before loading of LNG. BOG simulation results are validated using industrial data. The studied recovery options showed possible BOG recovery with energy usage less than 10% of the recovered energy from BOG. It is equivalent to more than 90% savings in energy when compared to complete flaring of BOG.

Since LNG loading is an intermittent and unsteady process, dynamic simulations are necessary to know the rate of change of BOG generation during loading, study the effect of BOG recycle on plant performance and controllability, and to study system behaviors with respect to changes in different parameters. In this paper, various factors were taken into consideration through dynamic simulations, viz., LNG composition and temperature, ambient temperature, jetty length, heat transfer coefficient of LNG tanks and pipelines. Further control strategy was developed to make system capable of tolerating disturbances and to improve the system performance. This study will help LNG plants choose a better option for BOG recovery based on individual plant environment and needs. Each LNG plant environment can be different in various ways like jetty length, LNG storage capacity, loading rate and frequency, available space for additional equipment. Similarly plants can have different demands of fuel gas, desired product specifications, controllability issues, which affects BOG recovery strategies.

Methane to higher hydrocarbons by a non-oxidative catalytic conversion

Shaima Nahreen, Julibeth M. Martinez de la Hoz*, Perla B. Balbuena* and Ram B. Gupta

Department of Chemical Engineering, Auburn University, Auburn, AL-36849

***Department of Chemical Engineering, Texas A&M University, College Station, TX -77843**

Abstract:

Discovery of large shale gas reserves in recent years effected in reduction of price of natural gas. For adding value to methane gas to be used as liquid fuel, many processes have been studied, among them steam reforming followed by Fischer-Tropsch synthesis is the most well-known. To convert methane in a direct and more energy efficient route, non-oxidative catalytic conversion is potentially an attractive option which includes activation of methane molecules at a lower temperature. Transition metal oxides such as Mo, Fe, V, W, Cr, Zn, and Cu have been studied for catalyzing the methane activation and conversion reaction, but the conversion is usually lower than 15% and operating temperature needed is above 800°C which causes coking and thus early catalyst deactivation. In this work, a noble transition metal, ruthenium, is chosen as the catalyst with the objective of lowering the methane activation temperature and also to have better conversion than other transition metal catalysts. The catalyst is prepared by 3 wt% ruthenium loading on ZSM-5, zeolite support. Methane is pumped in a tubular reactor through a packed bed of catalyst in a continuous process. After exchanging heat and phase separation the product is collected at room temperature and analyzed. The operating temperature is varied from 300° to 500°C. From GC analysis it is found that at reaction temperature 300°C almost no conversion of methane is taking place but at 400°C and 500°C operating temperatures, the product gas are comprised of 31% and 39% of higher hydrocarbon gases by volume respectively, but no unsaturated hydrocarbons are being formed. A catalyst bed of 3 wt% Ru/SiO_2 followed by pure ZSM-5 in series is also studied and the products are compared with that of 3 wt% Ru/ZSM-5 catalyst bed. In the series bed, the purpose of the first catalyst, that is 3 wt% Ru/SiO_2 is to activate methane and the pure ZSM-5 catalyst in series is for producing liquid aromatics from the activated methane. This series catalyst bed can produce liquid hydrocarbon directly from methane at much lower temperature than other transition metal catalysts. The origin of the lower-temperature activation effect on transition metals is examined with density functional theory methods. Preliminary results suggest that the confinement effect occurring in the zeolite environment may be lowering activation energies for methane decomposition on the transition metal sites.

388894 Measurements and Modeling of High-Pressure Gas Adsorption on Shales

Abstract:

In the coming years, recovery of natural gas from shale reservoirs is expected to increase significantly due to technological advancement in drilling and horizontal fracturing. Shale formation is categorized as unconventional gas reserve where a large amount of gas is adsorbed within the shale matrix. Therefore, knowledge of adsorption behavior is important in predicting reservoir gas content. However, availability of adsorption data on shales is very limited, especially at higher pressures, which hinders development of generalized predictions offered by theory-based adsorption models.

To enrich the gas adsorption data on shales, adsorption isotherms of methane, nitrogen and CO_2 were measured on several shale samples. The measurements were made at 328 K and pressures up to 12.4 MPa. The expected experimental uncertainties were estimated by propagating errors in the experimentally measured quantities. Further, the experimental design was re-optimized to minimize the expected experimental errors by minimizing the void space in the adsorption cell and injecting gas at pressures that reduced the uncertainty in the density.

A theory-based model such as the simplified local-density (SLD) model was applied to describe the adsorption data on diverse shales. The SLD model was found capable of representing the adsorption capacities on these shales with satisfactory precision. Further, several case studies were undertaken to investigate the predictive capability of the SLD model when applied to shales.

391329 Coalescence and Destruction of Nanoparticle Stabilized Foams after Hydraulic Fracturing Via Pressure Reduction

Abstract:

Foams, dispersions of gases in liquids, are great substitutes for conventional hydraulic fracturing fluids. Conventional methods use water, chemical additives and thickeners such as guar gum to increase viscosity of the fluids. This enables the pad to carry proppants into the fractures. The problem with hydraulic fracking processes appears during flowback of the fracking fluid. Although during injection step having high viscosity is essential to carry sand into the fracking zone, during flowback high viscosity of the pad can displace sand away from the fracture and clog the passages designed for the flow of shale gas and oil. Foams are considered as alternative smart pads. They significantly reduce water consumption during fracking jobs; exhibit high viscosity to move proppants into the fracking zone; and their physical properties are tunable either by changing macroscopic variables such as pressure or tuning microscopic chemistry at gas-liquid interface. Nanoparticle stabilized foams are generated at high pressures while their texture and stability appears to be a strong function of pressure. Moreover, addition of polymers and surfactants synergistically boosts foam generation. Gas bubbles are separated by liquid lamellas while disjoining pressure at liquid lamella is a function of interactions such as attractive van der Waals forces and repulsive electrostatic and oscillatory forces. The nanoparticles with diameters between 5-28 nm are adsorbed at gas-liquid interface to protect gas bubbles from coalescence while adsorption of the nanoparticles is essentially irreversible. During gas expansion the surface density of nanoparticles is reduced until they meet a threshold density at which the neighboring bubbles will coalesce into one and foam catastrophically collapse. Here we have developed a model to determine optimum pressure profiles to keep foams stable during fracture growth and break them during flowback.

370539 Hybrid Dynamical Modeling of Artificial-Lift Operation in Shale Gas Wells

Abstract:

Oil and gas reservoirs can be classified as conventional and unconventional depending on the ease of recovery and the quality of reservoirs. Shale gas – an unconventional resource – is natural gas found in shale rock formations, typically at depths of 10,000 ft or greater. Recent advances in horizontal drilling and hydraulic fracturing have made shale gas extraction economically viable. Hydraulic fluid consisting of water, proppants and other chemical components is pumped at high pressures to induce fractures in the reservoir and thereby increase the net permeability. About a third of the pumped fluid returns to the surface, while the remaining is produced along with natural gas during normal well operation. With time, liquids (water and condensates) accumulate at the well bottom of the well, hindering the gas flow to the well-head. This phenomenon is called liquid loading. *Artificial lift* techniques are employed at this stage to remove the accumulated fluids, to operate each individual well in an optimized manner to extract most of the useful resources.

This contribution focuses on computationally efficient modeling of a class of Artificial Lift techniques for dewatering of shale gas well. Specifically, artificial lift techniques that use well's own energy for de-liquefication of shale gas wells are considered. Examples of this include periodic well shut-in, plunger lift, gas lift, and a combination thereof. In the first two cases, the well is shut-in for a certain period of time, allowing the gas pressure to build up. Thereafter, a control valve is opened and the liquids accumulated at well-bottom are produced at the surface due to flow of the pressurized gas, either as a mist (periodic shut-in) or as a slug above a metal rod called plunger (plunger lift). In case of gas lift, some of the produced gas is pumped back into the well to augment well's energy for dewatering.

In general, artificial lift operations such as periodic shut-in and plunger lift, consist of several modes of operation based on whether the production valve is closed or open. Each of these modes follows different nonlinear dynamics. The system switches from one mode to another mode based on occurrence of various discrete events e.g., switching on/off production choke, plunger reaching at surface, etc. Thus, artificial lift model is inherently hybrid in nature.

We propose a generic hybrid dynamical model for the above-mentioned artificial lift operations for de-liquefication of shale gas well. The model consists of reservoir flow component that accounts for the flow of fluids from the reservoir to the well-bore; a component for gas dynamics; and a component for liquid loading. In particular, we identified various modes of operation and derived computationally efficient dynamic model for each of these modes. Various conditions responsible for switching system dynamics from one mode to another are also identified and integrated with the continuous dynamic of various modes.

CATALYTIC DIOXYGENATION OF CATECHOL BY NON-HEME IRON(III) COMPLEXES AS A SUSTAINABLE ROUTE TO DIMETHYL ADIPATE

Robin Jastrzebski, Emily J. van den Berg, Bert M. Weckhuysen,
Pieter C.A. Bruijnincx*

Inorganic Chemistry and Catalysis, Debye Institute for Nanomaterials Science, Utrecht University, Universiteitsweg 99, 3584 CG Utrecht, The Netherlands

The atom-efficient, direct conversion of functionalized biomass-derived platform molecules to bulk chemicals is an important challenge in the valorization of biomass.[1] Catechols constitute an interesting class of renewable platform molecules, as these functionalized aromatics may be obtained from biomass in multiple ways, including fermentation of glucose, pyrolysis of woody biomass or catalytic depolymerization of lignin. In nature, catechols are metabolized by intradiol catechol dioxygenase enzymes, to afford derivatives of *cis,cis*-muconic acid. Although bio-inspired non-heme iron(III) complexes have been investigated as models for this enzymatic reaction, the *catalytic* dioxygenation of pyrocatechol using such systems has not received much attention.[2] This reaction holds considerable promise, however, as simple hydrogenation of the obtained *cis,cis*-muconic acid would yield adipic acid, an important bulk chemical consumed on a billion kg/year scale for the production of nylons.

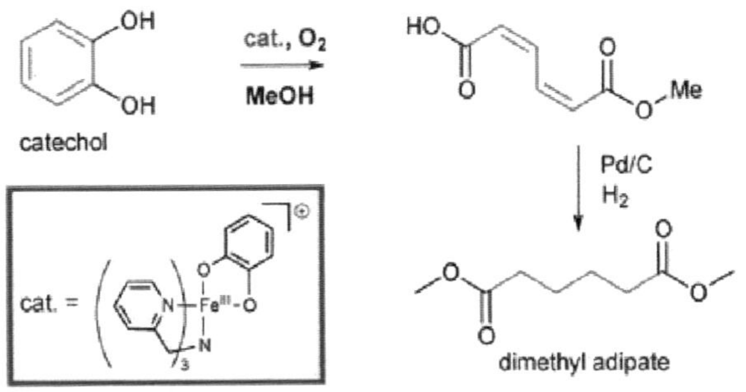

Scheme 1: The dioxygenation of pyrocatechol affords O-*methyl*-cis,cis-*muconic acid as the major product, which can be readily converted to dimethyl adipate.*

Here, we report on the first chemocatalytic, selective intradiol dioxygenation of pyrocatechol using iron(III) complexes of tris(2-pyridylmethyl)amine (tpa).[3] The active catalyst was readily prepared *in-situ* from simple iron(III) salts, tpa and a weak base. The reaction uses molecular oxygen as the oxidant and is run under moderate conditions in methanol (5 mol% Fe, 50 °C, 1 atm air). In the first step, the ring is selectively cleaved to yield a mixture of intradiol muconic acid products with *O*-methyl-*cis,cis*-muconic acid being the major product. Optimization of reaction conditions in a semi-batch autoclave, which enabled reactions at more severe conditions (up to 80 °C and 30 bar 5vol% O_2 in N_2), afforded a more than ten-fold increase in turnover numbers. The muconics acids are subsequently hydrogenated over commercial hydrogenation catalysts such as Pd/C and Raney Ni. A final acid-catalyzed trans-esterification step then fully resolved the complexity of the product mixture and gave dimethyl adipate, the preferred nylon precursor, as the single end product with high selectivity.

References:

1. Zakzeski, J.; Bruijnincx, P. C. A.; Jongerius, A. L.; Weckhuysen, B. M. *Chem. Rev.* **2010**, *110*, 3552–3599.
2. Yamahara, R.; Ogo, S.; Masuda, H.; Watanabe, Y. *J. Inorg. Biochem.* **2002**, *88*, 284–294.
3. Jastrzebski, R.; Weckhuysen, B. M.; Bruijnincx, P. C. A. *Chem. Commun.* **2013**, *49*, 6912–6914.

367347 Greening Paints, Plastics, and Fibers: Renewable Phthalic Anhydride from Biomass-Derived Sugars

Abstract:

Every year, 3 million tonnes of phthalic anhydride are manufactured worldwide for the production of paints, plasticizers, and fabrics. However, phthalic anhydride is currently produced by the catalytic oxidation of o-xylene, a chemical derived from petroleum. Here, we suggest a renewable route to phthalic anhydride from biomass-derived furan and maleic anhydride. Furan and maleic anhydride were converted to phthalic anhydride in two reaction steps: Diels–Alder cycloaddition followed by dehydration. Excellent yields for the Diels–Alder reaction between furan and maleic-anhydride were obtained at room temperature and solvent-free conditions (SFC) yielding 96% exo-4,10-dioxa-tricyclo[5.2.1.0]dec-8-ene-3,5-dione (oxanorbornene dicarboxylic anhydride) after 4 h of reaction. It is shown that this reaction is resistant to thermal runaway because of its reversibility and exothermicity. Subsequently, the dehydration of the oxanorbornene was investigated using mixed-sulfonic carboxylic anhydrides in methanesulfonic acid (MSA). An 80% selectivity to phthalic anhydride (87% selectivity to phthalic anhydride and phthalic acid) was obtained after running the reaction for 2 h at 298 K to form a stable intermediate followed by 4 h at 353 K to drive the reaction to completion. The structure of the intermediate was determined. This result is much better than the 11% selectivity obtained in neat MSA using similar reaction conditions.

Molybdenum based polyoxometalates as efficient catalysts for C-C rearrangement in the epimerization of carbohydrates

FeifeiJu[†], David VanderVelde[‡], and Eranda Nikolla*[†]

[†]Department of Chemical Engineering and Material Science, Wayne State University, Detroit, Michigan 48202, United States
[‡]Division of Chemistry and Chemical Engineering, California Institute of Technology, Pasadena, California 91125, United States

Polyoxometalates have been extensively studied for many years as catalysts for oxidation and photochemical processes[1]. While there exists a wealth of literature on the oxidative and photocatalytic properties of POMs, there are no reports on their ability to catalyze the C-C backbone rearrangement in cyclic molecules, such as in the case of the epimerization of aldose, under aqueous conditions. Herein, we demonstrate for the first time the activity and selectivity of molybdenum-based POMs in catalyzing the C-C rearrangement in the epimerization of aldoses over a broad range of reaction conditions[2]. Near-equilibrium conversions and high selectivities were obtained under aqueous conditions at relatively low temperatures. Our studies demonstrated that the molybdenum oxide octahedra on the surface of the Keggin structure POMs were the active sites for the aldose epimerization. Isotope labeling experiments along with ^{13}C NMR studies were conducted and showed that the epimerization of aldoses involves an intramolecular C1-C2 rearrangement, where the bond between C2 and C3 is cleaved and a new bond between C1 and C3 is formed leading to the formation of the epimer.

References:

1. Pope, M. T.; Muller A. Angew. Chem. Int. Ed. 30(1991), 34-48.
2. Ju, F.; VanderVelde, D.; Nikolla, E. ACS Catal. 4 (2014), 1358-1364.

SURFACE-LEVEL INVESTIGATIONS OF BENZYL ALCOHOL REACTION ON THIOLATE-COVERED SURFACES

Chih-Heng Lien and J. Will Medlin*

Department of Chemical and Biological Engineering, University of Colorado Boulder, Jennie Smoly Caruthers Biotechnology Building 598 UCB, Boulder, CO 80309-0596, USA

* Will.Medlin@colorado.edu

Abstract

Benzyl alcohol is a significant biomass-derived intermediate from lignin and a useful probe molecule for aromatic oxygenates. During reaction with hydrogen, benzyl alcohol can be converted to benzaldehyde, toluene and benzene. Recently, self-assembled monolayers-coated (SAM-coated) catalysts, which can improve the reaction selectivity through modifying the catalyst surface and blocking non-selective catalytic sites, have recently received attention[1]. Therefore, in this study, we explored the use of thiol SAMs to modify Pd/Al_2O_3 and enhance the selectivity of benzyl alcohol hydrodeoxygenation to toluene, and used surface science techniques to characterize the mechanism.

On uncoated Pd catalysts, after an initial deactivation period, the selectivity to toluene under steady state conditions was less than 35%, and the dominant product was benzene. After the SAM modification, the selectivity to toluene can reach 85% and the production of benzene is remarkably reduced, which shows that the SAMs are effective in enhancing the selectivity to toluene. Remarkably, on surfaces coated with adamantanethiol, both selectivity and overall catalytic activity were improved. Surface science techniques were employed to further investigate this chemistry, with Auger electron spectroscopy (AES) used to confirm sulfur coverage and temperature programmed desorption (TPD) to monitor the thermal chemistry of adsorbates. On octadecanethiol-coated surfaces, the benzene signal disappeared along with the hydrogen peak around 495K indicating reduced decarbonylation activity. Both of these observations indicate that the flat-lying conformation, which was the conformation found to generate benzene[2], was restricted and the production of benzene was thus decreased. In contrast, production of toluene was still observed on the thiol-coated surface, consistent with toluene formation occurring through a more upright conformation. Thus, the SAM modification strategy could be useful in making production of biomass-derived products more feasible and competitive by lower purification cost and less environmental impact.

Pore Structure and Bifunctional Catalyst Activity of Overlayers Applied by Atomic Layer Deposition on Copper Nanoparticles

Ana C. Alba-Rubio[1], Brandon J. O'Neill[1], Fengyuan Shi[2], Cem Akatay[3], Christian Canlas[4], Tao Li[5], Randall Winans[5], Jeffrey W. Elam[4], Eric A. Stach[6], Paul M. Voyles[2], and James A. Dumesic[1].

[1]Department of Chemical and Biological Engineering and [2]Materials Science Program, University of Wisconsin, Madison, Wisconsin 53706, United States.
[3]School of Materials Engineering, Purdue University, West Lafayette, Indiana 47907, United States.
[4]Energy Systems Division and [5]X-ray Sciences Division, Argonne National Laboratory, Argonne, Illinois 60439, United States.
[6]Center for Functional Materials, Brookhaven National Laboratory, Upton, New York 11973, United States.

For reasons of environmental and economic sustainability, it would be desirable to replace precious-metal catalyst (e.g. Pt) with base metals that are more earth-abundant (e.g. Cu). However, base metal catalysts are especially susceptible to irreversible deactivation through sintering or leaching, especially under liquid phase reaction conditions [1]. Recently, O'Neill et al [2] reported that atomic layer deposition (ALD) can be employed to stabilize a base-metal catalyst for liquid-phase catalytic processing. Herein, we present a model system that can be used to facilitate investigation of the ALD overcoat structure that leads to catalyst stabilization, and we extend the functionality of the ALD overcoated catalyst architecture to include bifunctional catalytic activity.

Cu nanoparticles were highly dispersed on non-porous silica spheres to facilitate characterization of the materials, followed by ALD overcoating with amorphous AlO_x followed by heat treatment to produce porosity. The very small (<1 nm) pores make the direct imaging of the porosity complicated and in this case not possible. To circumvent this complication, we designed a method to image the pores by scanning transmission electron microscopy and electron energy loss spectroscopy (STEM/EELS), which is facilitated by the spherical support. The porosity of the AlO_x overcoat was probed by deposition of NbO_x by ALD. Porosity was also confirmed by small-angle X-ray scattering (SAXS), and the catalyst performed comparably to previous work on conventional supports [2].

In addition to nanoparticle stabilization, we note that the use of ALD overcoating allows for the synthesis of novel and finely controlled catalytic structures. The ability to add bifunctionality with atomic precision is an important and promising future area in heterogeneous catalysis. In this proof of concept example, the addition of NbO_x into the pores of a silica-supported copper catalyst with an ALD-alumina overcoating leads to the formation of new acid sites capable of catalyzing the etherification of furfuryl alcohol with 1-butanol to form furfuryl butyl ether.

In summary, we have employed a model system based on the use of nonporous supports to facilitate the detailed characterization of supported metal catalysts stabilized by ALD, and we have developed a method to indirectly image the pores in the ALD overcoat. This strategy could be useful for better imaging and understanding of other phenomena in catalysis. In addition, we have demonstrated the use of ALD to synthesize novel bifunctional catalysts.

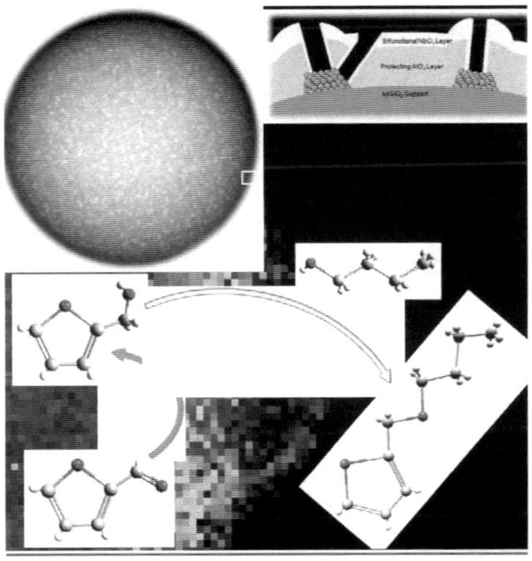

Schematic representation of the ALD layers and reactions that take place

[1] Bullock, R. M. In Catalysis without Precious Metals; Bullock, R. M., Ed.; Wiley-VCH Verlag GmbH & Co. KGaA: Richland, WA, 2010; pp I–XVIII.

[2] O'Neill, B. J.; Jackson, D. H. K.; Crisci, A. J.; Farberow, C. A.; Shi, F.; Alba-Rubio, A. C.; Lu, J.; Dietrich, P. J.; Gu, X.; Marshall, C. L.; Stair, P. C.; Elam, J. W.; Miller, J. T.; Ribeiro, F. H.; Voyles, P. M.; Greeley, J.; Mavrikakis, M.; Scott, S. L.; Kuech, T. F.; Dumesic, J. A. Angew. Chem., Int. Ed. 2013, 52, 13808–13812.

Aqueous-phase hydrogenation of levulinic acid (LA) is the most commonly proposed strategy for the production of γ-valerolactone (GVL) from biomass resources. The reaction is generally carried out over supported metals, most commonly Ru on activated carbon. Ru/C is both active and selective for GVL production; however, it exhibits poor stability in the aqueous phase such that steady state hydrogenation rates decay to a small fraction of their initial values[1].Here, we probe mechanisms leading to the rapid deactivation of Ru/C in this system and propose design criteria for the synthesis of catalytic materials that offer improved performance. Differential rates of LA hydrogenation obtained in packed bed reactors are combined with electron microscopy, X-Ray diffraction, elemental analysis, and physi- and chemisorption to identify two primary modes of catalyst deactivation in this system. First, irreversible loss of activity is attributed to the growth of Ru nanoparticles, which is observed directly by TEM. Although Ru/C appears sinter-resistant in gas phase treatments at significantly higher temperatures (e.g., 673K), metal clusters grow rapidly in aqueous media, even at near-ambient conditions (e.g., 323K). Second, Ru/C displays a reversible loss of activity during LA hydrogenation, and this is attributed to site-blocking via strongly chemisorbed intermediates. We have observed that these two modes of deactivation correlate with support properties, and both can be managed by judicious selection of host materials. For example, Ru/C and Ru/SiO_2 are relatively sinter-prone, while TiO_2 and γ-Al_2O_3 retain high Ru dispersions during extended operation. Active site blocking appears to correlate with the support's point of zero charge (PZC), and supports with lower PZCs, such as amorphous SiO_2, display little reversible loss of activity. The above insights establish a framework for designing stable supported metal catalysts for use in aqueous phase hydrogenation and hydrodeoxygenation, thus addressing a central need in biomass processing.

1 - O. A. Abdelrahman, A. Heyden and J. Q. Bond, *Analysis of Kinetics and Reaction Pathways in the Aqueous-Phase Hydrogenation of Levulinic Acid To Form γ-Valerolactone over Ru/C*, ACS Catal., 2014, 4, 1171–1181.

Chemicals from bioethanol: tailoring copper electronic properties to enhance its performance on the one-pot ethyl acetate synthesis

Isabel C. De Freitas,[1] Jean Marcel R. Gallo,[1] José Maria C. Bueno,[2] Clelia Mara P. Marques[1]

[1]Department of Chemistry, Federal University of São Carlos, São Carlos, Brazil

[2]Department of Chemical Engineering, Federal University of São Carlos, São Carlos, Brazil

The USA and Brazil are responsible for approximately 90 % of the biomass-derived ethanol produced worldwide. Therefore, these countries share the interest in the transformation of bioethanol into chemicals, such as ethyl acetate. This ester is widely used as a replacement for aromatic solvents in paints, inks and dyes; moreover, it can be used in the pharmaceutical, cosmetic, and food industry. Ethyl acetate is commercially obtained from non-renewable sources though multiple reaction processes. Recent studies have shown that ethyl acetate can be obtained from bioethanol though a one-pot reaction catalyzed by copper-based catalysts (for example, $CuCr_2O_4$ and Cu/ZrO_2 catalysts). The reaction mechanism involves the dehydrogenation of ethanol on the metal surface, followed by coupling of the newly formed aldehyde with an ethanol molecule (Step 1: $CH_3CH_2OH \rightleftharpoons CH_3CHO + H_2$; Step 2: $CH_3CHO + CH_3CH_2OH \rightleftharpoons CH_3CO_2CH_2CH_3 + H_2$). Our group has shown that Cu/ZrO_2 is a remarkable catalyst for this reaction and proposed that the reaction takes place in the interface between metal and support. It was also shown that the equilibrium Cu^0/Cu^+ plays a major role in the catalytic performance. However, the low specific surface area of zirconia does not favor the formation of a large extent of interface metal-support.

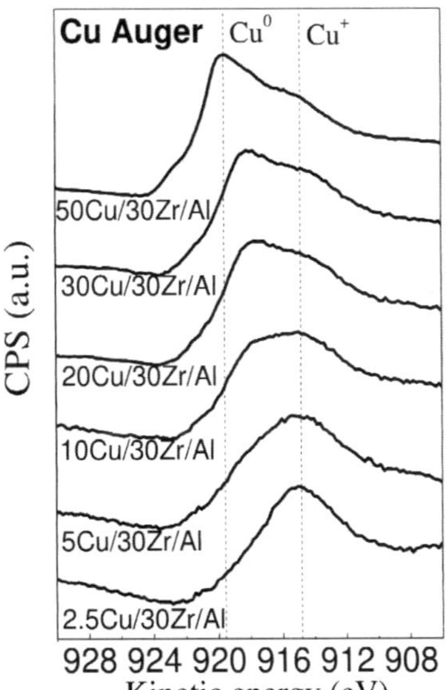

Fig 1. XPS of Cu Auger core level for reduced Cu supported on 30 wt. % Zirconia on Alumina.

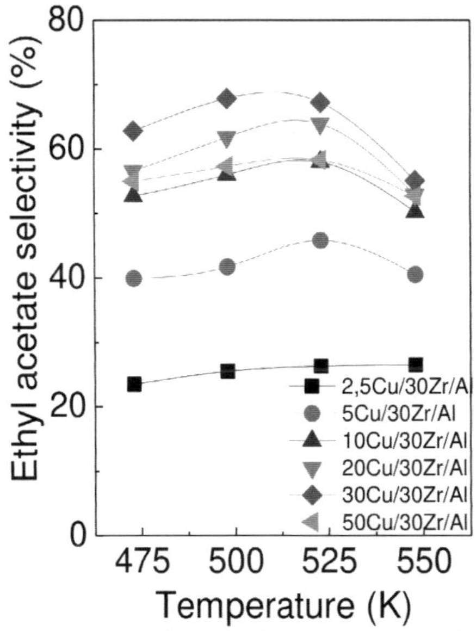

Fig 2. Evolution of the selectivity to ethyl acetate over xCu/30Zr/Al (where x is the wt. % of copper)

For catalysts syntheses, zirconia and copper were dispersed in a high surface area alumina support (370 m^2 g^{-1}) in order to increase the interface copper/zirconia. Different ratios alumina/zirconia/copper were studied aiming to find the optimal catalyst composition. The catalysts were studied by several characterization techniques, such as X-ray diffraction, nitrogen physisorption, and temperature programmed reduction. Analyses of infrared of the samples adsorbing carbon monoxide showed that copper is placed primarily on the zirconia surface, however, at high copper loadings a significant amount of copper is also observed in the alumina surface. Another important characterization technique used was X-ray photoelectron spectroscopy (XPS), which showed that increasing copper loading led to an higher Cu^0/Cu^+ ratio on the metal nanoparticle surface (as shown in Fig. 1). Comparing Fig. 1 and 2, it can be observed that the samples containing 20 and 30 % of copper (20Cu/30Zr/Al and 30Cu/30Zr/Al) were the ones with the optimum ratio of Cu^0 over Cu^+, since they displayed the highest selectivity for ethyl acetate (60-70 %). For the samples with lower copper loading (2.5Cu/30Zr/Al, 5Cu/30Zr/Al, and 10Cu/30Zr/Al), the higher Cu^+ concentration on the surface led to an increasing concentration of acetaldehyde on the metal surface and, therefore, facilitated the condensation reactions to heavy byproducts. Increasing the copper loading to 50 % (50Cu/30Zr/Al) led to the formation of bulk copper particles, which are selective to acetaldehyde formation, but lacks in the interface with zirconia, diminishing the formation of ethyl acetate.

It is important to mention, that Cu supported in alumina presents a low selectivity for ethyl acetate, and therefore zirconia is essential on providing active sites for the coupling reaction. If compared with the classical Cu/ZrO_2 catalyst, Cu/30Zr/Al (both at the same copper loading) has similar activity to ethanol conversion, however it shows significant improvement in the selectivity and reaction rate for ethyl acetate. This behavior can be attributed to the higher copper/zirconia interface on the Cu/30Zr/Al catalysts, once both zirconia and copper are highly dispersed in the alumina support.

Authors thank CNPq and FAPESP. JMR Gallo thanks FAPESP for the fellowship (process number 2013/15861-1)

Key parameters effecting the desalting of unconventional crude oil - Desalting Operation

Raghu ram. H[1], Utkarsh maheshwari[2]

[1]Department of chemical engineering, BITS, Pilani
[2]Department of chemical engineering, BITS, Pilani

[1]Email: raghuram6845@gmail.com

Abstract: Crude oils are traditionally characterized by API and sulphur. All refineries know that there is a difference between Arab Heavy having an API of 28 and a sulphur content of 3% and Arab Light with 33 and 2%. Refineries are designed to handle light or heavy crude oil, some are built to handle high sulphur and some can only cope with low sulphur. And there is also the acidity of the crude which needs to be considered. And that is not the end of the list of processing concerns when selecting opportunity crudes as refinery feedstock. Handling of heavy crudes for desalting process varies with conventional crudes. In this paper the important desalting process parameters of conventional and unconventional crude oils affecting the performance of the desalter operation is studied. This paper gives an overview of the extra challenges from heavy crude oils caused by being different in terms of viscosity, density, sulphur, water content, salt etc.

Keywords: desalting, conventional and unconventional crude oils, API, P_H.

1. Introduction: Oil produced in most of wells contains water and dissolved salts; mainly NaCl, $CaCl_2$ and $MgCl_2$, which can cause several problems in the refining processes. Salts and water are removed from crude oil in day-to-day operation for four reasons: corrosion, scale accumulation, lowering of activity of catalysts and plugging or fouling in pipelines. In a desalting process, dilution water mixes with the formation water. Then, water removes from crude oil. Settling, chemical, heat, mechanical, electrical coalescences used for crude oil desalting [1-2]. With the increasing regulations on effluent water and the ever increasing cost of producing a barrel of oil, the use of emulsion-treatment plants has become an important practice in crude oil processing. Treatment of emulsions has always ranged from simple methods such as gravity settlement to highly sophisticated methods such as tri-volted desalting and dehydration systems. The development of desalting systems has always been evaluated in terms of quantities of salt and water being removed. To reduce the amount of desalting required at the refinery, some oil purchasing contracts specify a maximum salt content as well as a maximum water content. A typical salt specification would be 10 Pounds per Thousand Barrels (10.0 PTB). To satisfy such purchase specifications, producers may be required to perform some oil desalting [3]. As the new unconventional crudes are discovered the refiners are accepting these challenging feed stocks as part of their crude slates to improve margins. Such crude oils may have a lower cost, but require more processing to yield end products. A significant production of heavy and extra-heavy oils in countries like Venezuela, a growing output of the nonconventional oils in Canada, the emerging production in countries like Colombia as well as the new developments to tap more into the 10 – 20 API type of oil reservoirs in many countries (Brazil, Angola, North Sea,...) have made that the so-called opportunity crudes have found their place on the oil market. Even so, more new refineries have been designed to handle at least a certain part - or even the whole part of their daily crude feed in the form of these "challenging or opportunity" crudes. But, getting these heavy crudes to the refinery site is not a simple operation and challenges are encountered at every part of the crude logistic chain, starting at the well, at the transport from the field by truck, pipeline, shuttle vessel or crude ship to the refinery storage tanks. As every field and its crude is unique in its own properties, the opportunities to use a specific crude will be different for each individual refinery. Therefore, each refiner will assess the possibilities to use heavy crudes in their system on an individual basis [4]. In this paper the key parameters that are effecting the desalting operation and the challenges involved while refining these unconventional crudes are discussed.

2. Literature survey:

Schramm et al discussed the basic principles of petroleum emulsions. The author noted that the factors that affect emulsion stability are: viscosity, density differential, water percentage, age of emulsion, control of emulsifying agents, and the agitation control. The author also noted that the success of chemical treatment methods is dependent upon the amount of the chemical, heating, and sufficient residence time for settling. In addition, Schramm described sampling and testing techniques that assist in characterizing a process stream's composition and thus in evaluating the effectiveness of a particular separation process [5]. Bartley described how heavy crudes present unique problems that require additional design considerations [6]. Chawla discussed the problem of wet crude in Kuwait and the need for desalting plants [7]. Al-Kandari reported that there are several problems generated from producing wet crude oil streams: corrosion, lowering the activities of catalysts, and scale development [8]. Anon noted that a static mixer could improve desalting efficiency. Burris designed a technique of using fresh dilution water to remove salt from crude oil. Baranov et al. recommended hydro cyclone vessels for use in preliminary or complete separation of inhomogeneous systems of the liquid-liquid type, such as water-oil emulsions. Taylor focused on the electrically enhanced phase separation of water-in-oil emulsions. He used AC, DC, pulsed DC, and combinations of these in the separation of water-oil emulsions [9-12].

There are many recent studies that seek to develop a comprehensive general framework for removing water from crude oil that requires additional processing beyond gravitational separation. For the best desalting operating conditions, Axsia suggested that there are a number of important design features to be considered [20]. Recent advances in the development of efficient desalting/dehydration processes have made it possible to reach an optimum set of factors. One method is the one-stage system where the desalting process is performed in disc-stack centrifuges that replace high-volume gravity desalter [21]. The benefits gained are to be found in the process simplification of turning the two-stage process into a single-stage process, resulting in less freshwater consumption. On the other hand, Agar developed a system of a minimum of two or a maximum of four energy absorption instruments: three located in the desalting vessel designated for service and one installed in the feed line.

3. Desalter theory

The Desalter removes contaminants from crude oil by first emulsifying the crude oil with chemicals and wash water to promote thorough contact of the water and oil. The salts containing some of the metals that can poison catalysts are dissolved in the water phase. After the oil has been washed and mixed as an emulsion of oil and water, demulsifying chemicals are then added and electrostatic fields are used to break the emulsion. Desalters are sized to allow the water and oil to settle and separate according to Stokes Law.

In addition, solids present in the crude will accumulate in the bottom of the desalter vessel. The desalter must be periodically washed to remove the accumulated solids [13].

A mud washing system is installed in the bottom of the vessel to periodically remove the solids. Mud washing consists of recycling a portion of the desalter effluent water to agitate the accumulated solids so that they are washed out into the effluent water. These solids are then routed to the wastewater system.

Desalter effluent is a combination or many things including the mud wash done at periodic intervals, produced water that came with the crude, and the brine wash water resulting from the dilution and removal of salts and other contaminants.

4. Key parameters in desalting heavy/ unconventional crudes

In general, the production of heavy and unconventional crudes are requiring an higher energy input compared to conventional production, this already due to the higher crude viscosity but also influenced by other reservoir properties, like temperature etc. Some of this extra energy is however off-set by the lack of gas that needs flaring off at the oil field, like it is often the case with lighter crude oils. Most crude units designed for light or even moderately heavy crudes require additional desalter volume to satisfactorily desalt heavy unconventional crude. Some refiners have tried to defer investment in additional desalting by operating the first- and second-stage desalters in parallel. The trade-off from converting a two-stage desalter to a single stage has been a high desalted crude salt content (3-6 ptb), high corrosion rates and poor reliability [26]. Controlling corrosion with single-stage desalting and caustic addition has proven difficult for heavy crude processing.

Due to the difficulties associated with desalting unconventional crude, it is equally important to pay special attention to other critical desalter parameters. The overall performance will depend on desalter size, as well as attention to these other variables:

- Operating temperature
- Amount and quality of water
- Water and oil mixing
- Mud washing to remove solids
- Brine cooling heat exchanger design
- Chemical treatment
- Desalter design

Crude preheat trains must have the flexibility to meet the desalter temperature required for optimum performance. The preheat train must have the flexibility to vary desalter temperature irrespective of crude blend changes. This means the ability to shift heat to and from the raw crude exchangers based on the desalter requirements.

4.1 Operating temperature

Optimum desalter temperature depends on the specific crude or crude blend, and it should be an Operating variable, not a consequence of the exchanger network design. Many existing crude units were designed for light or moderately heavy crude oils. It is not unusual to have desalter temperatures in the 220-240°F (104-115°C) range when these units process heavy crude because the exchanger network has poor flexibility or heat exchanger fouling is high. When the exchangers are clean or the crude is lighter, the desalter temperature may exceed the maximum temperature for the grid insulator bushings, causing shorting and damage of some or all of the grid system.

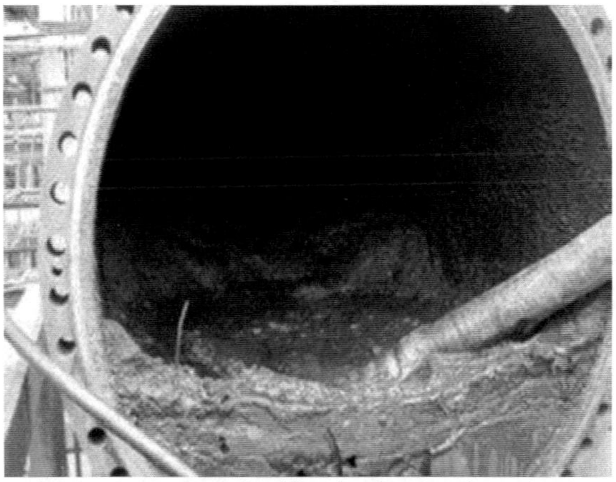

Figure 1: Severe fouling in the crude column top pump- around.

Optimum desalter temperature may be 280-290°F (138-143°C) for some heavy crude. However, with some heavy Canadian crudes, the optimum temperature may be as low as 240-260°F (115-127°C) to avoid excessive asphaltene precipitation and increases in conductivity that occur at higher temperatures [24]. When asphaltenes precipitate in the desalter they collect at the oil/ water interface and stabilise the emulsion. In some cases, rag layer removal headers can be used to remove hard-to-break emulsions from the desalter.

4.2 Amount and quality of water

Desalter water rate and make-up water quality must be maintained to properly desalt. With heavy crudes, the total desalter water rate should typically be 7-10 vol% of the crude, and some extreme cases may need as much as 14 vol%. In order to reduce the amount of effluent water, total wash water may consist of a portion of recycle with the make-up water when good-quality water is available. For heavy crudes, a higher water rate is needed to increase the number of water droplets, resulting in better oil and water contact, making the droplets closer together and therefore improving coalescing inside the electric field. This will also allow for larger droplets, thus reducing settling time.

Desalter water should be of good quality, which generally means low hardness, suspended solids and ammonia (pH). Ideally, the desalter makeup Water's pH should be maintained between 5 and 7 [25]. Since heavy crudes can have significant amounts of amines and stripped sour water is often used as make-up, water pH can be high, making it necessary to inject acid into the make-up water for pH control. High pH water, in conjunction with high naphthenic acid, results in the formation of soaps, harder-to-break emulsions and poor desalting.

4.3 Water and oil mixing

Proper oil and water mixing is essential. The Mix valve must create enough shears to produce a small enough droplet size to allow the water to contact the oil, permitting contaminants to be dissolved in the water. The objective is to try and make wash water droplets the same or similar size as the brine droplets, so that when coalescence occurs the brine will be removed with the make-up water. If the mix valve pressure drop is too low, the oil and water will not mix properly. Hence, salt, filterable solids and amines removal will be poor. Optimum mix valve pressure drop will vary, and it must be determined through adjustment and desalter performance monitoring [27].

4.4 Mud washing to remove solids

Filterable solid content varies significantly, depending on the specific crude source. It is not uncommon to have filterable solids above 100 ptb and sometimes higher than 300 ptb. These solids tend to stabilise the oil/water emulsion, leading to a large rag layer. The desalter size is often a function of the emulsion layer resolution rather than the droplet coalescing. The filterable solids that accumulate in the bottom of the desalter must be removed through intermittent or continuous mud washing. If the solids are not removed, they reduce residence time, leading to poor brine effluent quality. Mud washing is recommended in both first- and second-stage desalters.

Figure 2: Two stage desalter system

Make-up water must be heated to the desalter temperature and brine cooled sufficiently to allow further effluent treating. Since the amount of solids removed during mud washing fouls conventional heat exchangers, it is not unusual for the make-up water rate to be limited by the amount of brine that can be pressured from the system.

4.5 Brine cooling heat exchanger design

When the brine/make-up water and brine/CW exchangers foul, they can limit water make-up. Fouling-resistant spiral heat exchangers are needed in the brine service. Some crude units have too many exchanger services and excess surface area in front of the desalter, and others have too little. Unfortunately, CDU/VDU exchanger network designs are increasingly being done by pinch theorists with little understanding of critical operating parameters or challenges associated with heavy Canadian crude processing. The resulting designs do not have the robustness required to deliver the optimum desalter temperature required for specific crude blends.

4.6 Chemical treatment

Good chemical treatment is essential and the types of chemicals used are important. New chemicals are being developed to deal with the challenges of desalting heavy Canadian crude oils such as oil - soluble emulsion breakers and wetting agents. A thorough review of the type of chemicals used is required to process these crudes effectively.

4.7 Optimum desalter design

Each desalter should be specifically designed for the crude type or blend, based on operating conditions and process specifications. The most important factor in designing for heavy Canadian crude is controlling the interface emulsion. These crudes tend to be "emulsion sensitive", making resolution of the interface very difficult. Stokes Law does not factor this component into sizing calculations, and it can only be derived empirically. Most desalters are sized based on gravity and viscosity, typically making the desalter too small to resolve the interface emulsion at a rate as fast as it is being created. This leads to increased upsets, oily effluent water, increased chemical consumption and frequent use of interface draw-off, creating additional slop and associated operating costs to treat this material.

Along with vessel sizing, distribution of the oil can also be improved in the environment for resolving the interface emulsion. Having the proper distribution and the ability to control the residence time of the interface can have a huge impact on overall performance.

Due to the conductivity of these crude types, voltage gradient must also be optimum for steady-state operation [28]. Lower voltages tend to provide more flexibility than conventional desalter secondary outputs. A range of voltage outputs is selected so that optimisation for any particular crude can be realised.

5. Challenges facing while handling unconventional crudes

The most challenging feed stocks, from the standpoint of desalting, are the heavier crudes that contain high levels of Filterable solids, unstable asphaltenes or difficult-to remove chloride salts. Since the filterable solids, asphaltenes and salts are concentrated in the bottoms fractions, the more difficult feed stocks tend to be: Heavy, conventionally produced crude oils Oil sands bitumen's diluted with lighter hydrocarbons or synthetic crudes to meet pipeline gravity and viscosity specifications.

Most heavy feed stocks exported from this region range from 19°API to 22°API, which tends to create desalter challenges due to the resulting higher crude blend viscosities [28]. However, there are also lower density crudes coming from this region that can be problematic, owing to their high filterable solids contents. The large and growing number of different unconventional feed stocks produced, and the variability in quality observed, requires increased vigilance by refiners and service companies seeking to understand and to overcome the impact of these feed stocks on desalter operations.

5.1 Inorganic solids in feed stocks

As with any crude oil, the solids present in feed stocks originate in the formation from which they are produced, plus any sediment or corrosion scale picked up during transportation. These solids can stabilize desalter emulsions. Samples of desalter emulsions collected during upset conditions have been found to contain high concentrations of the same solids materials that are found in raw crude samples. Crude blend filterable solids levels significantly above 100 PTB (using a 0.45μ filter pad) can cause high desalter electrode current draw, high desalted crude oil basic sediment and water (BS&W) content, emulsion layer build-up and oily desalter effluent water [29].

5.2 Asphaltenes increase emulsion stability

Several refineries processing high percentages of unconventional feed stocks have experienced desalter problems related to asphaltenes' instability in the crude oil blends. Asphaltenes are known to stabilize water-in oil emulsions, perhaps due to their concentration from the oil phase to the oil/water interface. BS&W carry over into desalted crude oil and the appearance of asphaltene particulates in the desalter effluent water. Asphaltene instability can also promote sludge accumulation in crude storage tanks,

and can cause accelerated crude preheat exchanger fouling and pre-flash or atmospheric column foaming.
Several methods have been used to assess crude blend incompatibility, including SARA (saturates, aromatics, resins and asphaltenes) testing, asphaltene to aromatic ratio and various precipitation tests. Proprietary screening tests also have been developed to better predict the asphaltene instability of hydrocarbon feed stocks. Test results can be used to predict problem crude blends, as well as to set up crude blending schemes and to select proper treatment programs to control problems associated with incompatible blends.

6. Conclusion:

Heavy crude oils will continue to be processed, both today and in the future. Providing a proactive approach to managing desalter impacts affords a refinery with reliable and predictable desalting performance. Desalting heavy crude is proven to be very challenging. Effective desalting requires large desalters and attention to all variables associated with good desalter performance. One tool or solution does not fit all. Specially tailored desalter chemicals may be needed to further improve salt removal and/or effluent water quality. In addition to the problems they create in the desalter, tramp amines and H_2S scavengers contained in heavy un-conventional crude pose additional concerns about reliability in the top of the crude column and its overhead system. It is becoming clear that these sections of the unit must be robustly designed to minimise the laydown of high salt point amine chlorides in order to achieve reliability targets. The process requires diligence, innovation and chemistries for success.

References:

[1] Gholam Reza Check a, , Dariush Mowla, (2013),Theoretical and experimental investigation of desalting and dehydration of crude oil by assistance of ultrasonic irradiation, Ultrasonics Sonochemistry 20, 378–385.

[2] Neda N, Salem B, (2011), Effect of Mixing Efficiency in Dilution Water Consumption in a Crude Oil Desalting Plant, IPCBEE, vol.20.

[3] M al-otaibi, A elkamel, T al-sahhaf, A S ahmed, (2003) Experimental investigation of crude oil desalting and dehydration Chem. Eng. Comm, 190: 65_82.

[4] Bieber.S, Kremer.L, (2008) Rethink desalting strategies when handling heavy feed stocks www.hydrocarbon processing .com /Article/ 2596989, September.

[5] Schramm, L.L. ed., (1992) Emulsion fundamentals and Applications in the Petroleum Industry, American Chemical Society, Washington, 1-45, 51-74, 295-312, 313-338, 341-382, D.C.

[6] Bartley.D, (1982) Heavy crude stocks pose desalting problems, Oil Gas J. 80(5), 117_124.

[7] Chawla, M. L., (1987) Field desalting of wet crude in Kuwait, Paper 15711 presented at SPE Middle East Show, Manama, Bahrain, Oct. 7-10.

[8] Al-Kandari J, (1997) Desalting crude oil,Engineers, 55,2225.

[9] Al-Otaibi M, (1999) Experimental Investigation of Kuwaiti Crude Oil Desalting-Dehydration Process, M.S. thesis, Department of Chemical Engineering, Kuwait University.

[10] Anon, A,(1983) Static mixer improves desalting efficiency, Oil Gas J.81(42), 128-129.

[11] Axsia.,(2000) Oil, gas and water Process treatment, www.offshore-technology.com / contractors / separation /axsia/ Axsia.

[12] Baranov, D. A., Kutepov, A. M., and Ternovskii, I. G,(1986) Separation of lube oil emulsions in hydrocyclone vessels, Khimiya i Tekhnologiya Topliv i Masel, Moscow Institute of Chemical Machinary Construction (MIKHM), No. 3, 16-18.

[13] M.A.Fahim etal.,(2010), Fundamentals of petroleum refining, Elsevier publications, first edition,.

[14] Elgibaly A., Nashawi I., Tantawy M., and Elkamel A, (1999) Flow properties of

weathered crude oils and their emulsions,J. Dispersion Sci. Technol, 20(3), 857-882.

[15] Kuwait Oil Company, Production Personnel Development Division, Production and Export Operations Group.(1987) Dehydration-Desalting Principles: A Training Guide.

[16] Lohne, K.,(1994) Separation of solids from produced water using hydro cyclone technology, Trans. IChemE, 72, A (March), 526-539.

[17] Sun, R. and Shook, C. A,(1996) Inversion of heavy crude oil-in-brine emulsions, J. Pet. Sci. Eng.14, 169-182.

[18] TPL Company, Desalters Phase IV (GCs 1, 3, 6, 8 and 21), (1992) Project Report for Kuwait Oil Company, Contract 92G026.

[19] Taylor, S. E, (1996) Theory and practice of electrically enhanced phase separation of water-in-oil emulsions, Trans. IChemE, 74, A- July, 526-539.

[20] Agar, Agar's solution to desalting systems. (2000)http:/ www.agarcorp.com/AppNote3.htm System 3.

[21] Alfa Laval, Crude oil desalting,(2000) http:/www.Alfalaval.com/separation customers/ sebhc1r.htm.

[22] Gary R. Brierley, Visnja A. Gembicki and Tim M., (2006) Changing Refinery Configuration for Heavy and Synthetic Crude Processing," UOP LLC, Des Plaines, IL, NPRA Annual Meeting, Paper, AM-06-16.

[23] Joseph L. Stark etal., (2002)Crude Stability as Related to Desalter Upsets, Baker Petrolite, AIChE Spring National Meeting, New Orleans, LA, March.

[24] White. P, Planning Ahead for Effective Canadian Crude Processing, (2010) www.bakerhughes.com /opps crudes.

[25] Hart Energy Consulting, (2010)'Heavy Crude Oil: A Global Analysis and Outlook to 2030', November.

[26] Tom Collins, Tony Barletta, (2012) Desalting heavy Canadian crudes, www.digitalrefining .com /article / 1000566,.

[27] Burris, D. R,(1978) How to design an efficient desalting system, World Oil, 186(7), 150-156,.

[28] Kremer, L. N., (2000)"Challenges to Desalting Heavy Crude Oil," International Conference on Refinery Processing, 2000 AIChE Spring National Meeting, March 5–9, Atlanta.

[29] VAUTRAIN, J.; (2011) Challenged Crude Issues., challenged and Opportunity Crude Conference, Singapore.

[30] Bridjanian, h., Khadem samini, a., (2011) Bottom of the barrel- an important challenge of the petroleum refining industry, Petroleum & Coal 53 (1) 13-21.

386273 Ionic Liquids As Viscosity Modifiers for Heavy and Extra-Heavy Crude Oils

Abstract:

In this paper, the potential of using various types of ionic liquids to reduce the viscosity of heavy oils and extra-heavy oils has been explored. Various properties of the ionic liquids such as type of head group, length of tail group, and the charge density of counter-ions have been studied. The results indicate that the addition of ionic liquids, in very small quantities, reduces the viscosity of heavy oils. The results are attributed to the interaction between the ionic liquids and the asphaltene molecules. These results set the stage for further research on the viscosity reduction of heavy oils by the use of ionic liquids.

Molecular Dynamics Simulations of Nanoparticle-Assisted Bitumen Recovery from Oil Sands

Oleg A. Mazyar and Valery N. Khabashesku

Center for Technology Innovation, Baker Hughes Incorporated, Houston, TX, 77040

Introduction

Crude oil is an important source of energy and one of the most important chemical feed stocks. World oil consumption parallels global social and economic development and continues to increase. To meet the demand for crude oil, resources can be mobilized through discovery of new oil deposits, improved recovery of oil in existing reservoirs, and exploitation of unconventional hydrocarbon resources. Unconventional resources include, among others, oil sands containing immobile natural bitumen with an API degree of less than 10 and an *in-situ* viscosity greater than 10,000 cP.[1]

Oil sands deposited at a depth of less than 100 m can be extracted using mining technology. The material is crushed, mixed with hot water and caustic soda, and then transported by pipeline to the processing plant. At the plant, air is added to the slurry and bitumen is recovered in the flotation process as froth at the top of the separation vessel.[2] Bitumen liberation from sand grains occurs during vigorous mechanical mixing of the oil sand slurry in the hydro-transport lines and the flotation vessels.

The effectiveness of bitumen liberation is critical for the overall success of bitumen recovery process. Consequently, many studies have focused on optimization of bitumen detachment from oil sand particles via fine-tuning of slurry temperature, shear rate, chemical additives, pH, and electrolyte content in the slurry.[3,4] Optimizing the bitumen separation becomes even more important as the availability of high-quality oil sands becomes increasingly limited, due to progressively poorer oil recoveries from decreasing grades of oil sands.[4]

This study focused on facilitating the detachment of bitumen from oil sand grains by admixing a colloidal suspension of nanoparticles to stirred oil sand slurries. Suspensions of nanometer-sized particles or supermolecules, such as surfactant micelles, have been previously suggested for enhanced oil recovery and oily soil removal.[5] It was found that solid-like ordering of nanoparticles inside the confined geometry of the wedge film formed between an oil drop and a hydrophilic solid surface immersed in an aqueous nanofluid results in disjoining pressure separating the oil drop from the solid surface.[5] Clearly, such colloidal ordering can progress only at conditions close to stationary. At stirred conditions, the mechanism of bitumen liberation is likely associated with rupture of the bitumen continuum surrounding the sand grain by the abrasive nanoparticles and bitumen recession on the sand surface driven by the moving nanoparticles. This study used molecular dynamics (MD) simulations to investigate the process of bitumen liberation from hydrophilic silica surfaces immersed in a stirred aqueous nanoparticle suspension at a molecular level. The effects of temperature and shear rate on bitumen detachment are considered.

Simulation Details

The model used to study nanoparticle-assisted bitumen detachment is shown in Figure 1. It consists of a 5 nm-diameter spherical OH-terminated silica

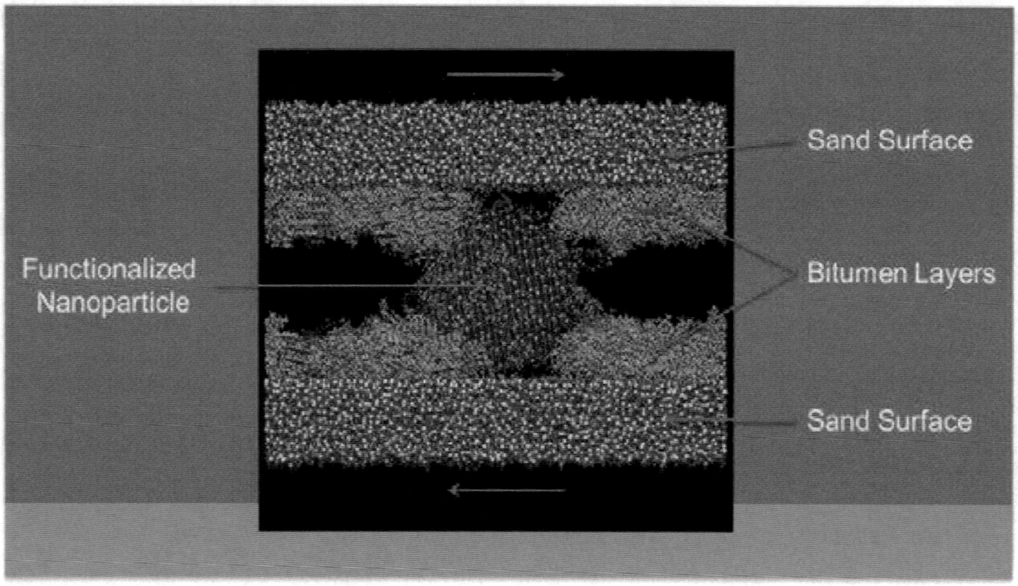

Figure 1. *Model used for nanoparticle-assisted bitumen liberation studies. Water molecules are not shown.*

nanoparticle that is functionalized with 35 hydrophilic (-O-CH_2)$_7$-CH_2-NH_2 groups and two OH-terminated silica surfaces sparsely coated with model bitumen layers, each composed of 243 asphaltene molecules shown in Figure 2. Each surface has

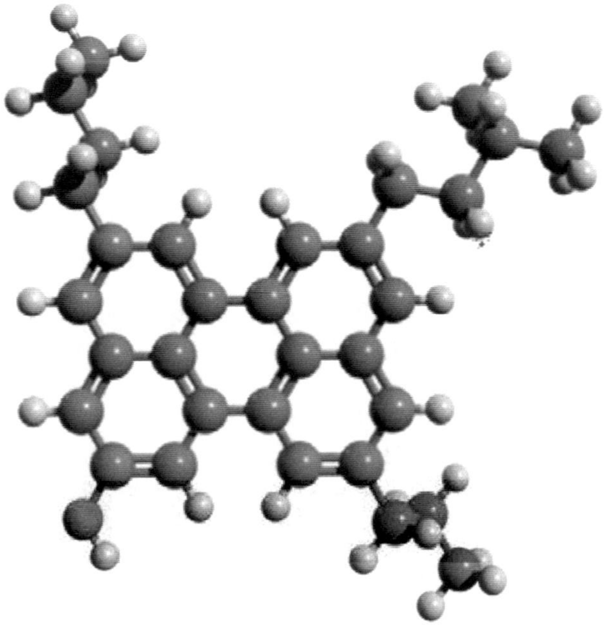

Figure 2. *Asphaltene molecule used for modeling of bitumen layers. Carbon atoms are dark grey, hydrogen atoms are light grey, and oxygen atom is red.*

an amorphous structure, a height of 25.4 Å, and an area of 150.5 × 146.3 Å^2 at 25 °C. The amorphous silica surfaces were obtained from thermal equilibration of an ideal β-cristobalite[6,7] with a lattice constant of 5.07 Å at 300 K and are composed of 13 Si atom and 13 O atom layers arranged in a hexagonal lattice of SiO_4 tetrahedra, alternatively pointing up and down. The lower model surface is obtained by cutting the β-cristobalite crystal above the layer of O atoms at the top of the upward SiO_4 tetrahedra. Each O atom of this layer is bonded to one H atom. The upper silica surface represents a mirror image of the lower surface, with the OH-terminated interface oriented toward the lower surface and the nanoparticle. The nanoparticle was placed in the grooves formed by the model bitumen layers to maintain contact with both silica surfaces and the bitumen layers. Finally, 29224 water molecules were added to fill the voids between the two silica surfaces. The system, which has a height of 113.6 Å, was placed in a rectangular box of 150.5 × 146.3 × 165.0 $Å^3$ to exclude the long-distance interactions between the upper and the lower surfaces under the 3D-periodic boundary conditions.

In this study, the CLAYFF force field[8] suitable for the simulation of mineral systems and their interfaces with aqueous solutions is used to describe the interactions within the OH-terminated silica surfaces and the core of the nanoparticle. The OLPS all-atom force field[9] is used to describe the bitumen layers and the $(-O-CH_2)_7-CH_2-NH_2$ functional groups. Water is simulated using the SPC/E water model.[10] The long-range intramolecular interactions and interactions between the nanoparticle, molecules, and the surfaces are represented according to the OPLS all atom force field rules.[9]

MD simulations were performed using the LAMMPS software package.[11] Application of the SHAKE algorithm[12] to constrain valence bonds and angles involving H atoms enabled the integration of the equations of motion using the velocity Verlet method with a larger 0.5-fs time step. A cutoff radius of 10.0 Å was used to compute short-range Coulombic and Lennard-Jones interactions. Long-range electrostatic interactions were computed using a particle-particle–particle-mesh solver.[13] 3D periodic boundary conditions were employed during the simulations. All atoms were coupled to a 1-atm Nosé-Hoover barostat and thermostat[14, 15] during an initial 2-ns period to obtain the systems equilibrated at 25 and 80 °C. Subsequently, seven outermost atomic layers of the upper and the lower silica surfaces were held rigid and moved in opposite directions at a constant velocity of 2.5, 5.0, and 7.5 m/s to simulate stirred conditions at different shear rates at 25 and 80 °C. The production runs were terminated when each surface moved ~ 20 nm.

Results and Discussion

It is important to optimize the temperature of bitumen recovery process and the shear rate of oil sand slurry in the separation vessel, because these factors strongly affect the energy consumption of the oil-recovery process. MD simulations can provide molecular-level insight into the dynamics of nanoparticle-assisted bitumen liberation. This insight is sometimes not easy to obtain from experiments, and is not available to guide engineers in designing more-efficient bitumen extraction processes.

To understand the dynamics of nanoparticle-assisted bitumen liberation, the energies of interactions between the bitumen layers and the silica surfaces were monitored as a function of distance traversed by the silica surfaces at different shear rates, modeled by varying the sliding velocities of the surfaces and temperatures.

As the silica surfaces move in opposite directions (see Figure 1), the nanoparticle exhibits complex stick-slip behavior. However, the center of mass of the particle does not displace significantly from its original position during the simulation run, while each of the silica surfaces moves ~20 nm. The nanoparticle obstructs the motion of bitumen layers adhered to the silica surfaces and disturbs the arrangement of molecules in them. At higher shear rates, when the relative velocity of the particle with respect to the bitumen layers is high, the moving nanoparticle can supply a significant amount of kinetic energy to the impacted bitumen molecules, overcoming the interactions with the adjacent bitumen molecules. In this case, the bitumen layers disintegrate. A small fraction of bitumen molecules start moving along with the nanoparticle, while the rest of the bitumen layers remain adhered to the silica surfaces for an extended period of time. At lower shear rates, the kinetic energy supplied to the impacted molecules can be redistributed between the adjacent bitumen molecules. In this case, the integrity of bitumen layers remains intact. The bitumen layers, obstructed by the nanoparticle, effectively separate from the flat silica surfaces and continue moving along with the nanoparticle. As shown in Figure 3, at the end of the simulations conducted at 25 °C, when the surfaces in both systems traversed ~20 nm, the separation of bitumen from the silica surfaces is not as complete at a sliding velocity of 7.5 m/s as at 2.5 m/s.

Increasing the temperature from 25 to 80 °C improved the efficiency of bitumen detachment. As can be seen in Figure 4, the separation of bitumen layers from the silica surfaces is almost as complete at a sliding velocity of 7.5 m/s as at 2.5 m/s. The observed different behavior at a sliding velocity of 7.5 m/s is associated

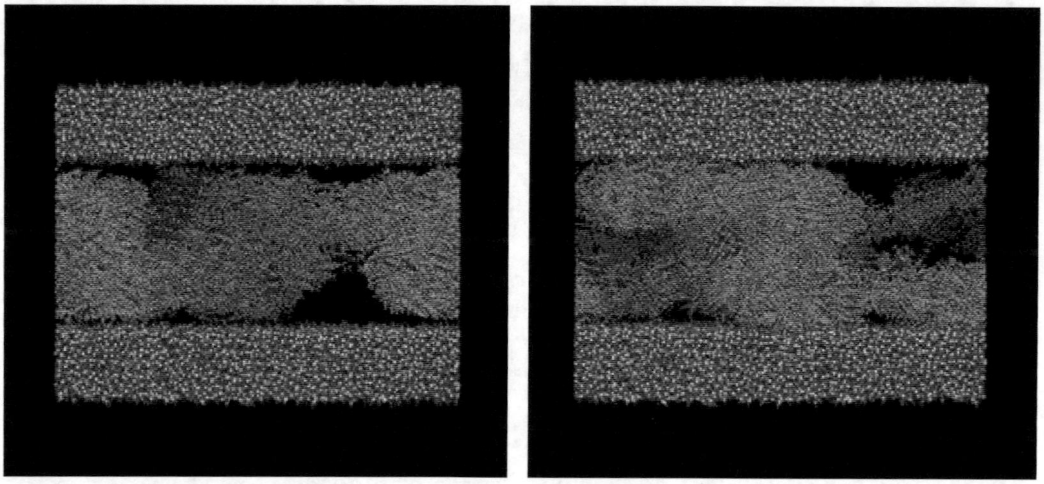

Figure 3. *Illustrations of the system shown in Figure 1 at the end of simulations at 25 °C and a sliding velocity of 2.5 m/s (left image) and 7.5 m/s (right image).*

with the difference in mechanical properties of the bitumen layers at 25 and 80 °C, as the model bitumen layers were found to be brittle at 25 °C, but appear to be viscoelastic at 80 °C.

The role of water flow is complex, and it is not fully understood. At the beginning of the simulations, when the bitumen layers are largely undisturbed, the

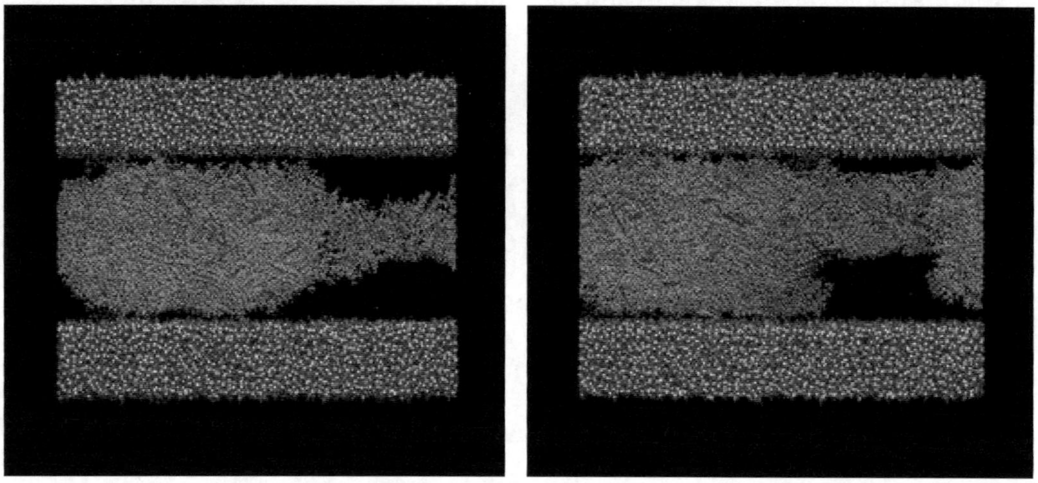

Figure 4. *Illustrations of the system shown in Figure 1 at the end of simulations at 80 °C and sliding velocity of 2.5 m/s (left image) and 7.5 m/s (right image).*

model bitumen molecules maintain stacked structures where the polyaromatic cores are oriented mostly parallel to the silica surfaces. As the advancing nanoparticle forces some bitumen molecules to deviate from the orientation parallel to the surface, the water flow may either separate the molecule from the bitumen layer or return the molecule to its original position in the layer, depending on the tilt angle of the molecule. By the end of simulations, the more-intensive flow promotes mixing of the upper and the lower bitumen layers, while the less-intensive flow allows for the formation of compact droplet-like bitumen aggregates.

Interaction energy between the silica surfaces and the bitumen layers is a convenient measure of bitumen liberation because the energy is low when bitumen adheres to the surfaces and increases as bitumen separates. These energies of interactions are shown in Figure 5 as a function of sliding distance traversed by the silica surfaces at sliding velocities of 2.5, 5.0, and 7.5 m/s at 25 and 80 °C. As can be

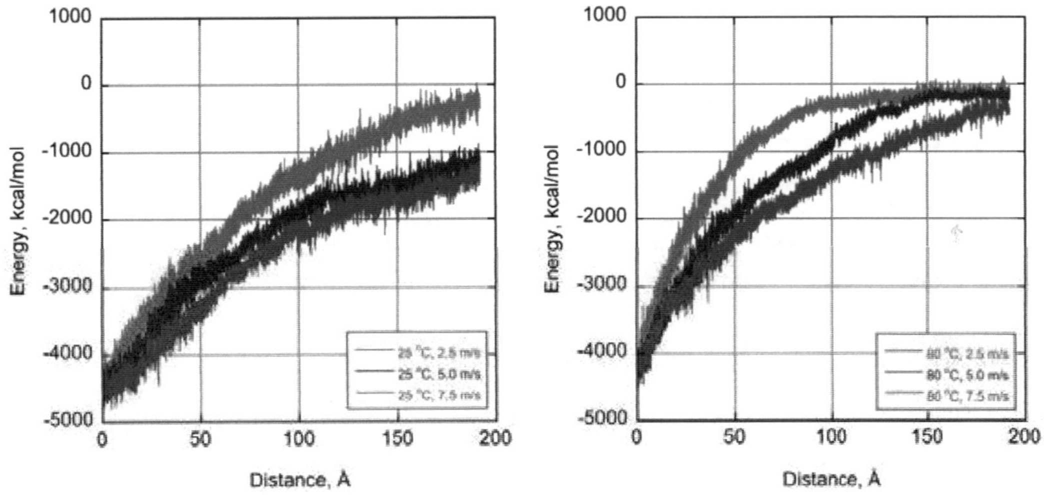

Figure 5. *Energies of interactions between the silica surfaces and bitumen layers as a function of sliding distance at sliding velocities of 2.5, 5.0, and 7.5 m/s at 25 (left image) and 80 °C (right image).*

seen in Figure 5, the efficiency of bitumen detachment increases as the shear rate decreases both at 25 and 80 °C in the studied range of shear rates, with the bitumen liberation at a sliding velocity of 2.5 m/s occurring almost as effectively at 25 °C as at 80 °C.

Thus, this work illustrates that MD simulations can provide valuable qualitative and quantitative data and guidance for optimization of important petroleum production processes.

References

1. Saniere, A. Definitions and Specifications. In *Heavy Crude Oils: From Geology to Upgrading: An Overview*; Huc, A.-Y., Ed.; Editions Technip: Paris, 2011; pp 11-23.
2. Delamaide, E. Oil Sands: Mining and Processing. In *Heavy Crude Oils: From Geology to Upgrading: An Overview*; Huc, A.-Y., Ed.; Editions Technip: Paris, 2011; pp 83-91.
3. Masliyah, J. H.; Czarnecki, J.; Xu, Z. *Handbook on Theory and Practice of Bitumen Recovery from Athabasca Oil Sands*; Kingsley Knowledge Publishing: Canada, 2011; Vol. 1, pp 282-294.
4. Schramm, L. L.; Stasiuk, E. N.; MacKinnon, M. Surfactants in Athabasca Oil Sands Slurry Conditioning, Flotation Recovery, and Tailings Processes. In *Surfactants: Fundamentals and Applications in the Petroleum Industry*; Schramm, L. L., Ed.; Cambridge University Press: Cambridge, 2000; pp 365-430.
5. Nikolov, A. D.; Wasan, D. T. Spreading of Nanofluids on Solids. *Nature* **2003**, *423*, pp 156-159.
6. Villars, P.; Calvert, L. D. *Pearson's Handbook of Crystallographic Data for Intermetallic Phases*, 2nd ed.; ASM International: Materials Park, Ohio, 1991.
7. Wyckoff, R. W. G., Ed. *Crystal Structures*; John Wiley & Sons: New York, London, 1963; Vol. 1, pp 318-319.
8. Cygan, R. T.; Liang, J.-J.; Kalinichev, A. G. Molecular Models of Hydroxide, Oxyhydroxide, and Clay Phases and the Development of a General Force Field. *J. Phys. Chem. B* **2004**, *108*, pp 1255-1266.
9. Jorgensen, W. L.; Maxwell, D. S.; Tirado-Rives, J. Development and Testing of the OPLS All-Atom Force Field on Conformational Energetics and Properties of Organic Liquids. *J. Am. Chem. Soc.* **1996**, *118*, pp 11225-11236.
10. Berendsen, H. J. C.; Grigera, J. R.; Straatsma, T. P. The missing term in effective pair potentials. *J. Phys. Chem.* **1987**, *91*, pp 6269–6271.
11. Plimpton, S. Fast Parallel Algorithms for Short-Range Molecular Dynamics. *J. Comp. Phys.* **1995**, *117*, pp. 1-19.
12. Ryckaert, J.-P.; Ciccotti, G.; and Berendsen, H. J. C. Numerical integration of the cartesian equations of motion of a system with constraints: molecular dynamics of *n*-alkanes. *J. Comput. Phys.* **1977**, *23*, pp 327–341.
13. Hockney, R.W.; Eastwood, J.W. *Computer Simulation Using Particles*; Taylor & Francis Group: New York, NY, 1988; pp 267-304.
14. Nosé, S. A unified formulation of the constant temperature molecular dynamics methods. *J. Chem. Phys.* **1984**, *81*, pp 511-519.
15. Hoover, W. G. Canonical dynamics: Equilibrium phase-space distributions. *Phys. Rev. A* **1985**, *31*, pp 1695-1697.

Selective Removal of Nitrogen Species from Bitumen Derived Gas Oil Using π-Acceptor Immobilized Polymers: Efficient Approach Towards Improving Hydrotreatment Technology

Prachee Misra[1], Ajay K. Dalai[1*], John Adjaye[2]
[1]*Catalysis and Chemical Reaction Engineering Laboratories, Department of Chemical and Biological Engineering, University of Saskatchewan, Saskatoon, SK (Canada) S7N 5A9*
[2]*Syncrude Canada Ltd., Edmonton Research Centre, Edmonton, AB (Canada) T6N 1H4*
*ajay.dalai@usask.ca

Increasing demand for liquid transportation fuel has intensified the exploitation of heavy oil and bitumen as refinery feedstock. Heavy crude oil is emerging as one of the alternative source to meet the present and future energy demands. However, bitumen derived heavy gas oil (HGO) contains high sulfur (approx. 4 wt.%) and nitrogen (approx. 0.3 wt.%) content as compared to those present in conventional crude oil (sulfur: approx. 2.5 wt.% and nitrogen: approx. 0.13 wt.%) which makes the hydrotreating processes challenging [1]. There is a critical need to improve the hydrotreatment technology in order to meet the stringent environmental regulations and produce ultra low sulphur fuel. In this regard, the improvement of the hydrodesulfurization (HDS) process by the removal of hydrotreating catalyst inhibitors, like heterocyclic nitrogen compounds (N-compounds) is worth considering. N-compounds deactivate hydrotreating catalysts used in the downstream secondary processing (such as fluid catalytic cracking and hydrocracking) of petroleum feedstock at a much faster rate than expected because of poisoning and inhibition by these nitrogen bearing species [2]. Removal of nitrogen compounds from HGO will restrain their irreversible adsorption onto the acidic sites of HDS catalyst, thus improving sulfur elimination. Removal of refractory sulfur and nitrogen compounds by non-catalytic processes is emerging as a new trend in hydrotreating of HGO because it reduces the hydrogen consumption during the process and can be achieved at less severe operating conditions. The present state of art in denitrogenation of gas oils has shown promising results by using an adsorbent material such as, π-acceptor immobilized polymers for the selective adsorption of refractory nitrogen species from gas oil. These refractory N-compounds are electron rich which are capable of forming charge transfer complexes (CTC) with electron deficient π-acceptors. Also, these π-acceptors can be immobilized on polymer supports to prevent the leaching of π-acceptors in gas oil resulting into their easier regeneration and reusability.

A novel pretreatment process employing π-acceptors, 2,7-Dinitro-9-fluorenone (DNF); 2,4,7-Trinitro-9-fluorenone (TriNF) and 2,4,5,7-tetranitro-9-fluorenone (TENF) immobilized on the polymer supports, poly(glycidyl methacrylate-co-ethylene glycol dimethacrylate) referred to as PGMA and polyHIPE (PH) via hydroxylamine (ON) linker is developed for the removal of nitrogen species from bitumen derived gas oil; in order to produce highly denitrogenated feed for hydrotreating applications. SEM, BET, CHNOS elemental analysis and TG/DTA were employed for determining the chemical and physical properties of the synthesized compounds. Characterization by FT-IR revealed the immobilization of π-acceptor on the polymers. Decrease in total nitrogen content of gas oil is a direct consequence of the formation of charge transfer complexes between π-acceptors immobilized PGMA and heterocyclic nitrogen species present in gas oil. Furthermore, spectroscopic features of the charge transfer interactions between π-acceptors and nitrogen species were confirmed by UV-VIS spectrophotometer. Total Nitrogen/Sulfur analyzer revealed the decrease in total nitrogen and sulfur content of bitumen derived gas oil after pre-treatment with immobilized polymers. ^{13}C NMR was used to calculate the decrease in aromatic content after treatment. As shown in Figure 1, the optimized polymers were successful in decreasing total nitrogen content in a single contact using 25 wt% of the immobilized polymers with respect to the gas oil while small amount of sulfur moieties were removed.

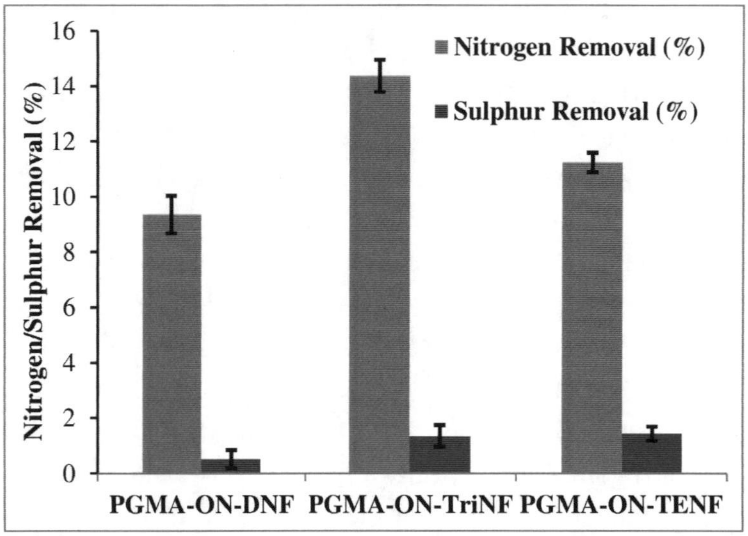

Figure 1: Percent nitrogen and sulfur removed during the first contact with gas oil.

All the immobilized polymers were successfully regenerated by washing multiple times with toluene and up to 7% of nitrogen removal was observed after the second contact of the fresh gas oil and regenerated polymers. Polymer synthesis procedures and process optimization studies for the three critical parameters (temperature, time and polymer loading) for the removal of sulfur and nitrogen species will be presented.

References:

[1] D. Ferdous, A.K. Dalai, J. Adjaye. *Energy & Fuels*, 17 (2003) 164–171.

[2] S.K. Bej, A.K. Dalai, J. Adjaye. *Energy & Fuels*, 15 (2001) 1103–1109.

CHARACTERIZATION OF VACUUM RESIDUES TYPICAL OF COLOMBIA WITH SUPERCRITICAL HEXANE

Adan Y. León. [1], **Martha Parra.** [2], **Alexander Guzmán.** [2], **Luis J. Hoyos.** [2], **Dionisio Laverde**[1] **and Cesar A. Mahecha**[2]

[1]*Escuela de Ingeniería Química UIS, Bucaramanga, Colombia*
[2]*Instituto Colombiano del Petróleo ICP– Ecopetrol.*

INTRODUCTON

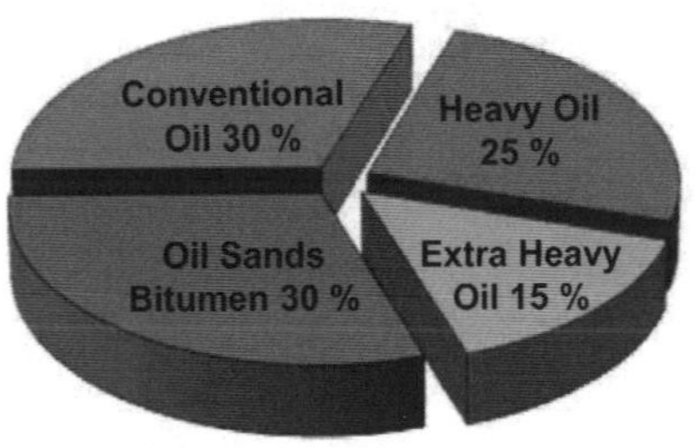

Figure 1. Total World Oil reserves

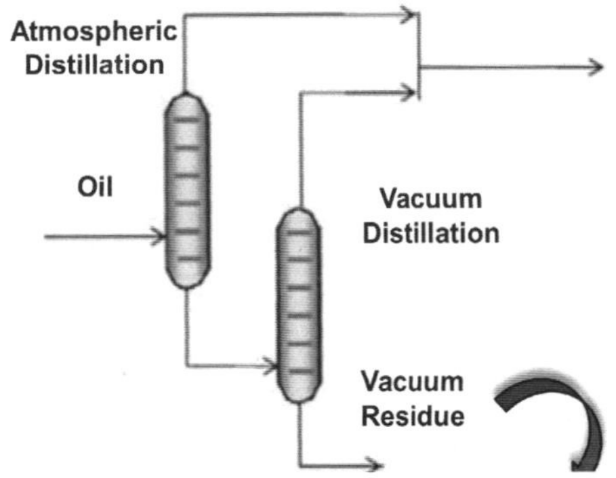

The processes must be adjusted with new diets oil and the production of high amounts of vacuum residue

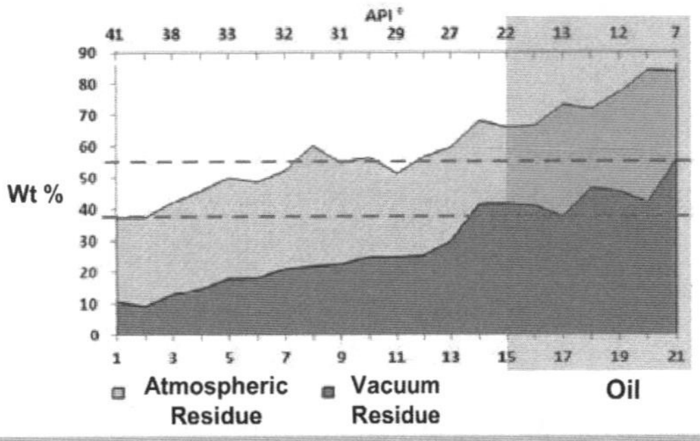

Figure 2. Contents of vacuum residue in crudes around world

- Alboudwarej H. Oilfield Review . 2006; 18 (2): 38-59.
- D. Stratiev et al. Petroleum & Coal 52(1) 35-43, 2010

ecopetrol

Heavy oil fractions

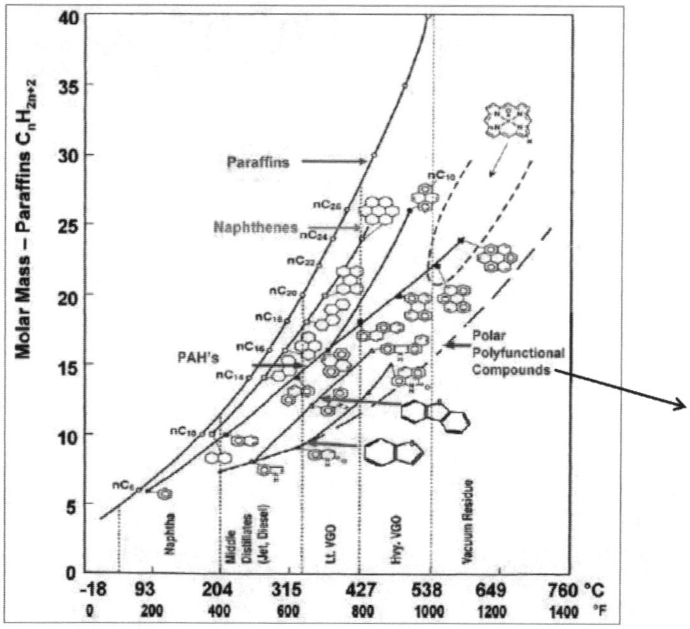

Thermal sensitivity

ecopetrol

Scheme of Characterization

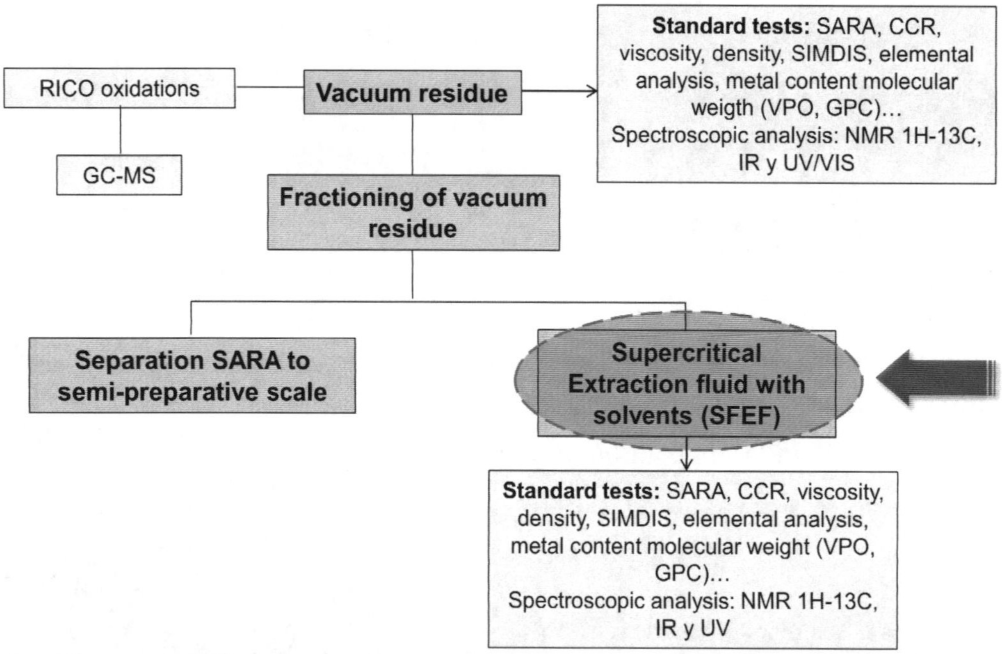

Extraction and Fractionation with Supercritical Fluids

Table 1. Operating Conditions. (Ecopetrol S. A Pilot Plant).

Temperatura °C	Presión, psi	Solvente	Características de la carga
250 - 280	450 a 1250	Hexano	Caracterización de DMO y Fondos de vacío: 0.7 ≤ API ≤ 15.3

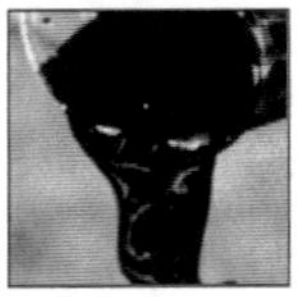

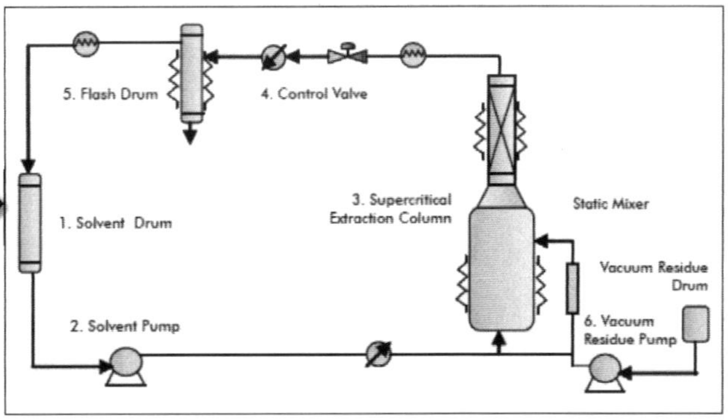

Process variables: Temperature, Pressure, Flow Solvent and extraction time

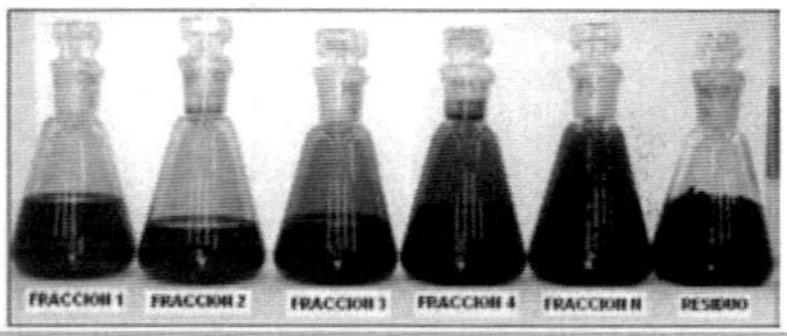

This technique facilitates the study of the composition of a vacuum residue from its fractions.

Materials and methods

Extraction and Fractionation with Supercritical Fluids

- Temperature: 250 - 265 °C
- Pressure range; 550 – 1250 psi.
- Solvent: n-hexane (P_C = 436.9 psi, T_C = 234.7 ° C).

Analytics: The vacuum residues and their fractions obtained by method of supercritical fluids extraction were characterized by the standard ASTM methods:

Table 2. Characterization of vacuum residues

Analysis	Vacuum Residue A	Vacuum Residue B
Sulfur (wt %) - ASTM D-4294	1,548	1,85
Nitrogen (wt %) - ASTM D-3228	0.168	0.195
API - ASTM D-4052 / 70	10.2	8.4
MCR (wt %) - ASTM D-4530	16,13	19,8
Viscosity (80 °C, Cp) - ASTM D-445	5960	14100
Metals – Norm B5863		
Nickel (mg/kg)	45,96	102,1
Vanadium (mg/kg)	140,02	196,4
SARA - ASTM D- 2007		
Saturates (wt %)	19,3	13,46
Aromatics (wt %)	47,5	49,98
Resins (wt %)	28,4	25,76
Asphaltenes (wt%)	4,8	10,8

INTRODUCTION | EXPERIMENTAL METHODOLOGY | RESULTS | CONCLUSIONS | REFERENCES

RESULTS

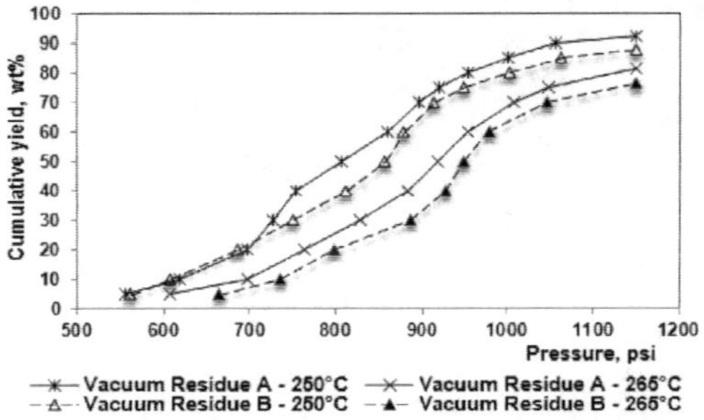

Influence of temperature and pressure

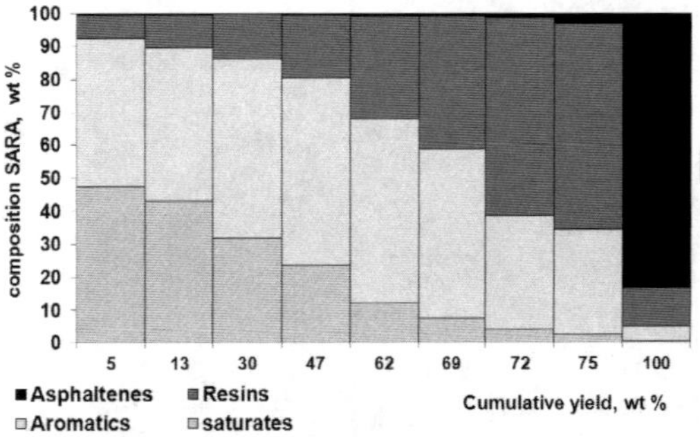

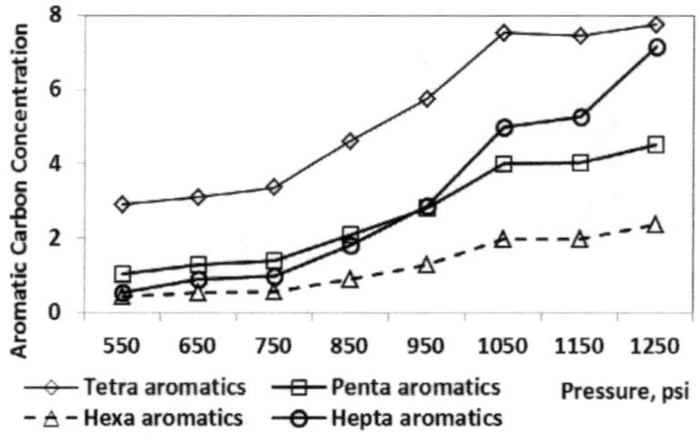

Figure 4. a- Accumulated Yield (wt %), **b-** Distribution SARA, **c-** Distribution aromatic carbon of vacuum residue B (265 °C)

7

INTRODUCTION | EXPERIMENTAL METHODOLOGY | RESULTS | CONCLUSIONS | REFERENCES

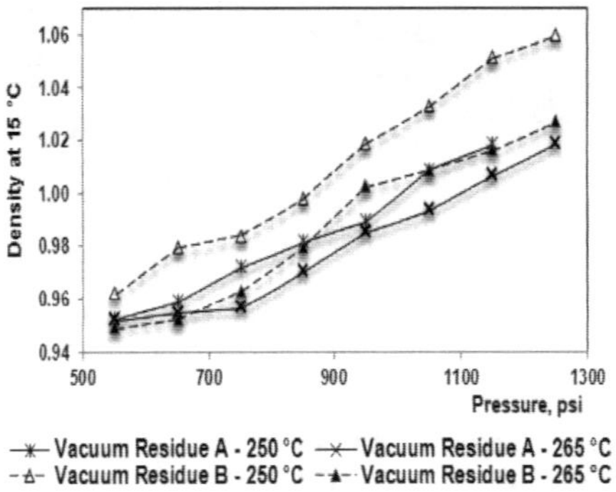

Profile of properties by type of vacuum residue.

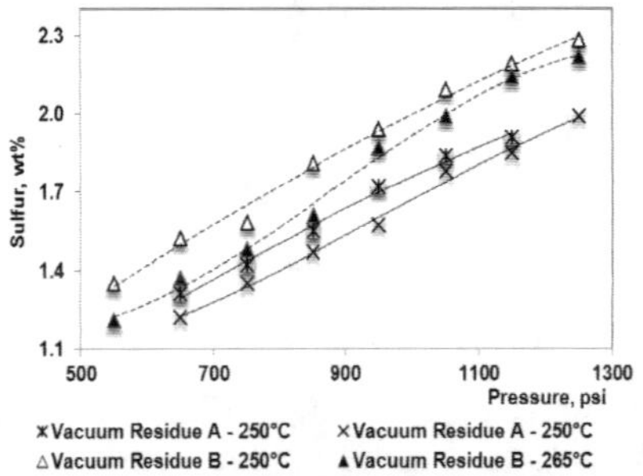

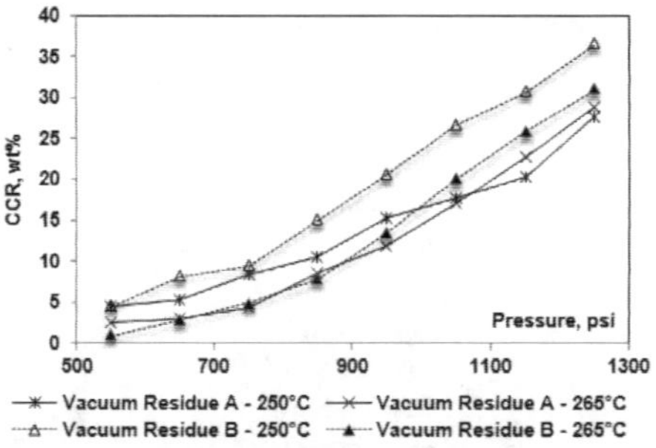

Figure 4. a- Density, **b-** Sulphur wt %, **c-** CCR wt % .

8

INTRODUCTION | EXPERIMENTAL METHODOLOGY | RESULTS | CONCLUSIONS | REFERENCES

ecoPETROL

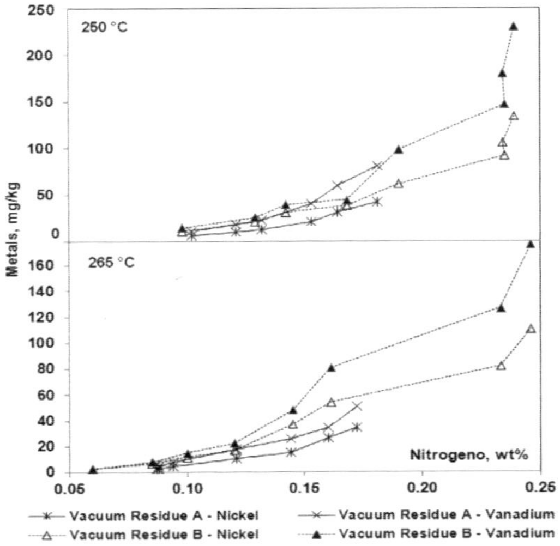

Figure 9. Distribution of nitrogen and metals in vacuum residues.

The extension of the distillation curves of the fractions obtained in this study were adjusted with the probability function proposed by Riazi.

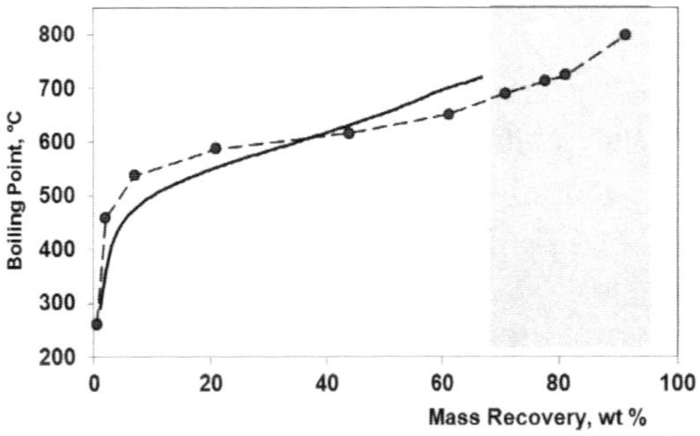

Figure 5. Distillation curve ASTM D-6532 vs average boiling temperature of fractions vacuum residue B (265 °C).

ecoPETROL

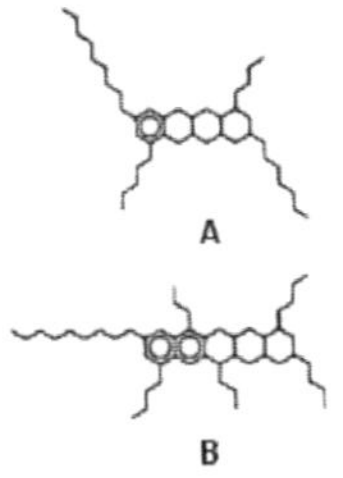

First fractions

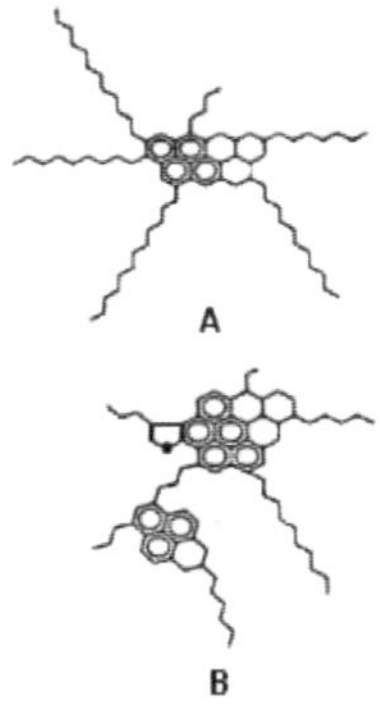

Middle fractions

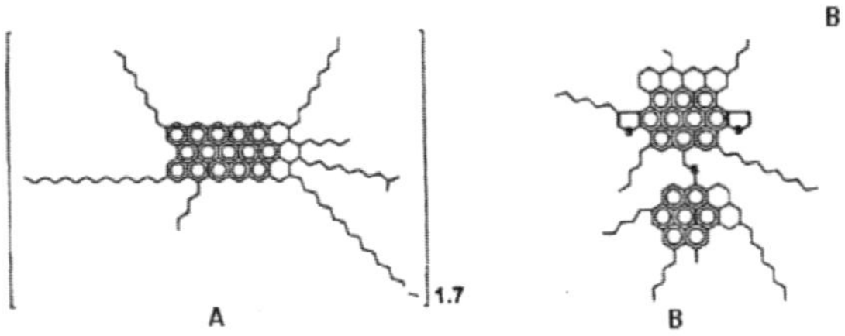

End fractions

The technique of extraction and fractionation with supercritical fluids is an alternative tool to improve and understand the refining processes of heavier petroleum fractions

- Suoqui Zhao. Fuel 84 (2005) 635–645

CONCLUSIONS

- The technique of extraction with supercritical fluids is a tool valuable to obtain significant amounts of fractions from vacuum residues.
- The distribution profile of density, average boiling point and CCR of the fractions obtained from the vacuum bottoms A and B, let show that the fractions become heavier as increases the pressure at temperature constant.
- The increase of the content of sulfur, nitrogen, nickel and vanadium is associated with the increase in complexity of the fractions obtained from a vacuum residue.
- The characterization results indicate that it is possible to fractionate the vacuum residues in cuts of different nature with equivalent atmospheric boiling points (AEBP) above 550 ° C.

REFERENCES

1. Tie-Pan shi, Yun-Xiang Hu, Tong su, and Ren An Wang. Ind. Eng. Chem. Res. 1997, Vol. 36, p 3988-3992.
2. Guangua yang, Ren An Wang. Jounal of petroleum science & engineering. 1999, Vol. 22, p 47-52.
3. Suoqi Zhao, Zhiming Xu, Chunming Xu, Keng H. Chung, Renan Wang.. Fuel 2005, Vol. 84, p 635-645.
4. Mohan S. Rana, Vicente Sámano, Jorge Ancheyta, J.A.I. Diaz. Fuel 86 (2007) 1216–1231.
5. Peter Behrenbruch, Thivanka Dedigama. Journal of Petroleum Science and Engineering 57 (2007) 166–180.
6. Zhan Gang Zhang, Shaohui Guo, Suoqi Zhao, Guangxu Yan, Energy and Fuels. 2009, 23. P. 374-385.
7. Parra M. J, León A.Y, and Hoyos L.J. CT&F. Volumen 4 número 2 - Diciembre 2010.
8. Stratiev. D, Dinkov. R, Petkov. K, Stanulov. K.. Petroleum & Coal 52(1) 35-43, 2010.
9. Linzhou Zhang, Zhiming Xu,* Quan Shi,* Xuewen Sun, Na Zhang, Yahe Zhang, Keng H. Chung, Chunming Xu, and Suoqi Zhao. Energy Fuels 2012, 26, 5795−5803.

Coke formation during pyrolysis of a Colombian heavy crude oil under in situ combustion conditions

Laura Urán[a], Luisa Carvajal[a], Laura Gaviria[a], Ding Wang[b], Eric Eddings[b], Alejandro Molina[a]*

[a]*Universidad Nacional de Colombia-Sede Medellín, Facultad de Minas, Bioprocesos y Flujos reactivos, Medellín, Colombia*

[b]*University of Utah, Department of Chemical Engineering, Salt Lake City, USA*

GRUPO DE INVESTIGACIÓN BIOPROCESOS Y FLUJOS REACTIVOS

UNIVERSIDAD NACIONAL DE COLOMBIA SEDE MEDELLÍN

The ISC is a complex process that involves different physicochemical phenomena: oxidation, pyrolysis, heat and mass transfer

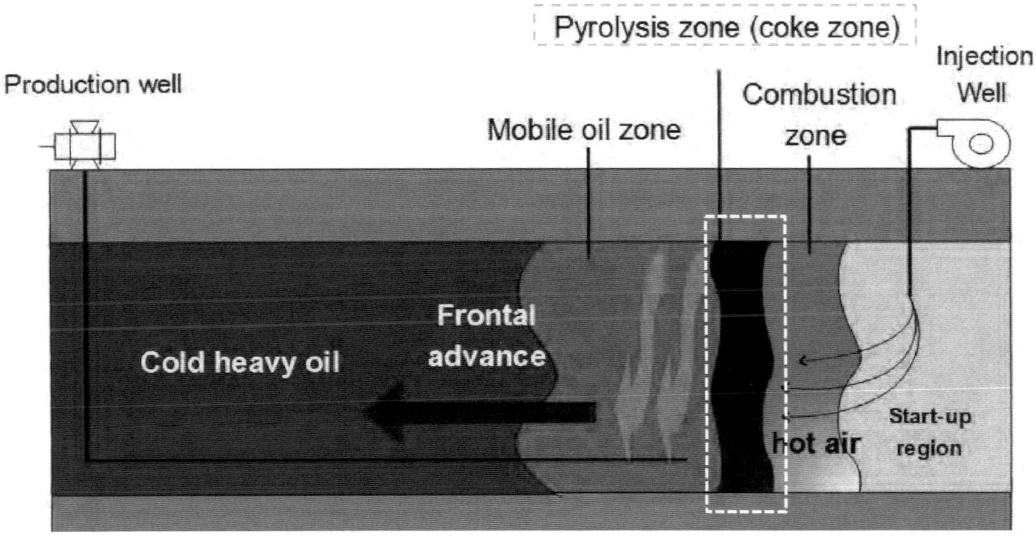

Coke forms under oxygen-free conditions and is of paramount importance because it fuels combustion and determines the recovery factor.

GRUPO DE INVESTIGACIÓN BIOPROCESOS Y FLUJOS REACTIVOS

UNIVERSIDAD NACIONAL DE COLOMBIA SEDE MEDELLÍN

The chemical composition of a crude oil can be described with SARA fractions

Saturates

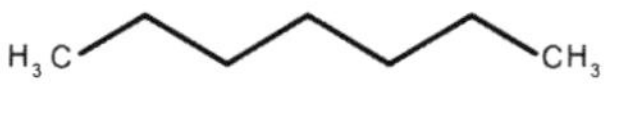

Aromatics

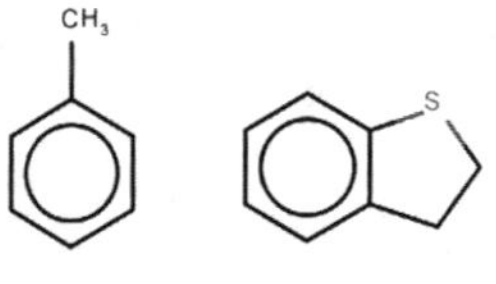

Resins

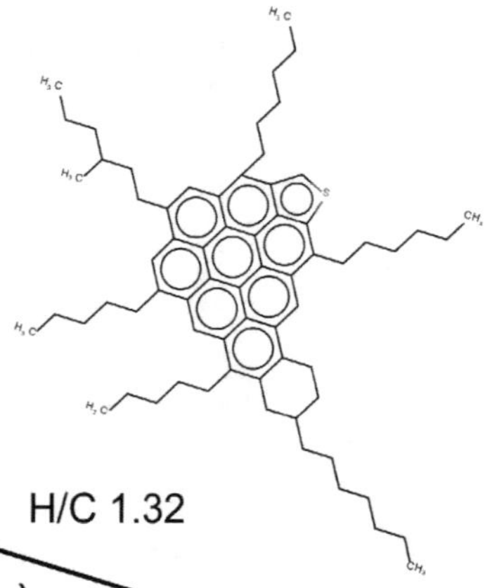

H/C 2.28

H/C 1.14

H/C 1.32

Maltenes (soluble fraction in n-heptane)

Maximum molecular weight about 1500 Dalton

Navarro et al. (2004)

GRUPO DE INVESTIGACIÓN BIOPROCESOS Y FLUJOS REACTIVOS

UNIVERSIDAD NACIONAL DE COLOMBIA SEDE MEDELLIN

3

Chemical composition of a crude oil can be described with SARA fractions

Asphaltenes (insoluble fraction in n-heptane)

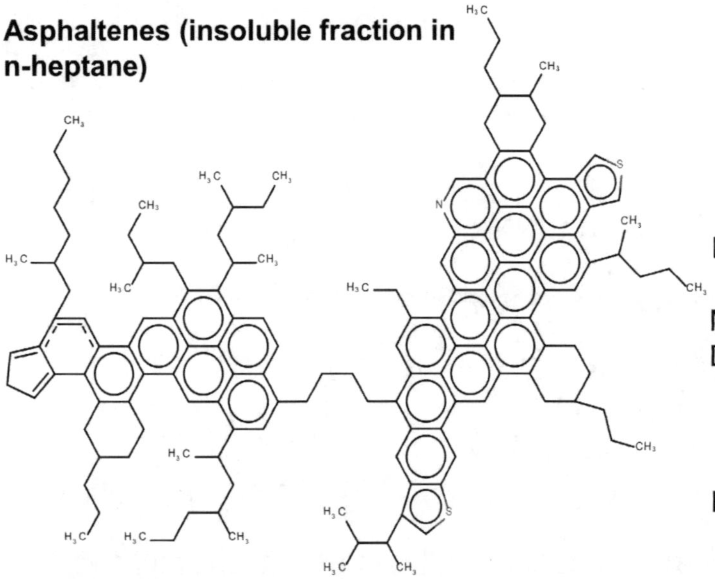

H/C 1.12

Molecular weight 2000 Dalton

Navarro et al. (2004)

GRUPO DE INVESTIGACIÓN BIOPROCESOS Y FLUJOS REACTIVOS

UNIVERSIDAD NACIONAL DE COLOMBIA SEDE MEDELLIN

4

The state of the art of the reaction mechanisms for pyrolysis during in situ combustion presents very general reaction mechanisms

Author / Reaction	Burger and Sahuquet (1971)	Belgrave et al (1993)	Martínez et al (1997)	Trejo et al (2014)	Urán et al. (2014)
Maltenes → Asphaltenes	✓	✓	☒	☒	
Maltenes → Coke	☒	☒	✓	☒	
Maltenes → Gas	☒	☒	☒	☒	
Maltenes → LMWM	☒	☒	☒	☒	
Maltenes → Maltenes*	☒	✓	☒	☒	
Asphaltenes → Maltenes	✓	☒	✓	✓	
Asphaltenes → Coke	✓	✓	✓	✓	
Asphaltenes → Gas	☒	✓	✓	✓	
Asphaltenes → LMWM	☒	☒	☒	☒	

We are proposing a new kinetic reaction mechanism that improves the understanding of the pyrolysis of a heavy crude oil

To establish the new reaction mechanism we perform experiments using the following materials:

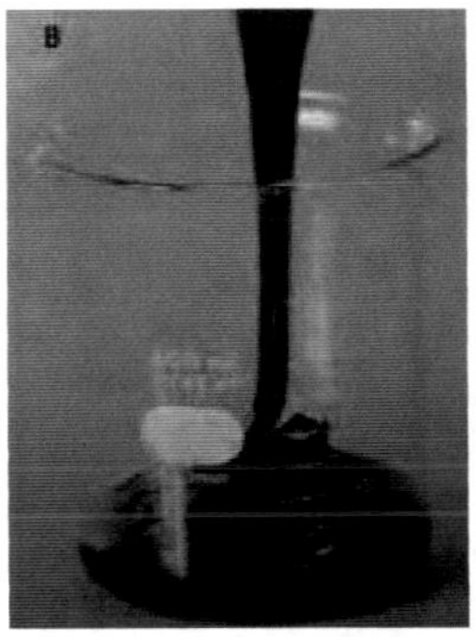

Castilla crude oil
- 81% maltenes 19% asphaltenes
- Density 0.98 g/cc at 15°C (API 13.4)
- Viscosity 971 cSt at 50°C

n-heptane

- soluble → **Maltenes**
- insoluble → **asphaltenic concentrated**

Methodology

TGA-DTG analysis (heating rate 2°C/min; T_{max} 500°C)

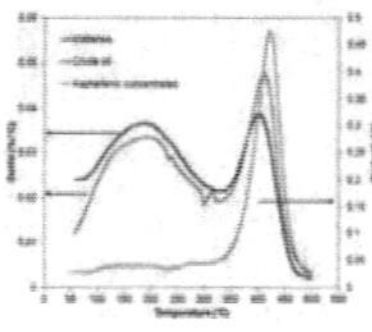

Pyrolysis experiments (heating rate 2°C/min; T_{max} 450°C)

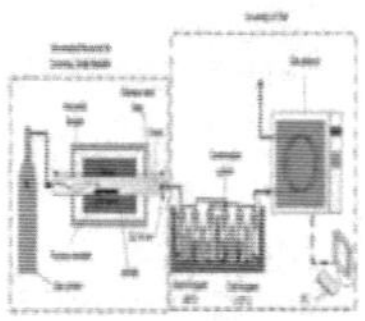

GRUPO DE INVESTIGACIÓN BIOPROCESOS Y FLUJOS REACTIVOS

UNIVERSIDAD NACIONAL DE COLOMBIA SEDE MEDELLIN

7

Pyrolysis experimental setup

Heating rate: 2 C/min

Temperatures: 300 C, 350 C, 400 C, 450 C

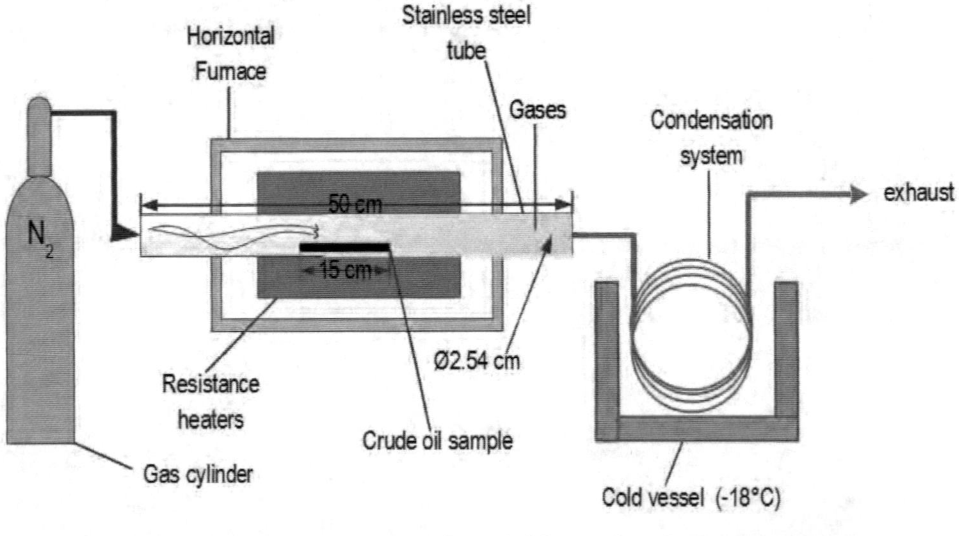

N_2 flow in the reactor: 1.1 sl/min

Gas residence time in the reactor 2.2 s

GRUPO DE INVESTIGACIÓN BIOPROCESOS Y FLUJOS REACTIVOS

UNIVERSIDAD NACIONAL DE COLOMBIA SEDE MEDELLIN

8

After pyrolysis the solid left in the furnace was analyzed to establish the yield of maltenes, asphaltenes and coke

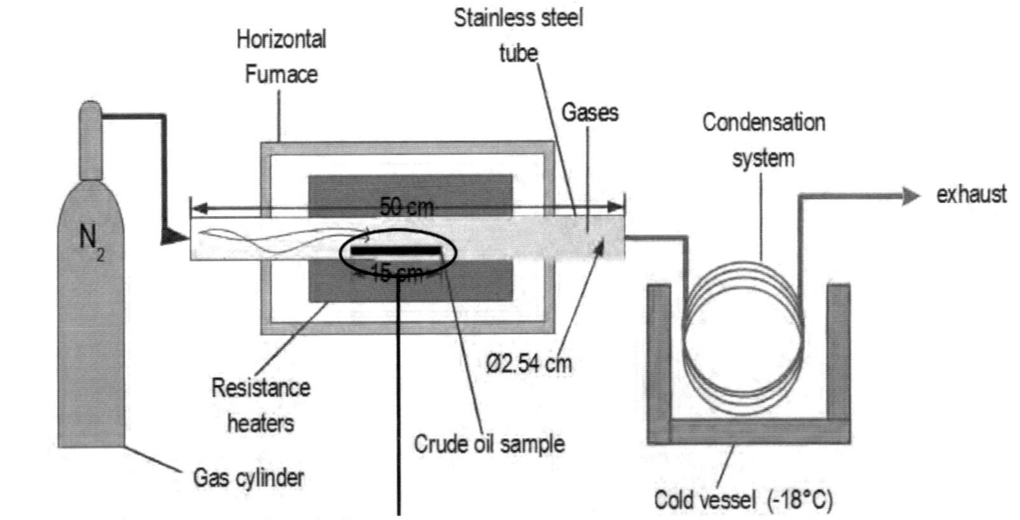

The yield of maltenes and asphaltenes was determined by test method IP 469

Coke was determined as the toluene-insoluble fraction

GRUPO DE INVESTIGACIÓN BIOPROCESOS Y FLUJOS REACTIVOS

UNIVERSIDAD NACIONAL DE COLOMBIA SEDE MEDELLÍN

9

University of Utah experiments allowed accurate analysis of the gas phase

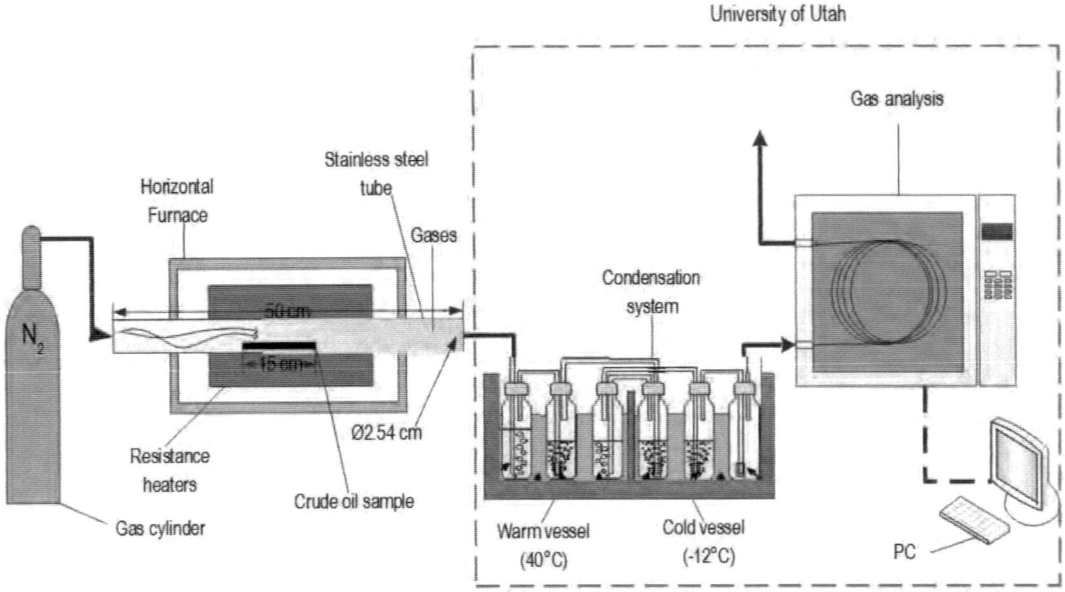

GRUPO DE INVESTIGACIÓN BIOPROCESOS Y FLUJOS REACTIVOS

UNIVERSIDAD NACIONAL DE COLOMBIA SEDE MEDELLÍN

10

Methodology

TGA-DTG analysis (heating rate 2°C/min; T_{max} 500°C)

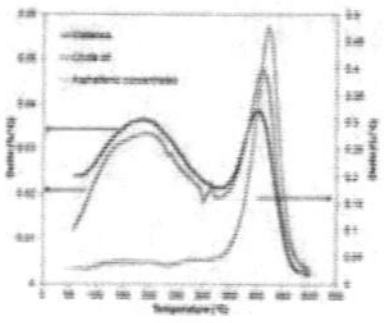

Pyrolysis experiments (heating rate 2°C/min; T_{max} 450°C)

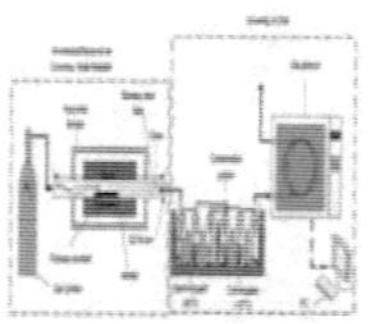

Reaction mechanism

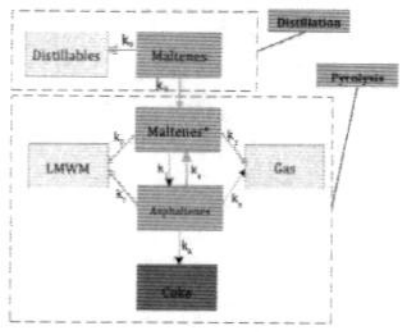

Determination of the kinetic parameter — **By minimizing the sum of square error (SSE)**

$$SSE = \sum_{i=1}^{n} \left(y_{i\exp} - y_{icalc} \right)^2$$

Results

TGA-DTG analysis: The first stage of the pyrolysis process is distillation, and the thermal cracking process follows

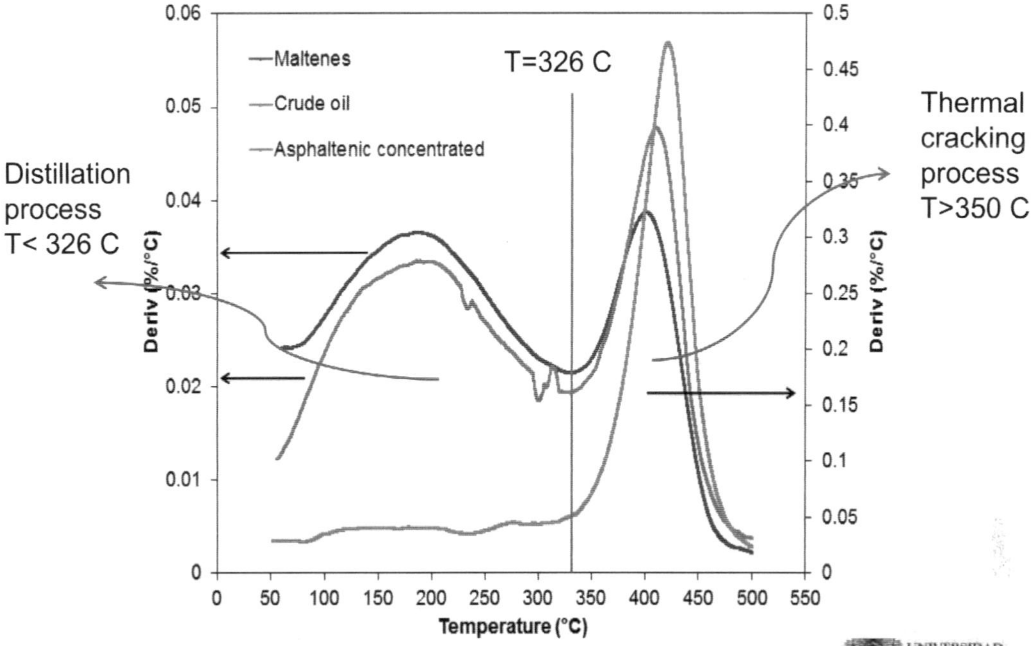

GRUPO DE INVESTIGACIÓN BIOPROCESOS Y FLUJOS REACTIVOS

UNIVERSIDAD NACIONAL DE COLOMBIA SEDE MEDELLIN 13

Pyrolysis experiments / crude oil

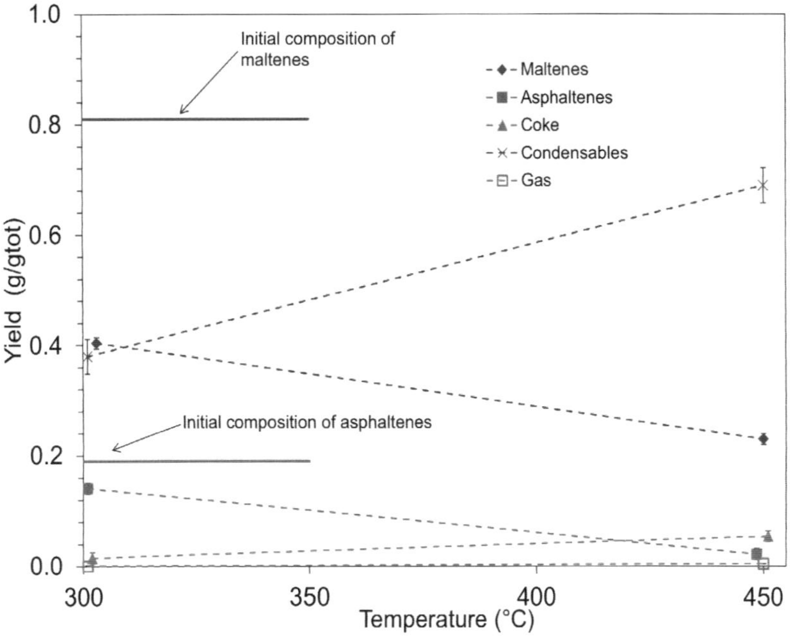

Temp 300C-450 C

Maltenes and asphaltenes yield mainly a condensable specie and, in a lower proportion, coke and gases.

3 repetitions.
The mass balance closes to within 94%

GRUPO DE INVESTIGACIÓN BIOPROCESOS Y FLUJOS REACTIVOS

UNIVERSIDAD NACIONAL DE COLOMBIA SEDE MEDELLIN 14

Pyrolysis experiments / maltenes

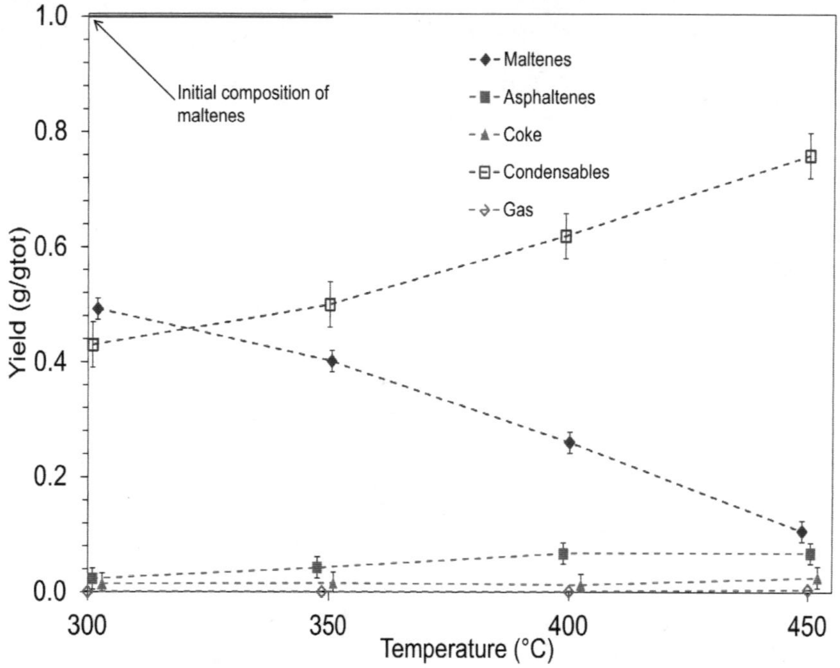

Maltenes yield mainly condensables and, in a lower proportion, coke and gases

The yield of asphaltenes becomes significant after 400C

Pyrolysis experiments / asphaltenic concentrate

Asphaltenes yielded species of lower-molecular weight like maltenes and gases below 400 C

After 400 C the formed maltenes yielded coke precursors because of secondary thermal cracking reactions

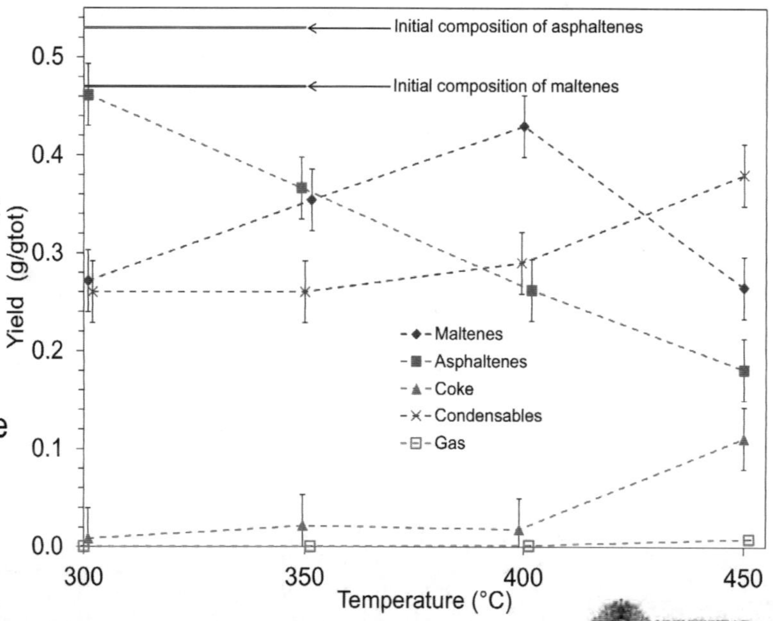

Part I. Proposed reaction mechanism

GRUPO DE INVESTIGACIÓN BIOPROCESOS Y FLUJOS REACTIVOS

UNIVERSIDAD NACIONAL DE COLOMBIA SEDE MEDELLÍN 17

A reaction mechanism that captures the chemistry of the process

The first stage of the pyrolysis process is distillation of maltenes. Thermal cracking follows.

Maltenes* and asphaltenes yield species of lower- and higher-molecular weight

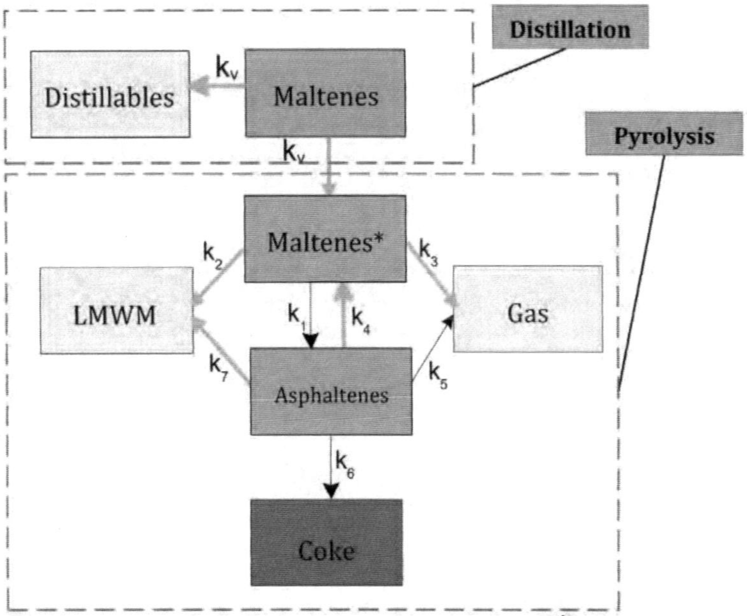

GRUPO DE INVESTIGACIÓN BIOPROCESOS Y FLUJOS REACTIVOS

UNIVERSIDAD NACIONAL DE COLOMBIA SEDE MEDELLÍN 18

We propose some new reactions compared with the state-of-the-art reaction mechanisms

Reaction \ Author	Burger and Sahuquet (1971)	Belgrave et al (1993)	Martínez et al (1997)	Trejo et al (2014)	Urán et al. (2014)
Maltenes → Asphaltenes	✓	✓	☒	☒	✓
Maltenes → Coke	☒	☒	✓	☒	☒
Maltenes → Gas	☒	☒	☒	☒	✓
Maltenes → LMWM	☒	☒	☒	☒	✓
Maltenes → Maltenes*	☒	✓	☒	☒	✓
Asphaltenes → Maltenes	✓	☒	✓	✓	✓
Asphaltenes → Coke	✓	✓	✓	✓	✓
Asphaltenes → Gas	☒	✓	✓	✓	✓
Asphaltenes → LMWM	☒	☒	☒	☒	✓

Part II. Kinetic analysis

The evaporation constant of the distillation process of Castilla crude oil was determined by regression

$$SSE = \sum_{i=1}^{n} \left(y_{i\exp} - y_{icalc} \right)^2$$

$$\frac{dy_{mal}}{dt} = -k_v y_{mal}$$

Experimental targets are from TGA of maltenes

kv = 7.46 × 10^{-5} (J/mol)

Fraction evaporated 0.42 g/gtot

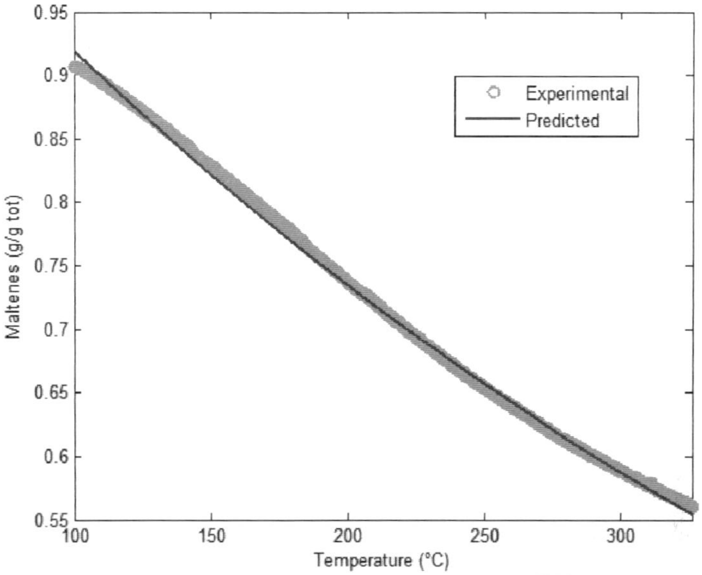

GRUPO DE INVESTIGACIÓN BIOPROCESOS Y FLUJOS REACTIVOS

UNIVERSIDAD NACIONAL DE COLOMBIA SEDE MEDELLÍN 21

We establish the kinetic parameters by minimizing the sum of square error (SSE) between the experimental and calculated yield for each pseudo-component

$$SSE = \sum_{i=1}^{n} \left(y_{i\exp} - y_{icalc} \right)^2$$

$$\frac{dy_{mal*}}{dt} = k_4 y_{asp} - (k_1 + k_2 + k_3) y_{mal*}$$

$$\frac{dy_{asp}}{dt} = k_1 y_{mal*} - (k_4 + k_5 + k_6 + k_7) y_{asp}$$

$$\frac{dy_{LMWM}}{dt} = k_2 y_{mal*} + k_7 y_{asp}$$

$$\frac{dy_{gas}}{dt} = k_3 y_{mal*} + k_5 y_{asp}$$

$$\frac{dy_{coke}}{dt} = k_6 y_{asp}$$

	k_1	k_2	k_3	k_4	k_5	k_6	k_7
Ea (J/mol)	7.10×10^4	8.20×10^4	1.27×10^5	4.34×10^4	2.28×10^5	1.58×10^5	1.14×10^4
k_0 (s^{-1})	3.59×10^{-3}	1.55	3.44×10^7	9.50×10^{-1}	3.0×10^{12}	2.09×10^8	1.0

GRUPO DE INVESTIGACIÓN BIOPROCESOS Y FLUJOS REACTIVOS

UNIVERSIDAD NACIONAL DE COLOMBIA SEDE MEDELLÍN 22

A parity plot shows good agreement (R^2=0.97) between experimental and calculated yields for pyrolysis of crude oil (CO), maltenes (M) and asphaltenic concentrate (AC)

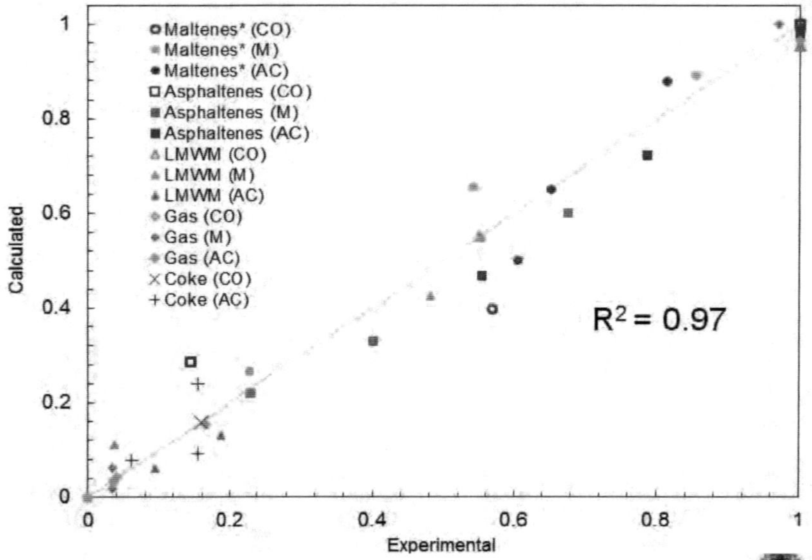

GRUPO DE INVESTIGACIÓN BIOPROCESOS Y FLUJOS REACTIVOS

UNIVERSIDAD NACIONAL DE COLOMBIA SEDE MEDELLÍN

23

The new reaction mechanism, considering a new pseudo-component (LMWM) and new reactions, gives a better prediction of the pyrolysis process of Castilla heavy crude oil

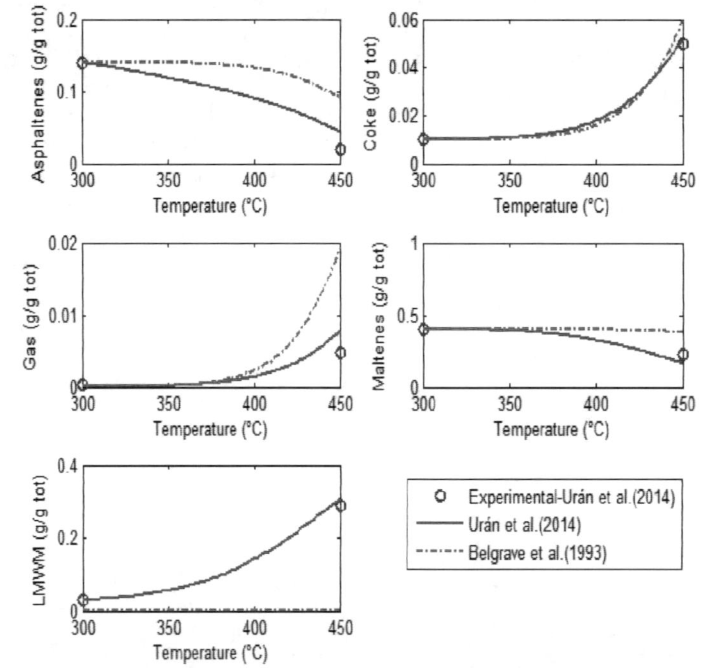

GRUPO DE INVESTIGACIÓN BIOPROCESOS Y FLUJOS REACTIVOS

UNIVERSIDAD NACIONAL DE COLOMBIA SEDE MEDELLÍN

24

Conclusions

1. The experiments, conducted under an oxygen-free atmosphere, suggest that the temperature needs to be higher than 400 C to produce coke during the pyrolysis of Castilla heavy crude oil.

2. Asphaltenes mainly yield low-molecular-weight fractions such as maltenes and LMWM below 400 C.

3. The proposed reaction mechanism includes, for the first time in the refereed literature, the formation of a low-molecular weight fraction (LMWM) and agrees with the experimental results described in this presentation.

GRUPO DE INVESTIGACIÓN BIOPROCESOS Y FLUJOS REACTIVOS

UNIVERSIDAD NACIONAL DE COLOMBIA SEDE MEDELLÍN 25

Acknowledgments

1. The authors acknowledge Colciencias and Ecopetrol for financial support of this research with the Contract RC. N° 0264-2013, project 1118-531-30561.

2. To the program ENLAZAMUNDOS for the partial financial support of the student Laura Urán during her research stay at the University of Utah.

GRUPO DE INVESTIGACIÓN BIOPROCESOS Y FLUJOS REACTIVOS

UNIVERSIDAD NACIONAL DE COLOMBIA SEDE MEDELLÍN 26

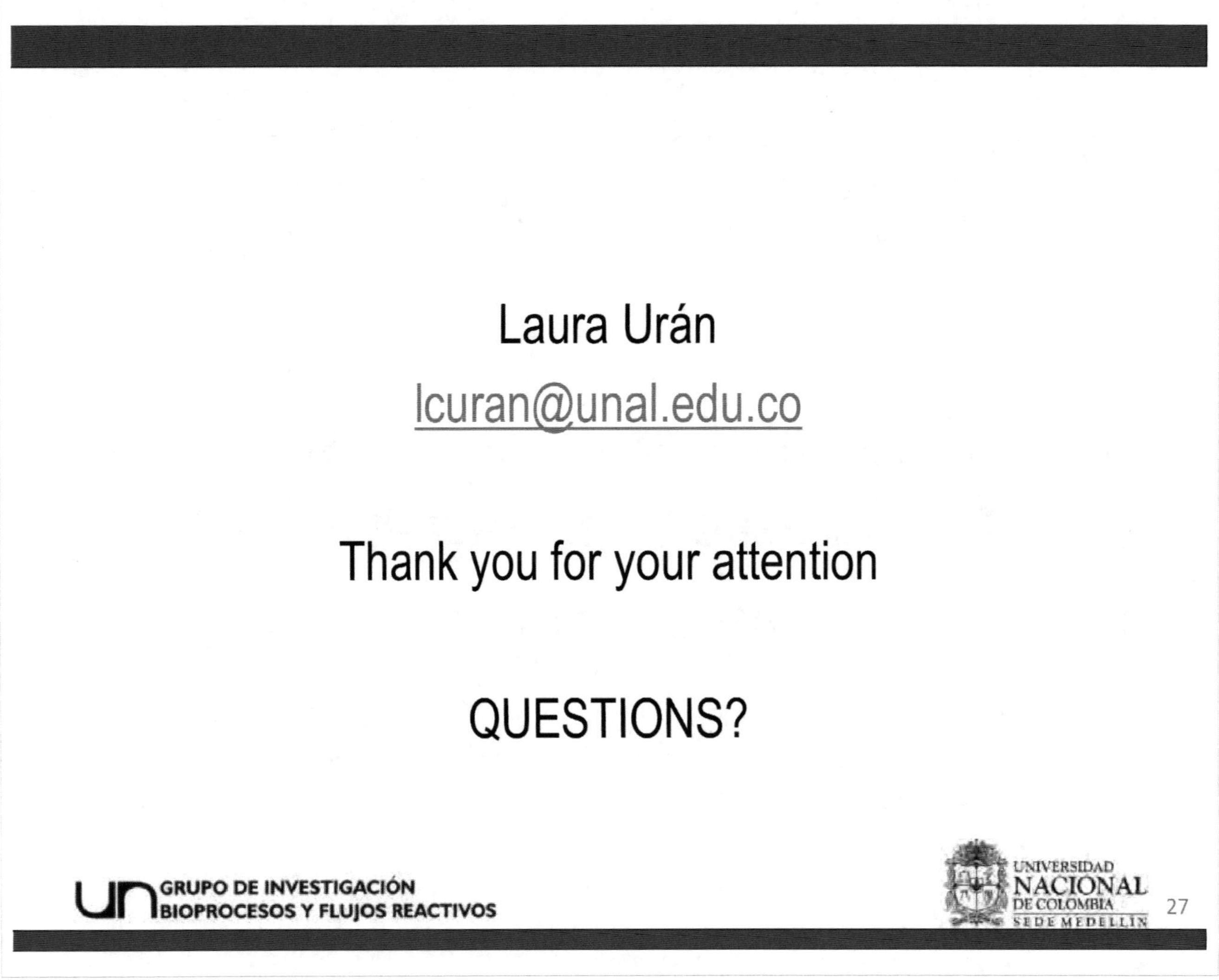
Laura Urán
lcuran@unal.edu.co
Thank you for your attention
QUESTIONS?
GRUPO DE INVESTIGACIÓN
BIOPROCESOS Y FLUJOS REACTIVOS
UNIVERSIDAD
NACIONAL
DE COLOMBIA
SEDE MEDELLÍN
27

A Mechanistic Investigation of Selective Glycerol Hydrogenolysis over Pt-Re Bimetallic Catalysts

Derek D Falcone (derekfalcone@virginia.edu)

John H Hack (jhh5za@virginia.edu)

Robert J Davis (rjd4f@virginia.edu)

Several chemical intermediates can be derived from renewable glycerol, a byproduct of the biodiesel transesterificaton process. In this work the selective hydrogenolysis of the secondary C-O bond in glycerol to produce 1,3-propanediol was studied over Pt-Re bimetallic catalysts in liquid water. In addition to shifting the selectivity of the reaction, Re also promoted the overall rate of glycerol hydrogenolysis by more than an order of magnitude compared to monometallic Pt. The role Re plays in the reaction as well as the oxidation state of Re is still a matter of debate in the literature. One possible role is that a Brønsted acid site is associated with the Re component in liquid water, which catalyzes the dehydration reaction at the secondary carbon of glycerol. In this work, XPS was used to evaluate the oxidation state of Pt and Re after in situ pretreatments in dihydrogen at 393 and 473 K. Infrared spectroscopy was also used to characterize the acidity of Pt-Re nanoparticles supported on silica using pyridine as an acid probe. In addition, batch reaction experiments showed that dilute NaOH substantially decreased the rate of glycerol hydrogenolysis, presumably by the neutralization of acid sites on the catalyst. Propyl acetate hydrolysis was also used as a probe reaction for acid catalysis and the Pt-Re bimetallic catalyst showed rates of hydrolysis comparable to rates in aqueous HCl. These results are consistent with the idea that Brønsted acid sites are present on the bimetallic catalyst in liquid water.

Glycerol hydrodeoxygenation catalyzed by Cu-Pd/Na-TiO_2: Structure of the catalysts

Alba N. Ardila, Erasmo Arriola, Gustavo A. Fuentes*
Dept. of Process Engineering, Universidad A. Metropolitana- Iztapalapa, A.P. 55-534; 09340 Mexico, DF; Mexico.
*- Email: gfuentes@xanum.uam.mx

There is much ongoing research to discover viable catalysts for the conversion of glycerol, a byproduct of biodiesel production, to different high added-value chemicals. On that line, we have developed a novel catalyst, Cu-Pd/Na-TiO_2, that shows very good characteristics during the aqueous phase hydrodeoxygenation of glycerol to 1,2- propanediol (PDO) under basic conditions. This catalyst requires low hydrogen pressure to accomplish that goal and at the same time has very low C-C hydrogenolysis activity, resulting in high yield and selectivity to the diol. It can also be reused a number of times with low decay in activity.

In this work we present results obtained using a combination of techniques, UV-Vis spectroscopy, H_2-TPR, XPS and high resolution STEM, coupled with reaction studies, to analyze the synergistic effect that occurs when Pd and Cu are combined on the same material.

Our results are consistent with the formation of Pd-Cu alloy nanoparticles. A consequence of alloying is the stabilization of Cu in a reduced state, what increases its dispersion and limits Cu leaching to the solution. Pd appears to be also responsible for partial reforming of glycerol to produce H_2 that is used in the hydrogenation of acetol in the absence of feed H_2. The role of the basic sites on the support is also discussed.

This class of materials may be a promising alternative for industrial applications, because the results obtained are better or comparable with those reported in the literature at much higher H_2 pressures.

Co-deposited electroless Cu-Ni-Mo electrocatalyst for glycerol oxidation

Oyidia Elendu, Yaw Yeboah and Egwu E. Kalu

Department of Chemical and Biomedical Engineering

FAMU-FSU College of Engineering

Tallahassee FL 32310

Cu, Ni and Mo have many properties that make them attractive as catalysts. Both the thermochemical catalytic and electrocatalytic activity of many metals are enhanced by the presence of other co-deposited metals. In addition, other synergies arise for co-deposited metals including enhanced deposit thickness and size, luster, wear, recyclability and other mechanical properties. Ni-Cu-Mo electrocatalysts were prepared by electroless deposition technique using a mixed reducing agent for the co-deposition. The electrocatalysts were characterized by XRD, SEM, EDAX, XPS and cyclic voltammetry techniques. The electrocatalysts were evaluated for the electrooxidation of glycerol in alkaline and acid media. The electrocatalytic activity of Ni-Cu-Mo catalysts on glycerol oxidation and product distribution will be presented and discussed.

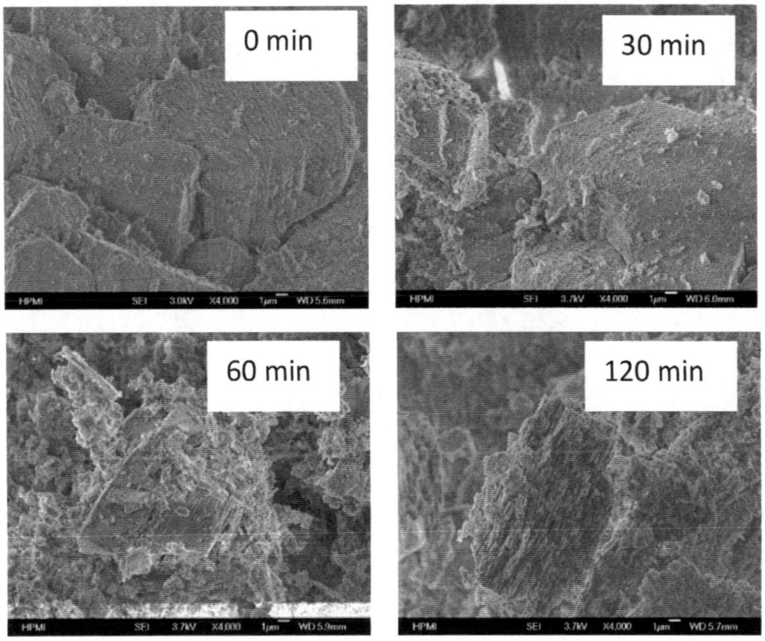

Fig. 1: SEM features of electroless deposited Ni-Cu-No electrocatalyst on Al_2O_3

389637 Base Free, One-Pot Synthesis of Lactic Acid from Glycerol Using a Bifunctional Pt/Sn-MFI Catalyst

Abstract:

Glycerol, an inevitable by-product from biodiesel production, is potentially obtained from non-edible biomass such as microalgae, cellulose and its derivatives. The conversion of glycerol to commodity chemicals not only reduces the chemical waste but also provides a sustainable and green pathway for our society. Lactic acid has been considered as one of the top 14 chemical building blocks produced from biomass conversion by US Department of Energy at 2010. Compared to glycerol, lactic acid (LA) and alkyl lactates have a higher value and offer important applications as green solvents, platform chemicals and precursors for the synthesis of polylactate or polylactic acid (PLA), the second most manufactured bioplastic of the world. LA can be produced from petroleum feedstocks in high yields using mineral acid catalysts. However, the use of toxic chemicals (e.g. hydrogen cyanide) and the high cost for product separation could lead to significant environmental issues. Therefore, currently LA is mainly manufactured by the fermentation of carbohydrates. However, the efficiency and productivity of the fermentation process are still low and need substantial improvements because of the high cost of enzyme catalysts and the need for precise control of operating conditions. There is thus a persistent need to develop efficient thermochemical pathways for the production of LA from new starting materials.

In this study, we report a base free one-step reaction pathway to selectively produce LA with high yields from the oxidation of glycerol over a bifunctional Pt/Sn-MFI catalyst under mild reaction conditions. 80.5% selectivity to LA was achieved at 89.8% conversion of glycerol[1]. In the cascade reaction route, the selective oxidation of glycerol into glyceraldehyde proceeds on Pt catalyst. Sn-MFI exhibits outstanding Lewis acidity for converting glyceraldehyde into dihydroxyacetone and dihydroxyacetone into the final product LA. The superior performance is mainly due to its microporous structure, accessible Sn site and Lewis acidity.

1. Cho, H. J.; Chang, C.-C.; Fan, W., "Base free, one-pot synthesis of lactic acid from glycerol using a bifunctional Pt/Sn-MFI catalyst". *Green Chem.* **2014**.

386098 The Catalytic Effect of H-Beta, Ce- and Ln-Doped Beta Zeolite on the Hydrothermal Transformation of Glycerol

Abstract:

The growth of the biodiesel industry over the last decade has resulted in a price drop of glycerol, making it an attractive feedstock for transformation into value-added chemicals [1]. The three hydroxyl groups of glycerol are highly reactive but their similar activation energies complicate their selective conversions. Among the processes tested for glycerol transformations, "hydrothermal technology" has gained attention due to its potential to overcome the difficulties associated with other methods such as long reaction times of fermentation [2]. High T/P water acts as a potent acid-base medium enabling short reaction times [2, 3] while the addition of a suitable catalyst may increase the selectivity to desired products [2].

Lewis acidic Sn-, Zr-, Ti- doped Beta zeolites exhibit high selectivity to the formation of lactic acid from trioses in both methanol and water at 80 °C and 125 °C, respectively [4]. In this study, the incorporation of even stronger Lewis acidic Ce and Ln into the BEA framework of zeolites was carried out using two different procedures: i) solid state ion exchange (SSIE) between H-Beta zeolite and a metal nitrate at 550 °C and ii) dealumination of H-Beta zeolite by strong mineral acid without collapsing the zeolite structures followed by SSIE [5].

Transformation of glycerol was investigated under hydrothermal conditions (270-360 °C, 55-186 bar, 5-180 min) using H-Beta, Ce- and Ln-doped Beta zeolites in batch tubular reactors. The catalysed glycerol transformation followed mainly two concomitant pathways in all systems: dehydration and/or oxidation. The former took place at faster reaction rates yielding ~17 %C yield of acrolein within the first 10 min. The latter primarily produced lactic acid having yield increased with reaction time; max ~20 %C. The metal-doped Beta zeolites prepared by simple SSIE showed a ~30% increase in selectivity towards lactic acid as compared to H-Beta zeolite. The incorporation of the metals resulted in lower Brønsted acidity of the zeolite surface but higher Lewis acidity which may play an important role in the transformation of pyruvaldehyde (PAL), the final intermediate product which transforms into lactic acid via Cannizzaro reaction as proposed by Bicker *et al.*[6]. The selectivity to lactic acid obtained was slightly higher in the reactions catalysed by Ce- (~30 %C) than those catalysed by Ln-doped Beta zeolite (~25 %C).

Herein, we report for the first time high yield of lactic acid (~ 20 %C, ~ 80 %mol glycerol conversion) achieved directly from glycerol under base-free hydrothermal conditions using high Brønsted-acidic H-Beta and Lewis acidic Ce- and Ln-doped Beta zeolites. The lactic acid yield achieved was competitive to those obtained in the reactions catalysed by Pd/TiO_2, Pt/TiO_2 or Au/TiO_2 at 90 °C in basic environment [7] and greater than in WO_3/TiO_2catalysed reactions under supercritical water conditions [8].

References

1. Yang, F., M. Hanna, and R. Sun, *Value-added uses for crude glycerol - a byproduct of biodiesel production.* Biotechnology for Biofuels, 2012. **5**(13).

2. Long, Y.-D. and Z. Fang, *Hydrothermal conversion of glycerol to chemicals and hydrogen: review and perspective.* Biofuels Bioproducts & Biorefining-Biofpr, 2012. **6**(6): p. 686-702.

3. Akiya, N. and P.E. Savage, *Roles of water for chemical reactions in high-temperature water.* Chemical Reviews, 2002. **102**(8): p. 2725-2750.

4. Taarning, E., et al., *Zeolite-catalyzed isomerization of triose sugars.* Chemsuschem, 2009. **2**(7): p. 625-627.

5. Hammond, C., S. Conrad, and I. Hermans, *Simple and scalable preparation of highly active Lewis acidic Sn-Beta.* Angewandte Chemie-International Edition, 2012. **51**(47): p. 11736-11739.

6. Bicker, M., et al., *Catalytical conversion of carbohydrates in subcritical water: A new chemical process for lactic acid production.* Journal of Molecular Catalysis A: Chemical, 2005. **239**(1–2): p. 151-157.

7. Shen, Y., et al., *Efficient Synthesis of Lactic Acid by Aerobic Oxidation of Glycerol on Au–Pt/TiO2 Catalysts.* Chemistry – A European Journal, 2010. **16**(25): p. 7368-7371.

8. Akizuki, M. and Y. Oshima, *Kinetics of Glycerol Dehydration with WO3/TiO2 in Supercritical Water.* Industrial & Engineering Chemistry Research, 2012. **51**(38): p. 12253-12257.

382616 Methane Conversion to Ethane and Ethylene Using Solid Oxide Membrane Reactors

Abstract:

Over the past few decades, natural gas production in the US has increased due to shale gas, and is projected to continue to increase. This increase in availability and production has driven prices down and prompted research into new ways to use methane, the main component in natural gas. Methane is currently reformed to produce synthesis gas, a mixture of CO, CO_2 and H_2, which is then used to make other chemicals and fuels. Although these processes are economical, they rely on large economies of scale, resulting in very large chemical plants and little use for stranded sources of natural gas. In contrast, direct conversion of methane to chemicals such as ethylene could create an economical use for natural gas that is currently flared.

Methane can be directly converted to C_2 products using oxidative coupling, where CH_3 radicals are formed using oxygen on a catalyst surface then couple to form ethane or ethylene. In this work membrane reactors are fabricated which can control oxygen anion flux through a membrane. These devices consist of three layers of solid oxide and perovskite materials. First a porous layer which dissociates oxygen from air, then a very thin solid layer that conducts oxygen anions, and finally another porous layer which facilitates methane activation. We will show conversion and selectivity results for these devices using several different membranes including: yttria stabilized zirconia (YSZ), scandia stabilized zirconia (ScSZ), and gadolinium doped ceria (GDC). We will also show results for factors that influence methane activation including the addition of transition metals.

357178 High Efficient and Recyclable Heterogeneous Catalysts (Ni-modified zeolites) for Reppe Carbonylation to Acrylic Acid

Abstract:

Reppe Carbonylation is undoubtedly one of the most important reactions in acetylene chemistry for yielding acrylic acid and its esters. Although this route had been replaced by propylene oxidation in the past decade owing to the global availability of inexpensive ethylene from petroleum, the desire for diversified feedstock for bulk chemical commodity is steadily on the increase. Because of the traditional homogeneous catalysts shows disadvantages in the catalyst -product separation and the scaling up of the process, more recent works were focused on the development of heterogeneous catalysts.

The current paper reports, for the first time, a highly efficient heterogeneous catalyst systems for the hydrocarboxylation of acetylene to acrylic acid. The nickel modified Y zeolites with carefully controlled concentrations of metallic and acidic sites were prepared, characterized and tested in acetylene carbonylation performed in a batch reactor. The catalysts were thoroughly characterized by ICP-AES, X-ray diffraction, Temperature-programmed Reduction, Extended X-ray absorption fine structure, Temperature-programmed desorption of NH_3, pyridine-FTIR, SEM, Raman and thermal gravimetric analysis. It was found that all of the exchanged Ni in zeolite were present as charge-compensating cations and remained this state through the overall reaction. The nickel sites play a determinative role in the carbonylation of acetylene, and the acidic sites would interact with acetylene and favor the protonation of acetylene to form active intermediates, but a too high density of acid sites would cause the formation of coke. Base on these findings, it is necessary to design a catalyst with a proper ratio between nickel ion and acid sites. In addition, promoters such as copper salts were required, and the $CuBr_2$/Ni-Y-zeolite has been proved to be a highly effective catalyst system in the hydrocarboxylation of acetylene with 1400 h^{-1} of average turn over frequency after six times of recycle at 235 and 3.6MPa of initial total pressure with 30min of reaction time. The amount of acrylic acids formed with these Ni-exchanged Y catalysts($4\sim5\times10^4$g/(molNi*h)) was far higher than those reported with Ni supported on silica gel (~3.4 g/(molNi*h)). Furthermore, the coke deposited on the zeolites, locating mainly on the external of external surface of the zeolite, was present as ordered filamentous nature, and much of them was separated from the zeolite crystals and thus keeps the catalysts with a remarkable reusability. The current heterogeneous catalysts system showes the best catalytic activity for synthesizing acrylic acid by carbonylation of acetylene with carbon monoxide and water to date.

375673 Neural Network Controller for Regulation of a Water-Cooled Fuel Cell Stack

Abstract:

Fuel cells are readily available for use as backup generators for residential applications in the 5 kW range. The advantages these generators have over conventional gasoline or natural gas powered generators include improved effciency, easy maintenance due to the absence of moving parts, and reduced noise during operation. The idea of using a fuel cell for dynamic power demand is relatively new and this application requires the development of a control system that keeps the system at optimal conditions during power demand uctuationss. In a previous paper, we developed a mathematical model for a novel water-cooled fuel-cell stack that provided power up to 5 kW for a residential application. In this paper, a backpropagation neural network is designed to control the temperature of a 5 kW hydrogen fuel cell stack. The power demand is obtained experimentally from a 3-bedroom house. The controller is initially trained to recognize the pattern of dynamic power demand from a consumption source. The power demand is the driving force in current production from the fuel cell, which in turn increases the temperature of the fuel cell stack. This temperature change is controlled by changing the flow rate of cooling water through the fuel cell stack. The results show that the neural network controller has excellent performance in maintaining the stack temperature at the desired set-point despite signicant fluctuations in power demand.

388461 Cathode Filtration System Design and Optimization for PEM Fuel Cells

Abstract:

Cathode contamination causes significant performance degradation of proton exchange membrane fuel cells, as a result of poisoning the membrane electrode assembly. Since the variety of contaminant species, the simplest solution to mitigate the contamination effect is using cathode filtration. However, the cathode filtration increases the parasitic power loss and hence reduces the net power output of fuel cells. Typically, the total system parasitic losses can reach up to 10-20% of net power output depending on the system design and optimization. This study aims to maximize the net power output and life-time of fuel cells in a polluted air environment by using novel filter media and optimizing the filtration design. A novel adsorptive sorbent media, activated carbon fiber enhanced microfibrous entrapped sorbent was developed and used as a polishing layer to improve the performance of traditional packed bed filter. The full size activated carbon filter (24×24") was tested in a full scale test rig which was constructed according to ANSI/ASHRAE Standard 145.2. The resulting breakthrough data was applied to determine the adsorptive filter parameters including the saturation capacity and adsorption rate constant. At last, a filter design methodology was developed through carefully considering the attributes of fuel cell, operating conditions (air flow rate, stoichiometric ratio, operating temperature and pressure, current density, etc.), and polluted air properties (contaminants types and concentration). The design methodology was successfully applied to a 60 kW fuel cell to simulate the relationships between net output and filtration with time. The simulation results show that the optimization of filtration design can significantly reduce the power consumption of compressor and hence improve the total efficiency of fuel cell system.

377357 Nickel-Based Anode with Microstructured MoO2 Internal Reformer for Liquid Hydrocarbon-Fueled SOFCs

Abstract:

The present paper describes the fabrication of a bilayer structured solid oxide fuel cell (SOFC). Its anode consisting of an internal MoO_2-based micro-reformer in the form of a porous thin film deposited over a Ni-based anode. Cell performance was measured by directly feeding a mixture of *n*-dodecane and air at different O_2/C ratios to the anode at 750°C. Our findings show that the bilayer structured SOFC operating at an O_2/C ratio of 0.64 leads to the highest initial cell performance with a maximum power density output > 4 W/cm^2. At a constant voltage of 0.7 V using an O_2/C ratio of 0.64, the bilayer structured SOFC showed a gradual increase in power density output over the first 2 h, followed by a stable output of 3.6 W/cm^2 for the next 10 h. The tested cell showed no indication of coking. When a conventional Ni-based SOFC, without the internal micro-reformer, was operated under similar conditions, its initial performance and long-term stability were found to be significantly lower than that of the bilayer structured SOFC. Hence, the present work demonstrates that a Ni-based SOFC with an integrated MoO_2 micro-reformer can effectively operate when fed directly with *n*-dodecane without an external reformer. These results open up new opportunities for efficiently generating electrical power from both existing transportation fuels and next generation biomass-derived liquid fuels using SOFCs with an integrated MoO_2 micro-reformer.

387241 Design Considerations of Deployable Solar-Hydrogen Generators

Abstract:

The generation of hydrogen from sunlight is an attractive alternative to directly capture and store solar-energy in the form of a fuel. Solar-fuel generators have the potential to significantly increase the share of renewable energy sources in our current energy portfolio. These devices need to be able to absorb light, generate and transport charge carriers to catalytic centers where water can be dissociated. To date, a significant amount of research has been devoted to the development of photoelectrochemical components (i.e. photovoltaics and electrocatalysts) that can split water in the presence of solar irradiation, but only limited attention has been given to understanding the key aspects that drive economically viable solar-fuel generators. In this study, we introduce a system agnostic approach to elucidate the economic factor behind the design of cost-effective solar-hydrogen generators. The methodology presented here makes use of electrochemical and cost models to compare different material combinations for light absorption and water splitting components on the basis of the hydrogen production cost (\$/Kg of H_2). Results arising from the analysis demonstrate that properly designed systems can lead to costs as low as \$2.90 per kilogram of hydrogen after compression and distribution (versus \$2.7-5.7 for hydrogen produced from fossil fuel). Moreover, the analysis demonstrates that the primary cost component in optimized solar-hydrogen generators corresponds to the light-absorbing units (up to 97% of the cost) while the materials selection for the electrolysis components has only small effects in the cost. These results arise from the fact that the photovoltaic components in optimized systems occupy areas that are 2 orders of magnitude larger than those required for electrolysis. Furthermore, physical models were developed to understand the cost effects of the incorporation of solar-concentrators and current invertors, and the optimal scenarios for their implementation will be discussed.

389466 Design and Construction of a Fluidized Bed Gasifier

Abstract:

political instability in producing regions, has stimulated the search for alternative and renewable fuels . In the hotbed of alternative energy at the Free University has been deepening in investigating the use of biomass for oil and power generation, highlighting the processes of pyrolysis and gasification . Gasification is an efficient method for the production of hydrocarbons and biomass energy has , which is a renewable resource and not edible , sometimes considered waste .

In Colombia palm oil producers are seeking a solution to what to do or where to deposit the daily waste production? This waste called pericarp or fart , is produced in considerable quantities every day , being used as a base material in internal way the plantations where the palm is grown , due to its high strength and light weight. For mechanical and chemical characteristics will be used as a source of energy and biofuel through fluidized bed gasification, with the process expected significant results that would be achieved vital for energy-environmental development in remote populations in the country where poor electrical wiring and fuel supply, as in the Meta, the Guaviare, the Amazon, the Orinoco, the Pacific Coast, these being the main producers of palm oil, a source of biomass. To achieve this end has experience in biomass thermochemical analysis and supported by expertise in gasification studies.

As these studies were obtained outstanding results through experience in fixed bed gasification. Looking to extend the field of investigation now designing a fluidized bed gasifier model shows how the gasification process , implemented in palm oil producing regions and geographically remote serve as a source of electrical energy, biofuels and in turn reducing waste , thus providing to improve the quality of life of these communities.

384211 Hydrodeoxygenation of Bio-Oil over Pt Supported Catalyst:: Importance of Mesopores and Acidity

Abstract:

Biofuel shows promising potentials to answer our current globule challenges on energy and environment. The key to its success lies on the conversion of biomass to high quality biofuel and/or chemical feedstocks. But unlike fossil fuel, biofuel is composed of many carbohydrates with high oxygen content and low hydrogen/carbon ratios, which leads to the low heat capacity, poor thermal and chemical stability of biofuel. Further hydrogenation and deoxygenation must be done before biofuel can serve as attractive replacements of the conventional fuel. We hypothesize that catalyst supported mesoporous zeolites (i.e., hierarchical zeolites) could provide superior activity and selectivity than other common support materials (e.g., alumina and silica): zeolites offer strong acidity for hydrogenation and good hydrothermal stability while the hierarchical structure improves the diffusion of large compounds (presenting with a high percentage in bio-oil), the reduction of pore blocking, and/or the lessening of deactivation of catalyst from coking. To demonstrate our concept, a novel dual-template synthesis approach was used to produce new mesoporous ZSM-5 zeolites and Pt catalyst was impregnated for catalytic hydrodeoxygenation of bio-oil. We found better catalytic activity of mesoporous ZSM-5 zeolites based catalysts than their conventional counterparts (i.e., microporous zeolites) and those on alumina or silica in the deoxygenation or hydrogenation of phenol-type bio-oil compounds. By testing on three model bio-oil compounds, it showed that the catalysts with strong acidity (e.g. ZSM-5) have high hydrodeoxygantion activity while mesopores of support can further help improve the catalytical performance of ZSM-5. Exploring on new synthesis strategies for more desired hierarchical zeolites (e.g., with large mesopores, strong acidity, or their combinations) and/or better catalytic performance of these new hierarchical zeolites are currently under investigation. Our efforts are expected to help make biomass/biofuel a truly renewable and environmental friendly energy source.

BIOFUELS FROM PYROLYSIS OIL: A TWO-STEP HYDRODEOXYGENATION PROCESS

Graz University of Technology

Nikolaus Schwaiger

Graz University of Technology
Institute of Chemical Engineering and Environmental Technology

BDI-BioEnergy International AG

Graz University of Technology

FFG

a3plus

Outline

- Introduction
  - Liquid phase pyrolysis (bioCRACK)
- Liquid phase pyrolysis oil
  - Specification
- Hydrodeoxygenation
  - Setup
  - Product Formation
  - Results
- Summary

Concept

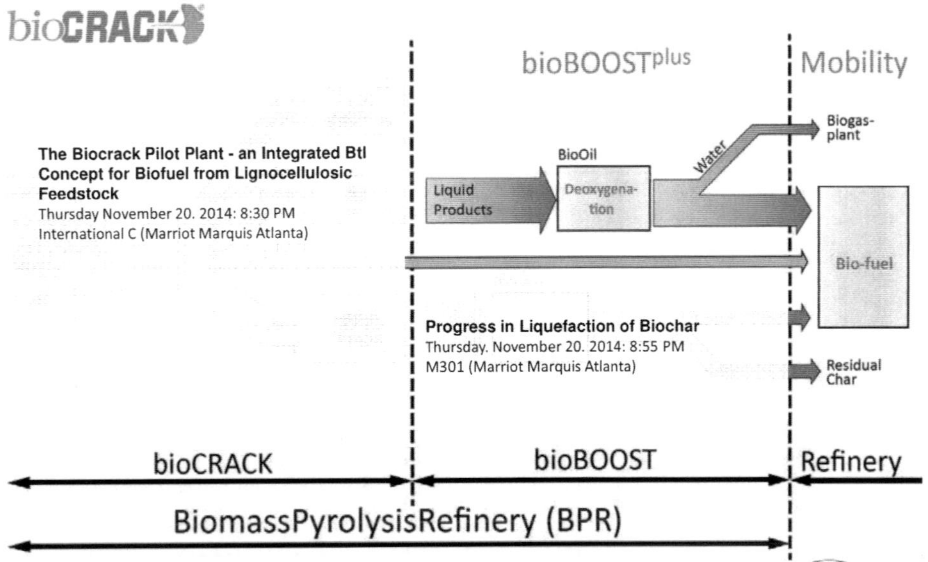

Liquid phase pyrolysis

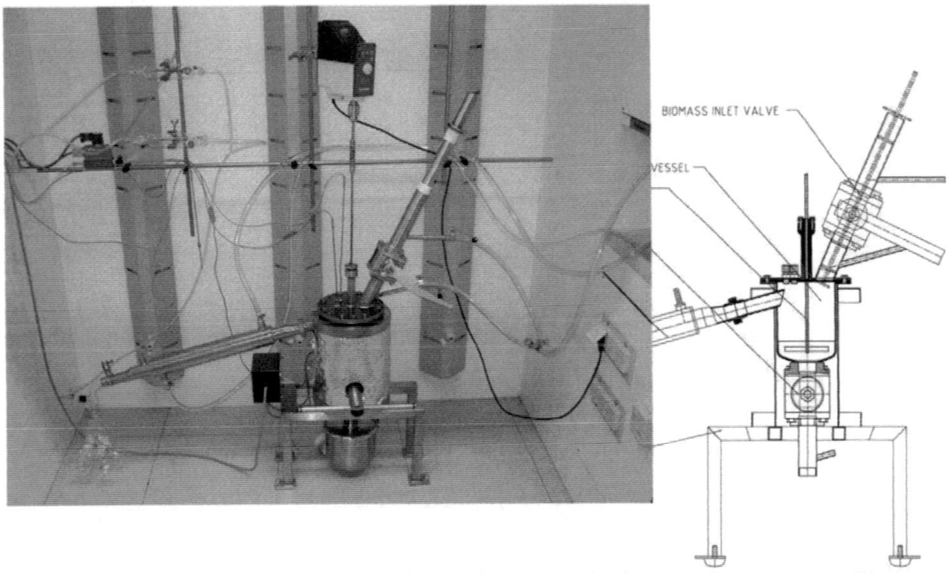

Liquid phase pyrolysis

100 kg/h Biomass

7 Mio. € Project

Pilot plant; OMV Vienna

BDI bioCRACK

OMV

klima+ energie fonds

CEET

Liquid phase pyrolysis in the bioCRACK pilot plant

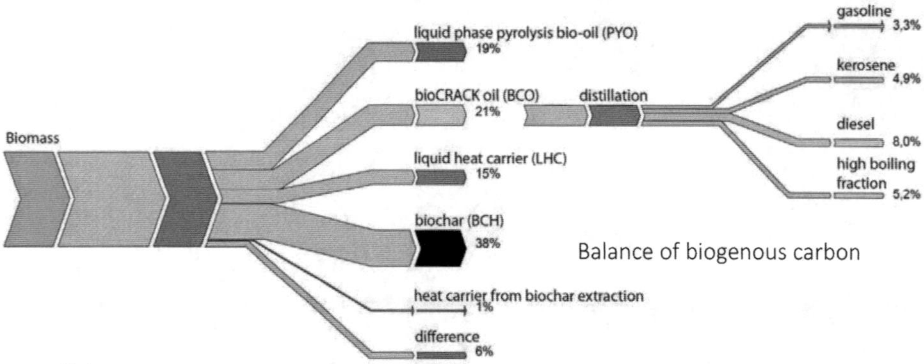

Balance of biogenous carbon

Setup:

- reaction temperature: 375°C
- duration of experiment: 38 h
- biomass feed rate 60 kg/h

CEET

Composition of bioCRACK oil and liquid phase pyrolysis oil

	Liquid phase pyrolysis oil	BCO
Watercontent [wt.%]	49,6	0,1
LHV [MJ/kg]	9,2	40,1
TAN [mg_{KOH}/g_{sample}]	82,2	-
Density [kg/m^3]	1,09	0,88
Viscosity [mPa s]	3,9	8,0
^{14}C-Content [wt.%]	100	9,8
	Elemental Composition	
Carbon content [wt.%]	25,6	83,6
Hydrogen content [wt.%]	9,2	11,8
Oxygen content [wt.%]	64,9	4,2
Nitrogen content [wt.%]	<1	<1

Concept

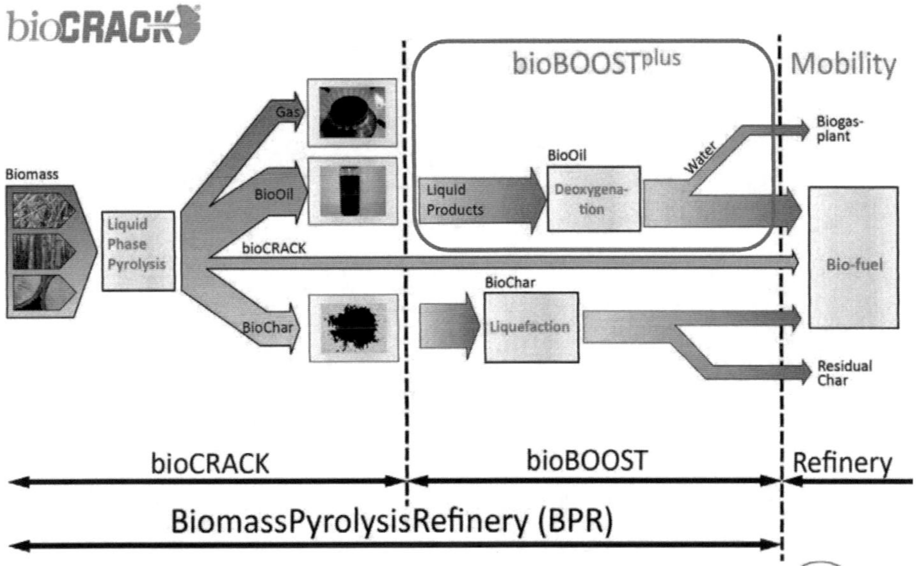

Liquid phase pyrolysis oil Hydrodeoxygenation

CEET

Setup

CEET

Setup

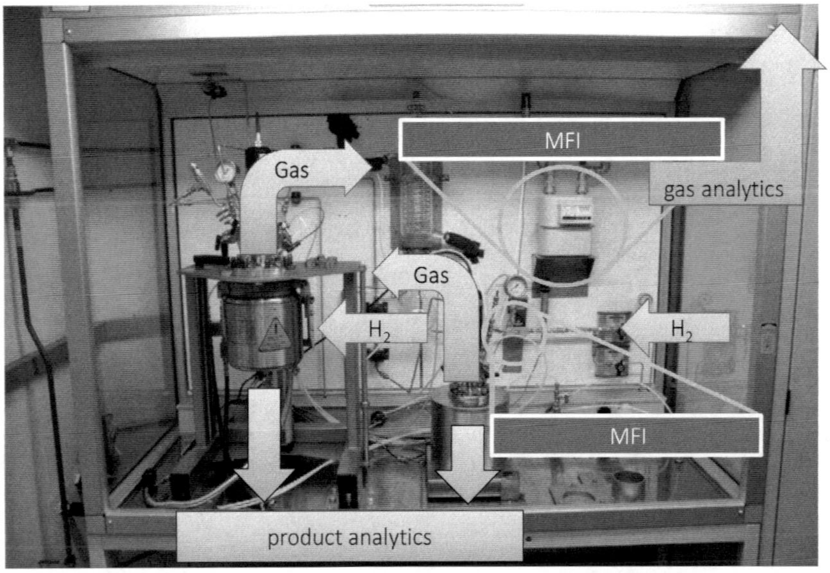

Product formation

Product formation

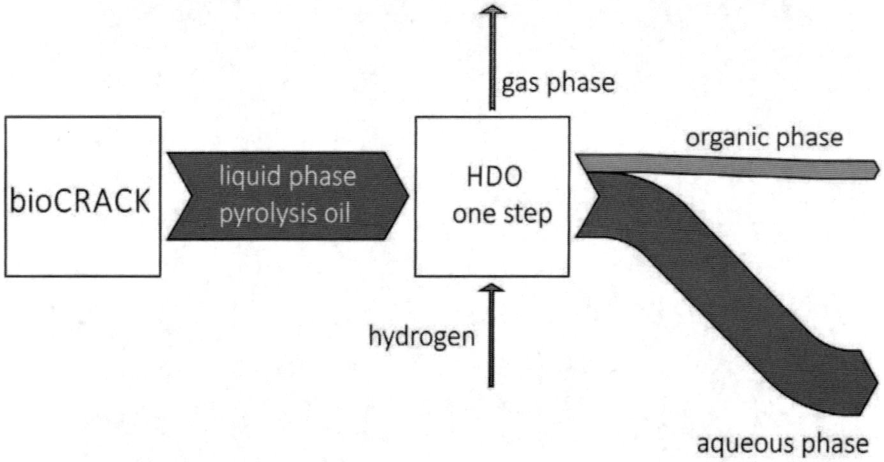

Reaction phase 1 = heating

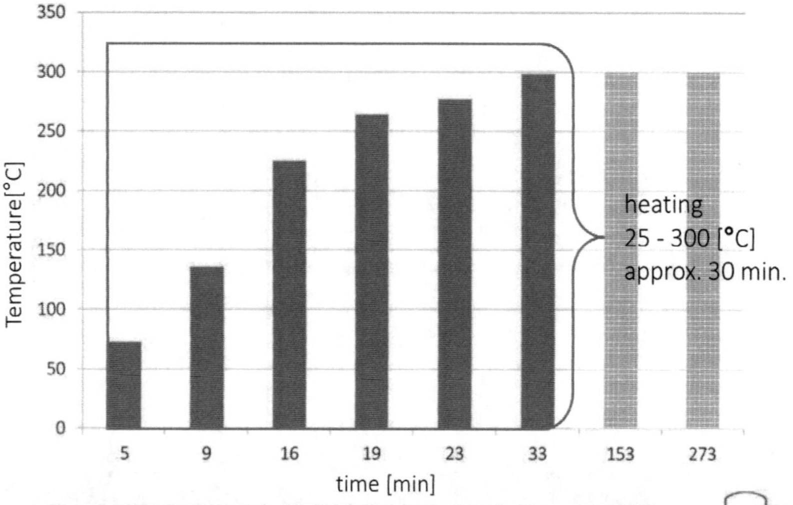

Reaction phase 2 = isothermal stage

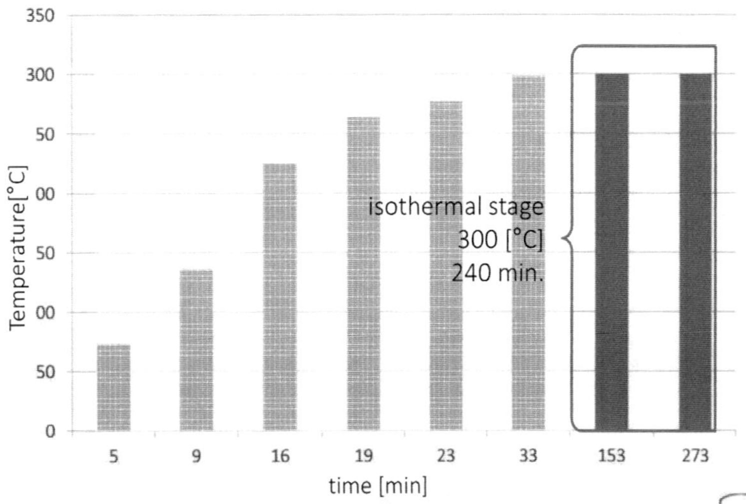

Product distribution

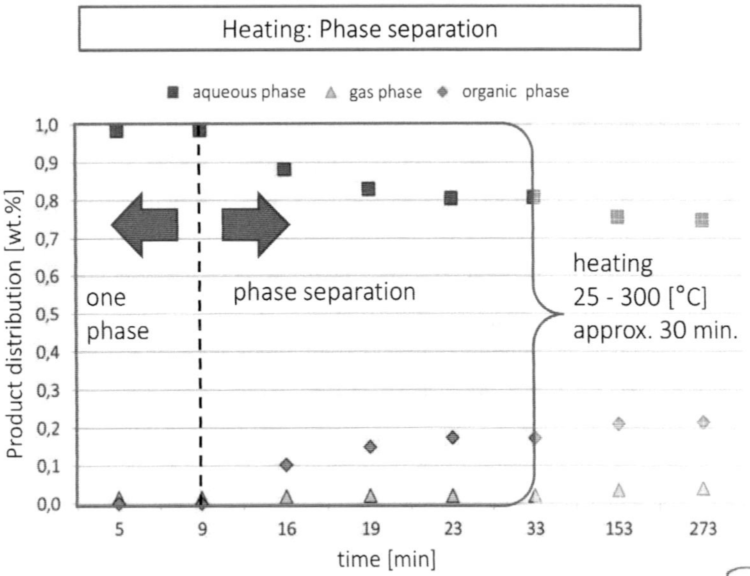

Product distribution

isothermal stage: minor changes

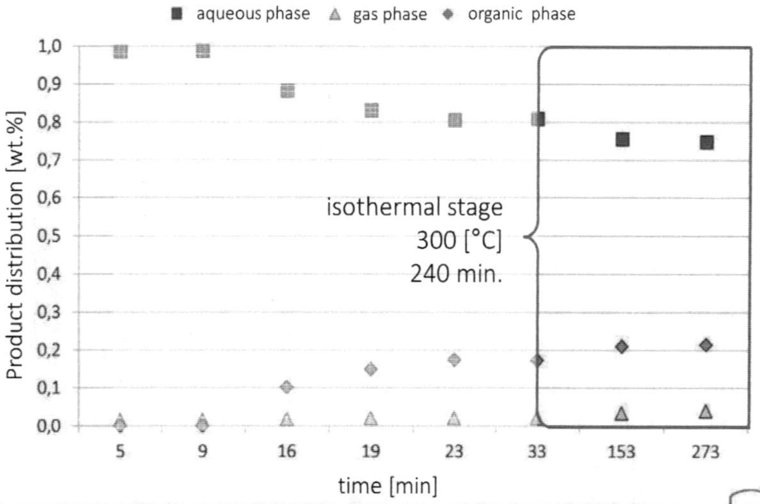

Elemental composition of product phase

phase separation leads to oxygen depletion

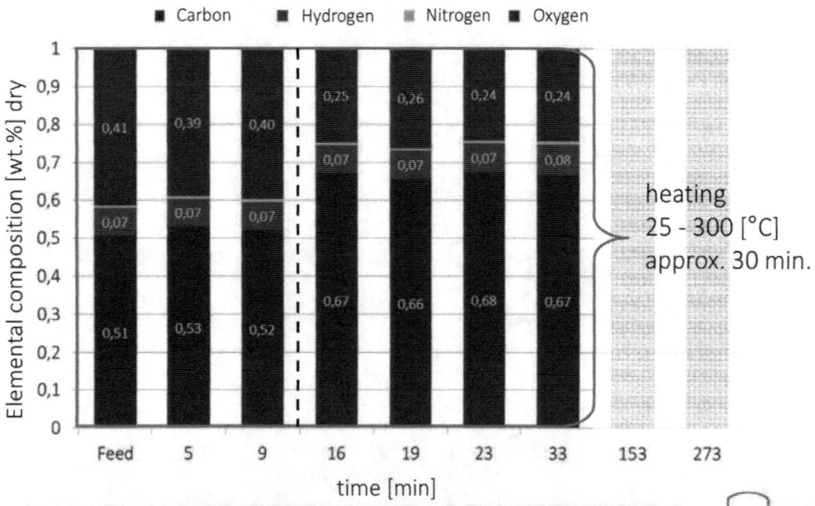

Elemental composition of product phase

HDO is performed in the heating stage

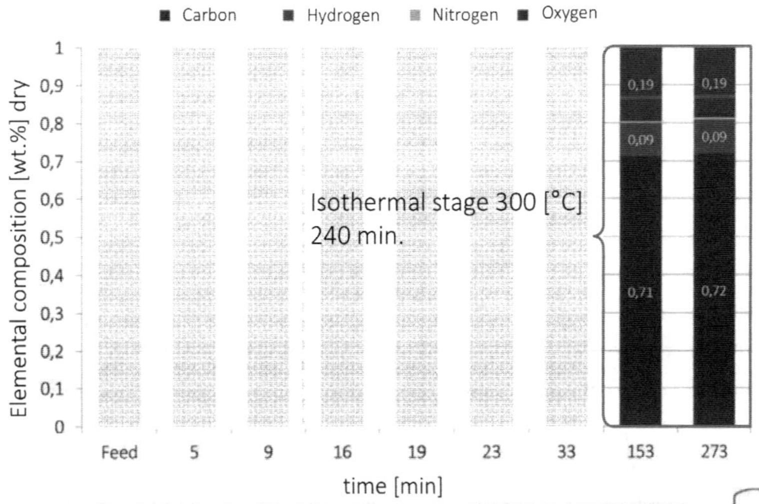

GPC Analysis

Molecular weight distribution during HDO

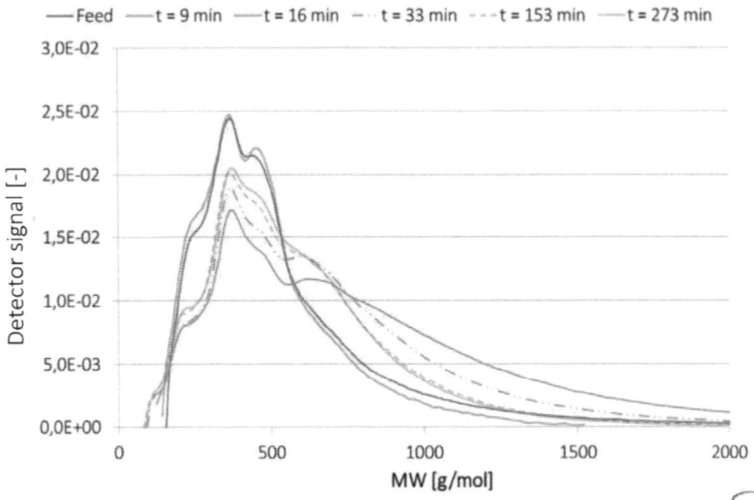

Van Krevelen Plot

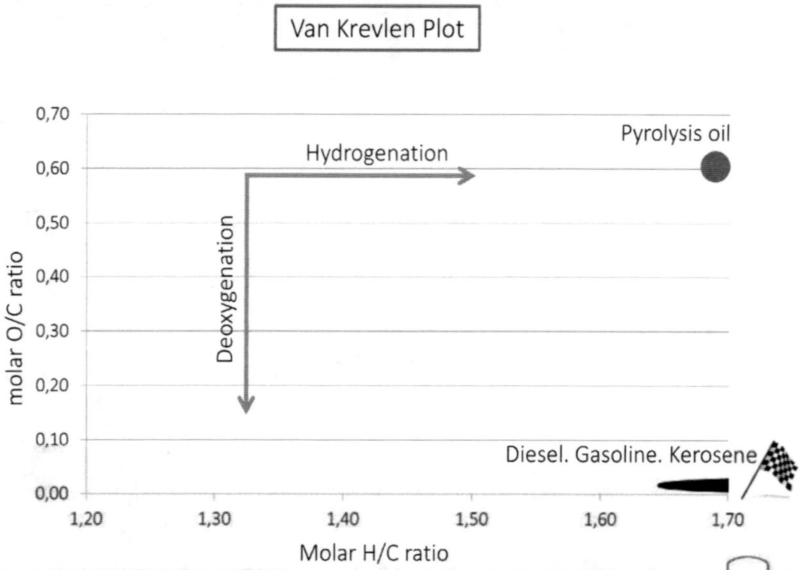

Van Krevelen Plot

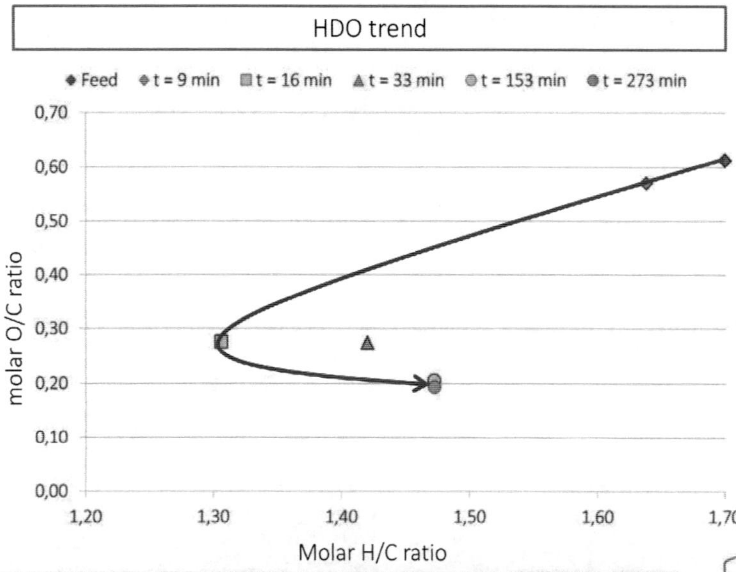

Two step process

Two step process

Process scheme

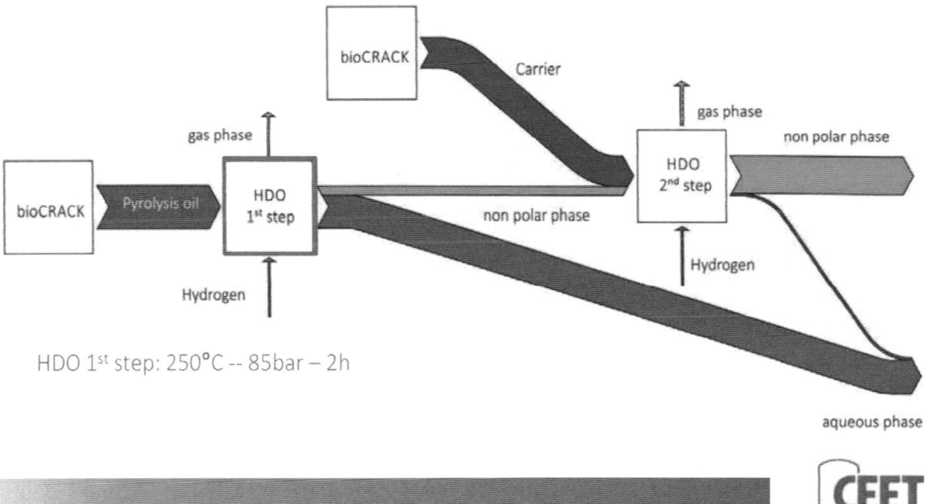

HDO 1st step: 250°C -- 85bar – 2h

Two step process

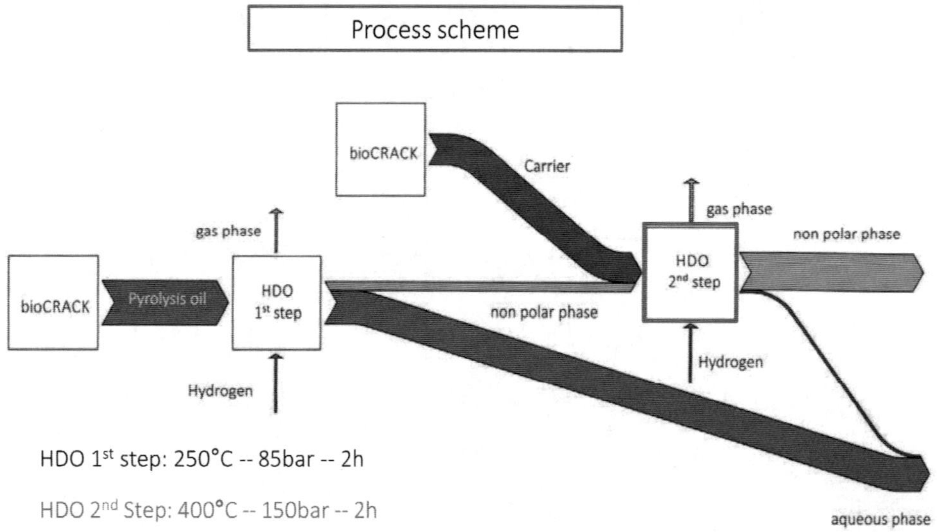

HDO 1st step: 250°C -- 85bar -- 2h

HDO 2nd Step: 400°C -- 150bar -- 2h

Van Krevelen plot – two step process

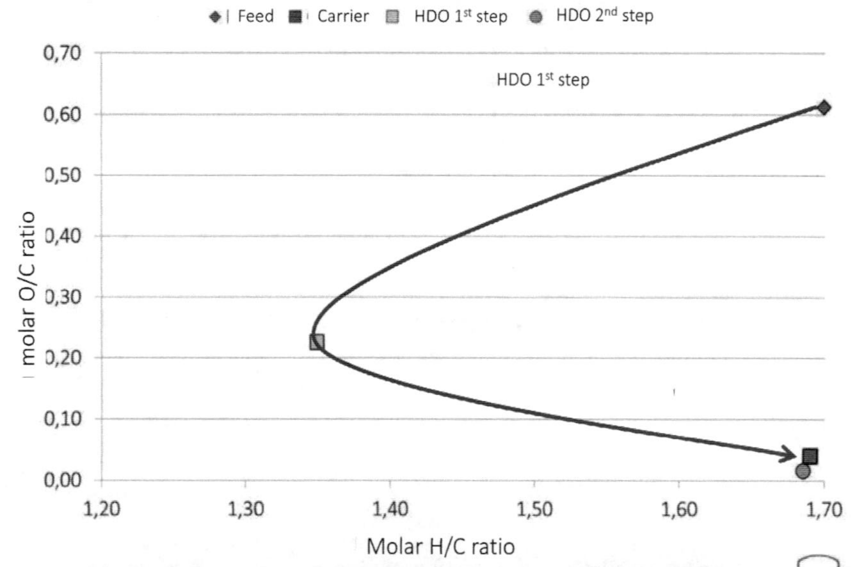

ATR-IR Spectra

Vanishing of oxygenated functional groups

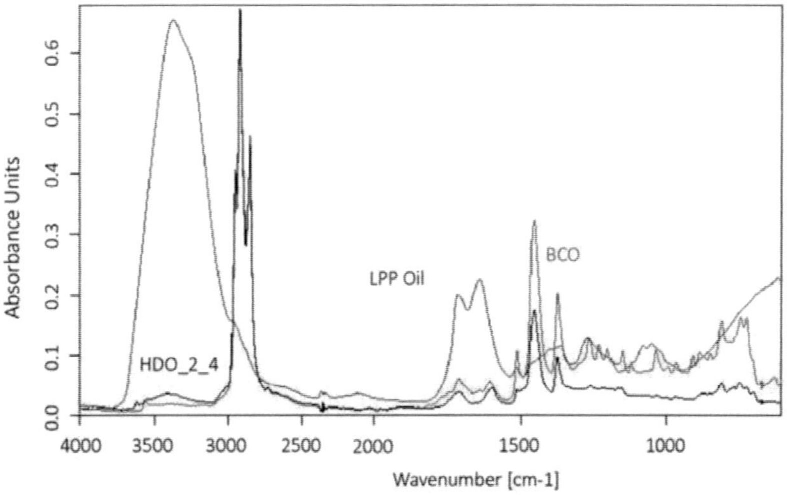

Chemical Engineering and Environmental Technology | www.icvt.tugraz.at 27

CEET

Boiling range

Boiling Range is comparable to Diesel

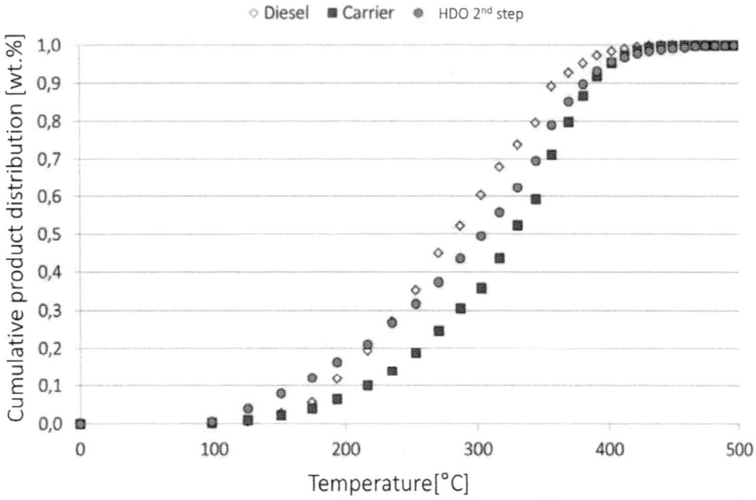

Chemical Engineering and Environmental Technology | www.icvt.tugraz.at 28

CEET

Conclusion

During the two step process 67% of chemical bonded energy is transferred from liquid phase pyrolysis oil into the fuel phase

	unit	Feed	HDO 1st step	HDO 2nd step	Diesel
Water content	[wt.%]	50	11	0.2	0.02
LHV	[MJ/kg]	8.7	26.4	41.2	42.5
Density	[kg/m^3]	1070	1100	870	845
Viscosity	[mPa s]	4	163	4.5	4.5
Carbon content	[wt.%]	25.6	62.5	85.5	85.9
Biogenous content	[wt.%]	100	100	28	<7
Hydrogen content	[wt.%]	9.2	8.3	12.1	13.3
Oxygen content	[wt.%]	64.9	28.7	1.9	<1
Nitrogen content	[wt.%]	<1	<1	<1	<1

Acknowledgement

Prof. Dr. Matthäus Siebenhofer
Dr. Hannes Pucher
Dr. Roland Feiner

Dr. Peter Pucher
Dipl.-Ing. Lisa Ellmaier

BIOFUELS FROM PYROLYSIS OIL: A TWO-STEP HYDRODEOXYGENATION PROCESS

Nikolaus Schwaiger

Graz University of Technology

Institute of Chemical Engineering and Environmental Technology
Inffeldgasse 25 C. 8010 Graz

Nikolaus.Schwaiger@tugraz.at

Tel: +43 316 873 7965

Ni-Cu catalyzed transfer hydrogenation of *p*-cresol and biomass pyrolysis oils

Hari Prasad Reddy Kannapu, Charles A. Mullen*, Akwasi A. Boateng

Eastern Regional Research Center, Agricultural Research Service, U. S. Department of Agriculture, 600 E. Mermaid Lane, Wyndmoor, PA 19038.
*charles.mullen@ars.usda.gov

Hydrodeoxygenation of biomass fast pyrolysis oils is an often cited potential pathway for production of drop in renewable fuels. However, the well documented negative properties of conventional fast pyrolysis oils, including incompatibility with hydrocarbons, thermal instability, acidity and corrosiveness result in unfeasible economic expense in the hydrotreating step due to high hydrogen consumption and short catalyst lifetimes. Reducing this cost by producing higher quality pyrolysis oil, developing more efficient upgrading technologies or both is vital to the economic feasibility of pyrolysis based biofuels. This has led researchers to seek advanced pyrolysis methods, those that produce stable, partially deoxygenated pyrolysis oils. Two such methods are catalytic fast pyrolysis (CFP) and tail gas reactive pyrolysis (TGRP), both of which can produce pyrolysis oils containing 10 wt% or less oxygen. A large fraction of the remaining oxygenated components in these stable pyrolysis oils are alkyl phenols. In this work, the hydrogenation and hydrodeoxygenation of *p*-cresol were studied to determine conditions for upgrading of these types of pyrolysis oils without the addition of external hydrogen. Ni-Cu bimetallic catalysts supported on both carbon and alumina in combination with isopropanol as a hydrogen donor molecule proved effective for hydrogenation to 4-methylcylcylohexanol with a small amount of the hydrodeoxygenation products toluene and 4-methylcyclohexane also produced. The optimized conditions were applied to authentic TGRP biomass pyrolysis oils where hydrogenation reactions that further stabilized the pyrolysis oil were observed. For authentic pyrolysis oils, carbon supported catalysts were preferred over the alumina supported catalysts because the later was highly selective for undesired condensation reactions between bio-oil components and isopropanol while the former were more selective for hydrogenation and hydrodeoxygenation reactions. The chemical and physical changes in bio-oil observed using this upgrading process will be discussed.

377016 Techno-Economics of Formate Assisted Pyrolysis

Abstract:

The purpose of this study is to determine the capital cost for producing liquid fuel from biomass by using Formate-Assisted Pyrolysis (FAsP) process. Many studies have been published about producing liquid fuel from biomass by using fast pyrolysis. Fast pyrolysis produces bio-oil with 60% mass yield; however, the bio-oil is unstable, has a high amount of oxygen 35 wt% and requires high capital cost to upgrade it. Formate-Assisted Pyrolysis (FAsP), a new method being developed at UMaine, produces a bio-oil that is more stable and has less amount oxygen. FAsP uses the same operating conditions as fast pyrolysis (500 °C and atmospheric pressure), but also integrates the thermal decomposition of formate salts. Typical FAsP oil has 15 wt% oxygen and 34% energy yield. Experiments were done using different calcium formate/biomass ratios.

An Aspen Plus® process model has been developed for a 1000 dry metric ton of biomass per day FAsP plant. The model includes feedstock sizing and drying, pyrolysis, hydrogen production and hydrotreatment of FAsP oils and recycle of calcium. The char is gasified to produce CO to regenerate calcium formate. Other gases from upgrading and pyrolysis are used to produce hydrogen which is necessary for bio-oil upgrading. After the bio-oil is upgraded, the bio-oil is treated to reach a final 13% gasoline/diesel mass yield and 33% energy yield based on the dry biomass fed.

380394 Fatty Acid Methyl Esters from Soybean Oil with Honeycomb Monolithic Catalysts

Abstract:

Viscous vegetable oils containing triglycerides of various fatty acids and free fatty acids are mostly converted to less viscous fatty acid methyl esters (FAME) through the methanol trans-esterification process, using homogeneous catalysts in batch reactors. FAME biofuels produced with homogeneous alkali catalysts through transesterifcation processes contain undesirable substances such as soap and homogeneous catalysts in addition to glycerol and water. Additional refining and purification processes are needed to remove these undesirable substances from FAME biofuels, generating wastewater. We converted viscous soybean oil to less viscous FAME biofuels with methanol and honeycomb monolithic catalysts in a flow reactor at moderate temperatures and pressures to develop monolithic catalysts for efficient methanol transesterification processes.

Active and stable monolithic catalysts were formulated with various active metals and honeycomb monolithic catalyst supports to produce FAME biofuels from soybean oil. Flow reactors with honeycomb monolithic catalysts have some advantages over flow reactors packed/fluidized with catalyst pellets for the transesterification of soybean oil. Pressure drops will be lower over flow reactors with honeycomb monolithic catalysts than those packed with pellet catalysts. High uniform mass transfer rates of methanol and soybean oil to porous surface of solid catalysts will occur in channels of monolithic catalysts in comparison with packed catalyst pellets. Formulated honeycomb monolithic catalysts are deactivated and regenerated during converting soybean oil in methanol to FAME biofuels in a vertical honeycomb monolithic catalyst flow reactor by showing conversion of soybean oil to FAME biofuels fluctuated, that is not observed with powdered/pellet catalysts in a packed or fluidized flow reactor.

Pacific Northwest NATIONAL LABORATORY
Proudly Operated by Battelle Since 1965

Bio-oil streams in the current fast pyrolysis – catalytic upgrading route of producing renewable liquid fuel from biomass

MV OLARTE, AH ZACHER, DC ELLIOTT, GG NEUENSCHWANDER, LJ ROTNESS, JR CORT, SD BURTON, AB PADMAPERUMA and C DRENNAN
Pacific Northwest National Laboratory, USA

J CHEN, C FAIRBRIDGE, R GIELECIAK and D HAGER
CanmetENERGY, Canada

CATALYTIC BIOFUELS REFINING
FALL 2014 AICHE MEETING
NOV. 18, 2014
ATLANTA, GEORGIA

U.S. DEPARTMENT OF ENERGY

PNNL-SA-106650

Outline of the talk

Pacific Northwest NATIONAL LABORATORY
Proudly Operated by Battelle Since 1965

- Introduction
  - Drivers for fast pyrolysis-catalytic upgrading process
  - Process description
  - The "plugging" challenge
  - Enabling advances to catalytic upgrading process
- Characterization of feed, intermediate and product streams
  - Pyrolysis oils
  - Pre-treated oils
  - Upgraded oils
- What we learned and path forward

2

Why produce renewable fuel from biomass?

Pacific Northwest NATIONAL LABORATORY
Proudly Operated by Battelle Since 1965

- Biomass Policy – *Energy Independence and Security Act (EISA) Renewable Fuels Standard*
  - By 2022, provide 36 billion gallons of renewable fuel (~20% of current transportation fuels)
  - Requires utilization of cellulosic biomass
  - Requires development of economical processes to **convert biomass to infrastructure ready fuels and chemical products**
  - Requires **integration** with existing infrastructure (oil refineries, transportation pipelines)

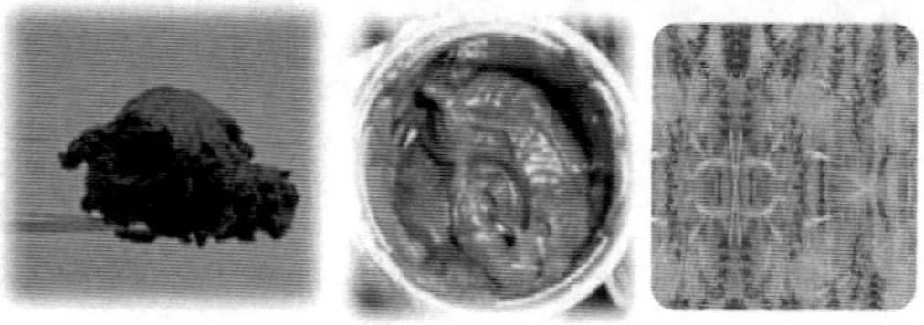

- Grand Challenges
  - Capability to understand fundamental chemistries and reaction pathways of hydrocarbon feedstocks to develop new conversion processes.
  - Enable efficient utilization and conversion of domestic hydrocarbon resources on a regional basis.
  - Capture current refinery capacity and consider regional, flexible mini-refineries (similar to "mini-mills" for steel industry).

Current BETO cost target by 2017: gasoline = $3/gge

3

Why produce renewable fuel from biomass?

Pacific Northwest NATIONAL LABORATORY
Proudly Operated by Battelle Since 1965

- Biomass – most accessible source of renewable liquid transportation fuel

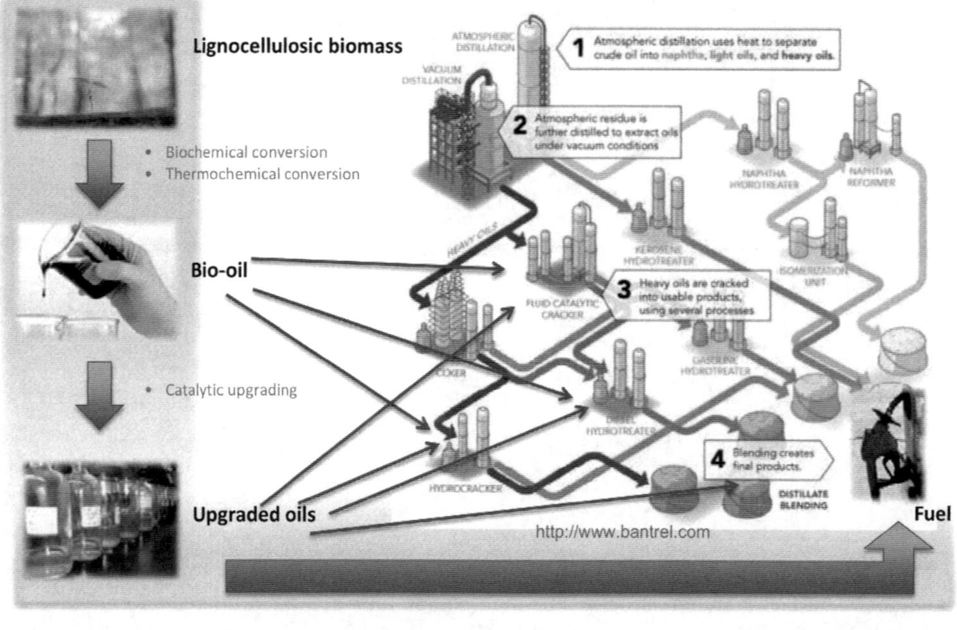

4

Fast pyrolysis process overview

Pacific Northwest NATIONAL LABORATORY
Proudly Operated by **Battelle** *Since 1965*

Treatment of size-reduced biomass at high temperature **(450-550°C)** under **oxygen-free atmosphere**, with a residence time **< 2 sec**.

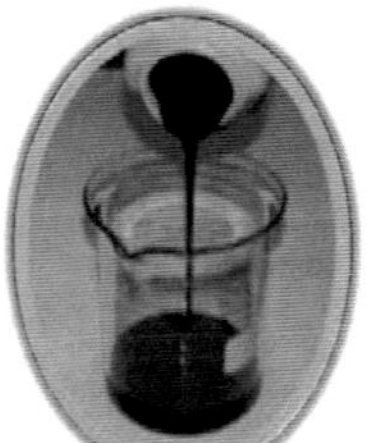

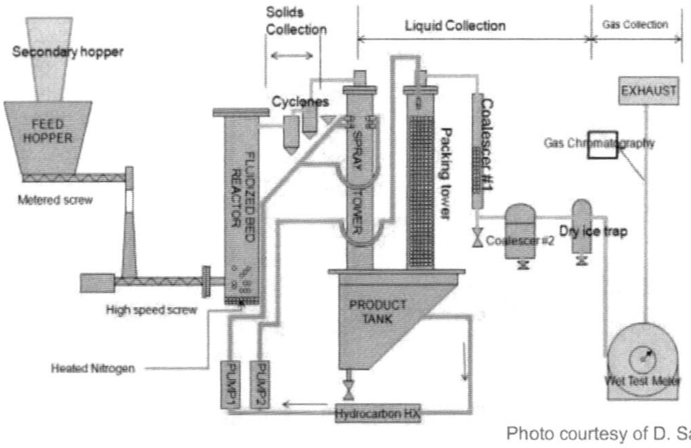

Photo courtesy of D. Santosa.

Hydrocarbon quench circulation flow

Recommended for processing of fairly dry biomass (≥ 90% solids)

5

Catalytic upgrading process overview

Pacific Northwest NATIONAL LABORATORY
Proudly Operated by **Battelle** *Since 1965*

Removal of O as CO_x but preferably as H_2O, under H_2 pressure (> 1200 psig) and with the use of catalysts, typically promoted (Ni, Co) sulfided Mo on catalyst supports

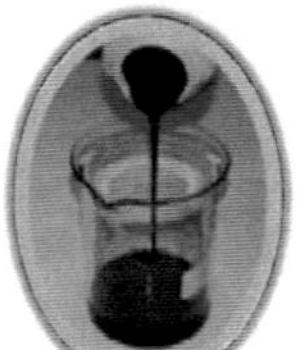

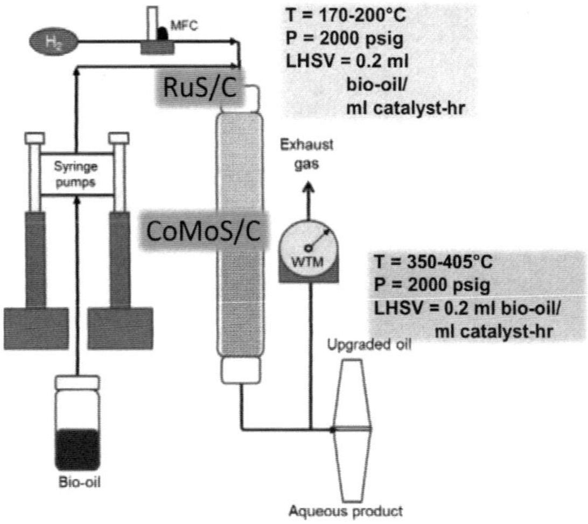

Hydrodeoxygenation/ Hydrocracking

6

PNNL catalytic upgrading at different scales

Pacific Northwest NATIONAL LABORATORY
Proudly Operated by Battelle Since 1965

~1.4 ml 8-reactor packed bed system

- ☐ Enables high-throughput testing

30/60 ml dual T zone packed bed reactor

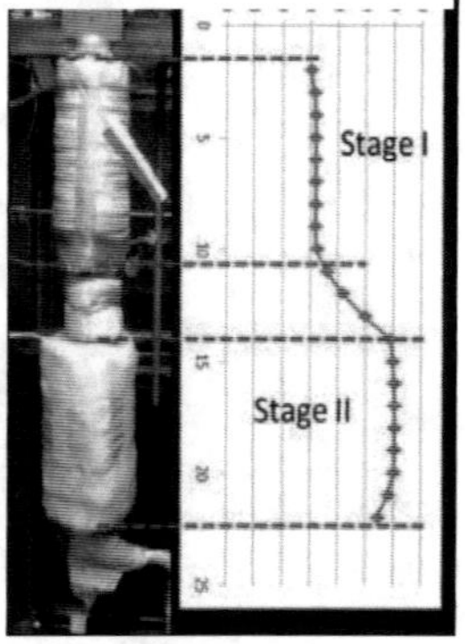

- ☐ Enables further screening and unattended catalyst lifetime studies

400/900 ml dual T zone packed bed reactors

- ☐ Enables scaled-up testing

7

Operational issues with FP-catalytic upgrading

Pacific Northwest NATIONAL LABORATORY
Proudly Operated by Battelle Since 1965

- ► Until ca. 2011: longest cumulative time on stream was about 150 hrs

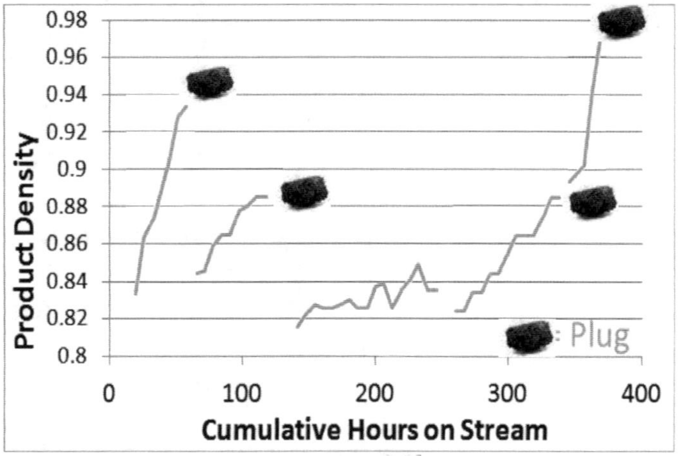

Stop experiment due to over-pressure

- ☐ Hydrotreating catalyst: CoMoS, 250°C-410°C
- ☐ Pre-hydrotreated bio-oils (unoptimized)
- ☐ 10% of the catalyst replaced when the bed plugged

8

Operational issue with FP- catalytic upgrading

Pacific Northwest NATIONAL LABORATORY
Proudly Operated by Battelle Since 1965

Possible causes of **reactor plugging:**

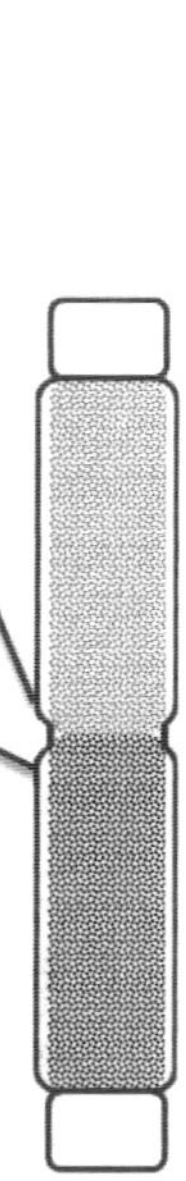

- Physical build-up of char/inorganic solids
  - Remedy: filtration of pyrolysis oil to remove char (<1% by 50:50 DCM:MeOH)
  - Inorganics typically associated with chars
- Catalyst deactivation/poisoning
  - Catalyst design
  - Implement processing and configuration changes
- Build up of polymeric carbons
  - Typically occurs at the transition zone between low and high temperature beds

9

Pre-treatment (stabilization) of bio-oil to improve catalyst lifetime

Pacific Northwest NATIONAL LABORATORY
Proudly Operated by Battelle Since 1965

Comparison of raw and stabilized feeds for catalytic upgrading:

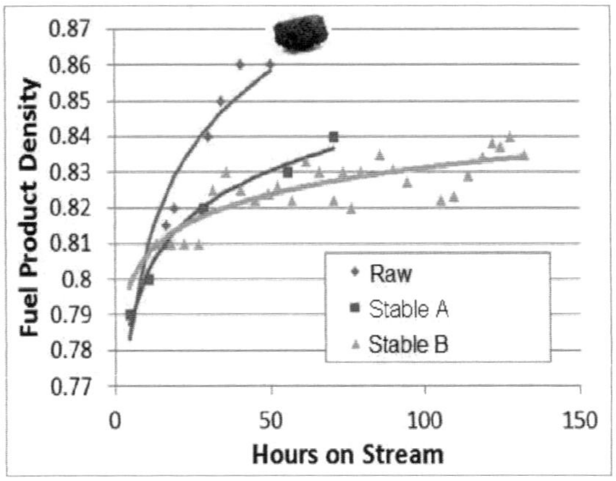

Feed	Raw	Stable A	Stable B
Sample time (hr)	48	75	141
WHSV	0.24	0.27	0.27
Stabilization	none	Ru	Ru
1st catalyst (RuS)	Commercial		Lab prep
2nd catalyst	Same in all tests		
Temp	170/402	171/406	170/429
Fuel density	0.86	0.84	0.84
Oil Yield	0.42	0.40	0.40
H used (g/g dry)	0.060	0.066	0.049
Oil C yield	60.7%	58.3%	59.6%
Aq C yield	0.1%	0.1%	0.4%
Outcome	**PLUG**	**clear**	**clear**

- ❑ Untreated bio-oil plugged after 48 hours TOS.
- ❑ Stabilized bio-oil was operated up to 141 hours without plugging.

Advanced catalytic upgrading: incorporation of pre-treatment process

Pacific Northwest NATIONAL LABORATORY
Proudly Operated by Battelle Since 1965

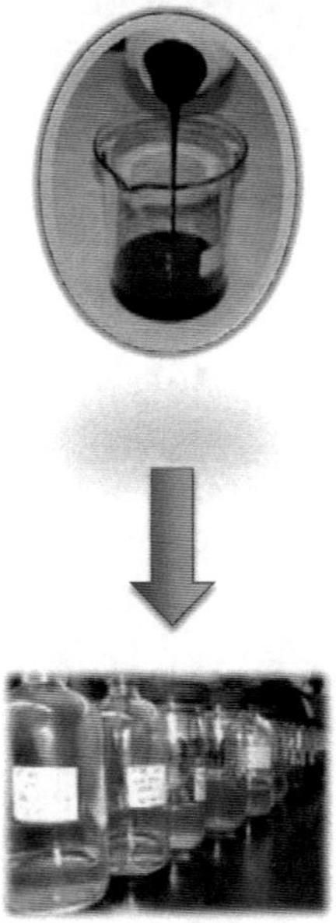

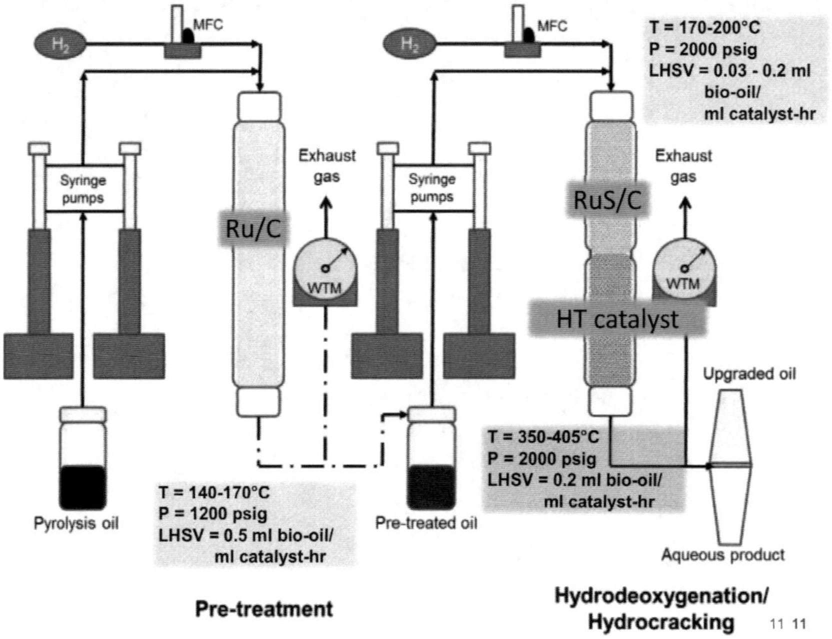

11 11

2013 run – long-term experiment

Pacific Northwest NATIONAL LABORATORY
Proudly Operated by Battelle Since 1965

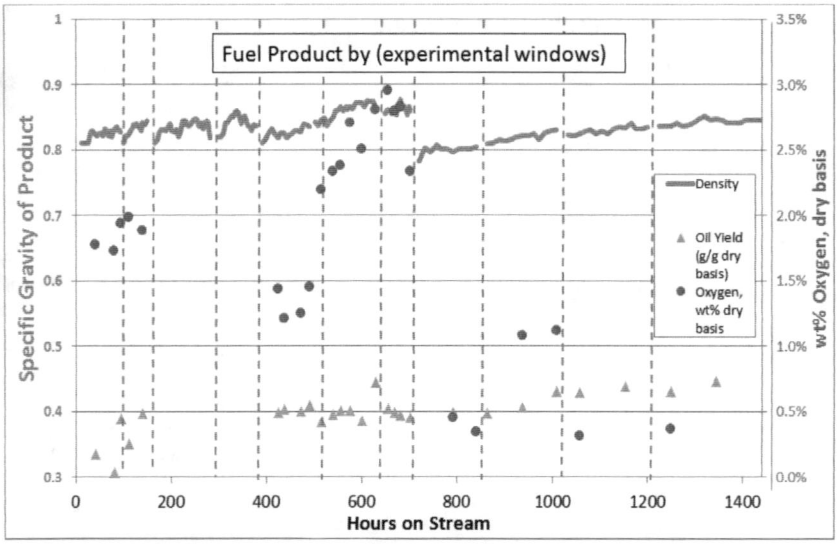

- The long-term experiment produced oils of oxygen contents below 3 wt%.
- Slow catalyst deactivation apparent with changes in product quality.

12

The stabilization challenge

Pacific Northwest NATIONAL LABORATORY Proudly Operated by Battelle Since 1965

- Addition of the pre-treatment bed, coupled with the more active hydrotreating catalyst, enabled the long-term processing.
- Catalyst deactivation still occurring.

There is a need to understand why

- ❑ Understand the quality that oil needs to be processed to enable successful operation at each process stage
  - ❑ For example: pre-treatment = no plug/appreciable delta P
  - ❑ Challenge: thorough and quantitative intermediate and product characterization
- ❑ Design better catalysts
  - ❑ Optimized activity for specific chemistries (i.e. very high activity isn't always desired)

13

Importance of characterization of product streams

Pacific Northwest NATIONAL LABORATORY Proudly Operated by Battelle Since 1965

❖ The compositional difference between crude oil and bio-oil presents significant risks to refineries.

Fast pyrolysis bio-oil is:

- ❑ corrosive
  - ❑ typical pH = 2 to 3
- ❑ thermally/chemically unstable
  - ❑ 47-116% increase in viscosity upon exposure to 80°C for 24 hours *(Elliott, et al. Energy Fuels 2012 (26) 3769 – 3776)*
- ❑ miscible with water
- ❑ low heating value

due to high oxygen content

Pyrolysis oil comprises different types of O functionalities:

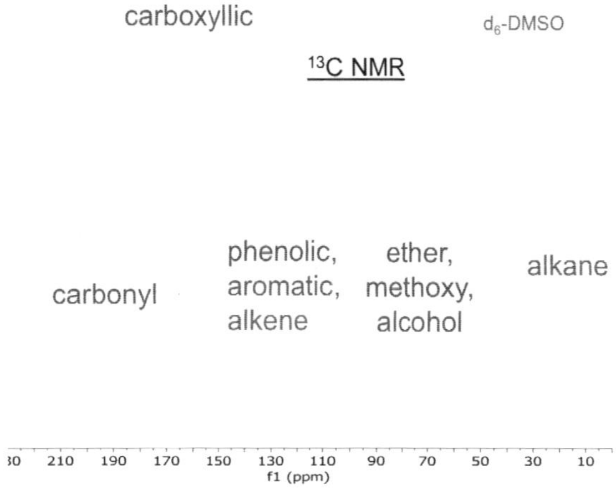

There is a need for catalytic upgrading to produce hydrocarbons.

Bio-oil characterization: O_2 content

Pacific Northwest NATIONAL LABORATORY
Proudly Operated by Battelle Since 1965

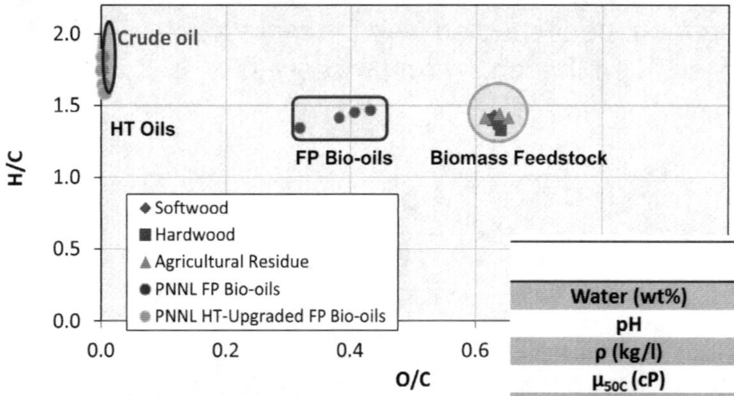

Objective:

- oxygen removal
- hydrogen content increase

	Bio-oil	Crude oil
Water (wt%)	15 – 30	0.1
pH	2.8 – 3.8	-
ρ (kg/l)	1.05 – 1.25	0.86
μ_{50C} (cP)	40 – 100	180
HHV (MJ/kg)	16 – 19	44
C (wt%)	55 – 65	83-86
H (wt%)	5 – 7	11-14
O (wt%)	28 – 40	<1
N (wt%)	<0.4	<1
S (wt%)	<0.05	<4
Ash (wt%)	<0.2	0.1
H/C	0.9 – 1.5	1.5 – 2.0
O/C	0.3 – 0.5	~0

Pyrolysis oils contain larger amounts of oxygen compared to crude oil.

Zacher, et al. 2014. Green Chem. 16: 491..

Typical oxygen content of each stream

Pacific Northwest NATIONAL LABORATORY
Proudly Operated by Battelle Since 1965

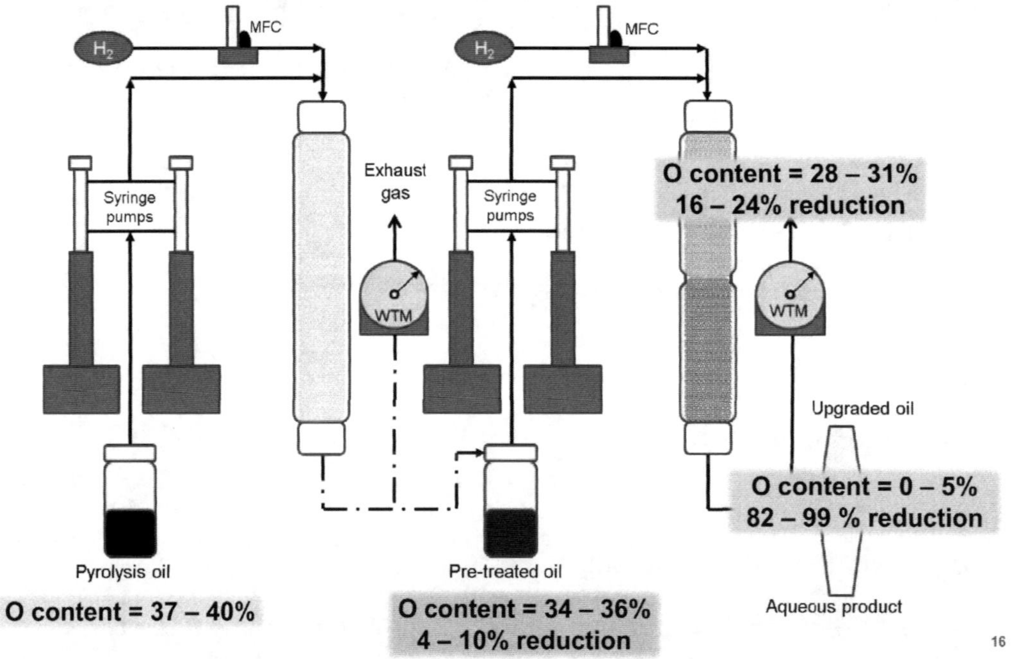

16

Not all oxygen is created equal.

Pacific Northwest
NATIONAL LABORATORY
Proudly Operated by Battelle Since 1965

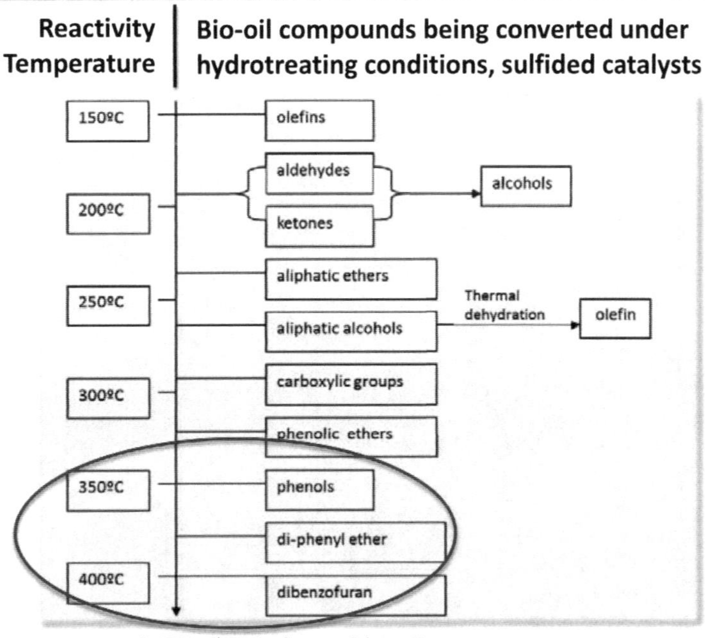

Laurent, et al. 1991. 6th EC Conference, Athens, Greece.

17

Pacific Northwest
NATIONAL LABORATORY
Proudly Operated by Battelle Since 1965

Characterization of pre-treated oils

18

Pre-treated oil: ^{13}C NMR analysis

Pacific Northwest NATIONAL LABORATORY
Proudly Operated by Battelle Since 1965

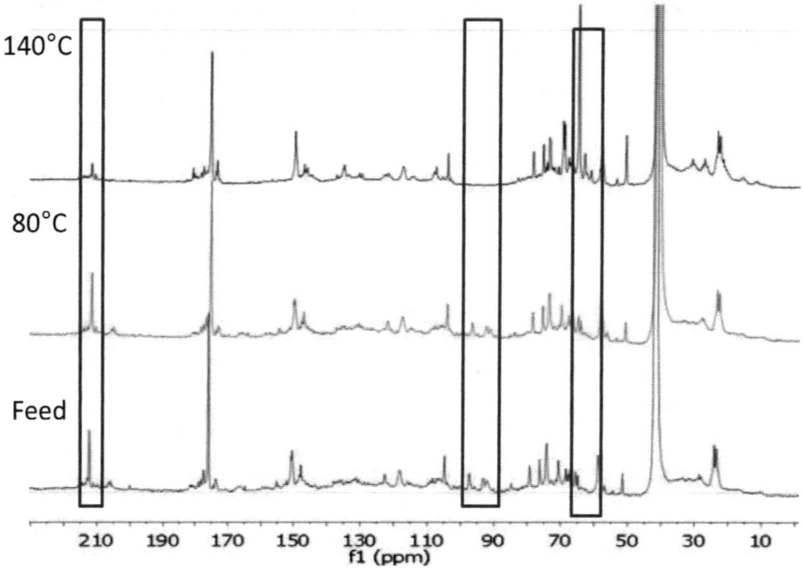

Changes in functional groups in the 140°C product oil are apparent.

Functional groups affected by pre-treatment

Pacific Northwest NATIONAL LABORATORY
Proudly Operated by Battelle Since 1965

Carbonyl region:

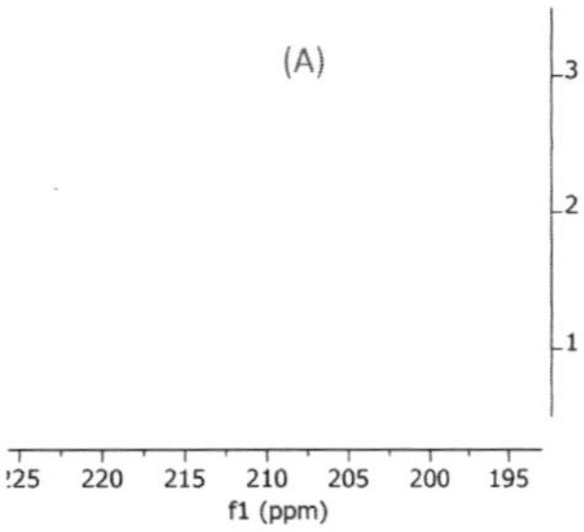

Alcohol/ether region:

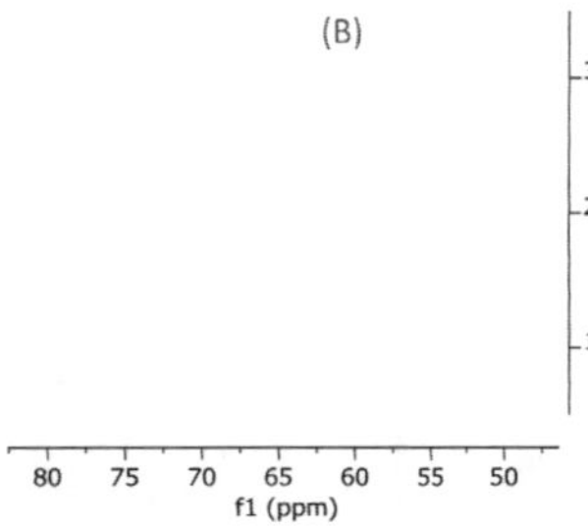

In the presence of **high pressure H_2** and **noble metal catalyst** (Ru/C), decrease in the carbonyl region and increase in the alcohol/ether region suggests **hydrogenation** of the carbonyls

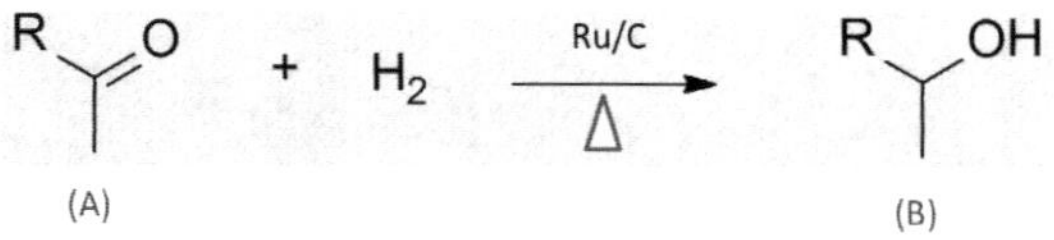

Region typically associated with anomeric C in carbohydrates:

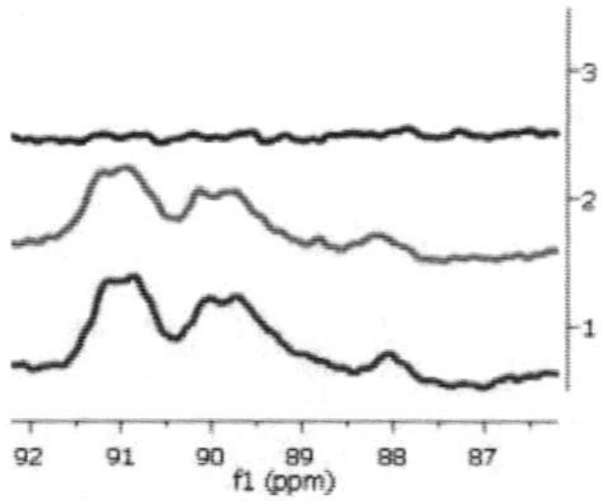

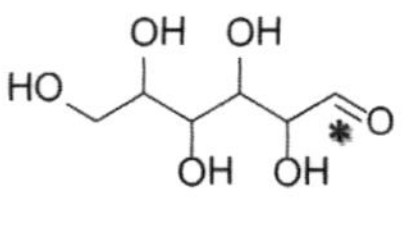

1 – Raw
2 – 80°C
3 – 140°C

2D-NMR Analysis of pre-treated oils

Pacific Northwest NATIONAL LABORATORY
Proudly Operated by Battelle Since 1965

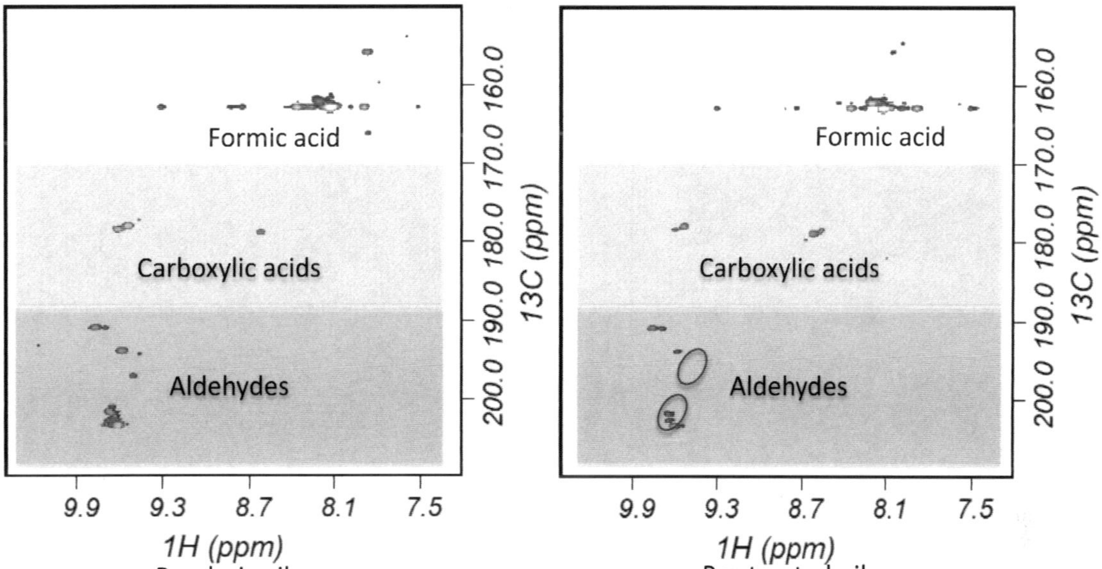

Pyrolysis oil | Pre-treated oil

Loss and decrease of certain aldehydes.

21

2D-NMR Analysis of pre-treated oils

Pacific Northwest NATIONAL LABORATORY
Proudly Operated by Battelle Since 1965

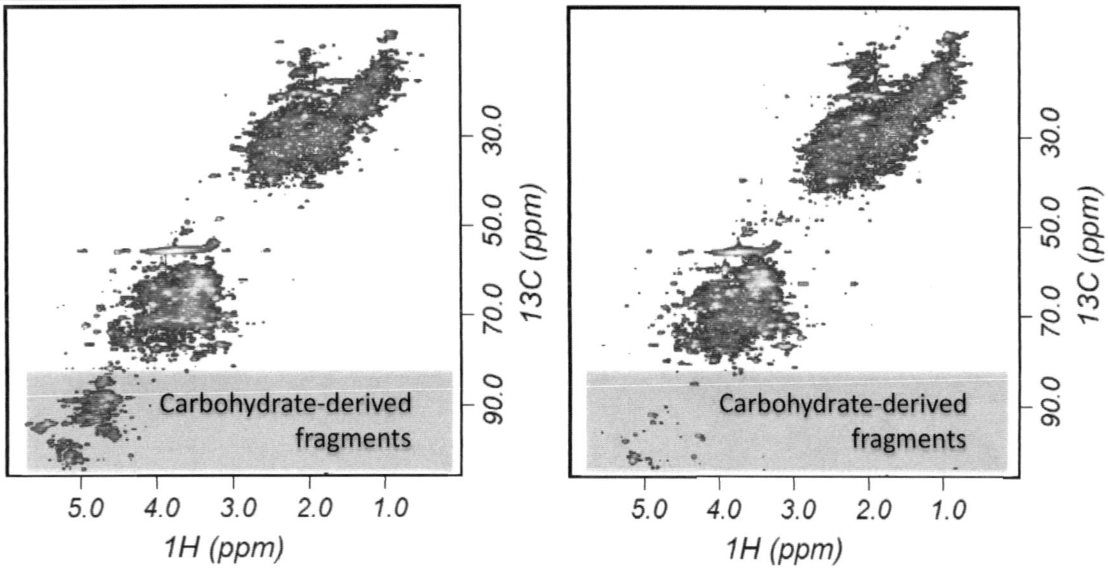

Pyrolysis oil | Pre-treated oil

Loss of anomeric C, potentially reduced to alcohols

22

2D-NMR Analysis of pre-treated oils

Pacific Northwest NATIONAL LABORATORY
Proudly Operated by Battelle Since 1965

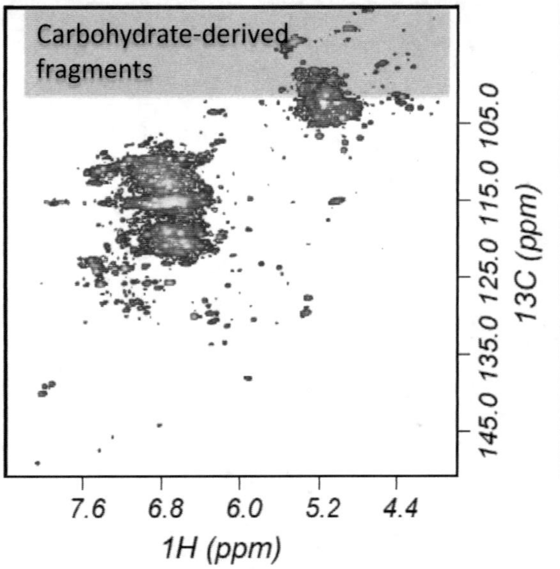

Pyrolysis oil

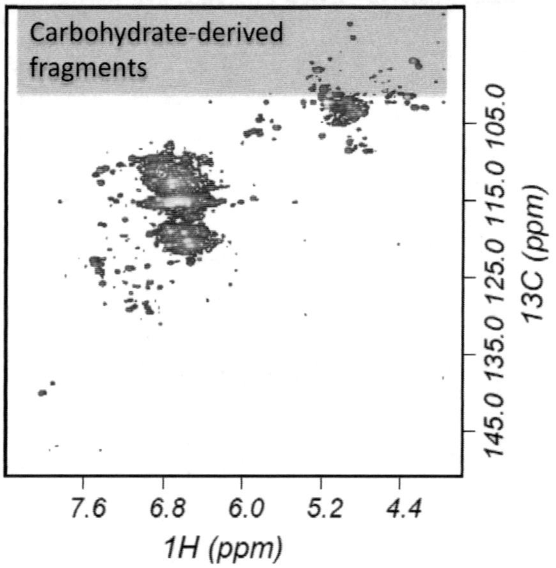

Pre-treated oil

Loss of anomeric C, potentially reduced to alcohols

23

Pacific Northwest NATIONAL LABORATORY
Proudly Operated by Battelle Since 1965

Characterization of upgraded oils

24

Upgraded oil: Simulated distillation

Pacific Northwest NATIONAL LABORATORY
Proudly Operated by Battelle Since 1965

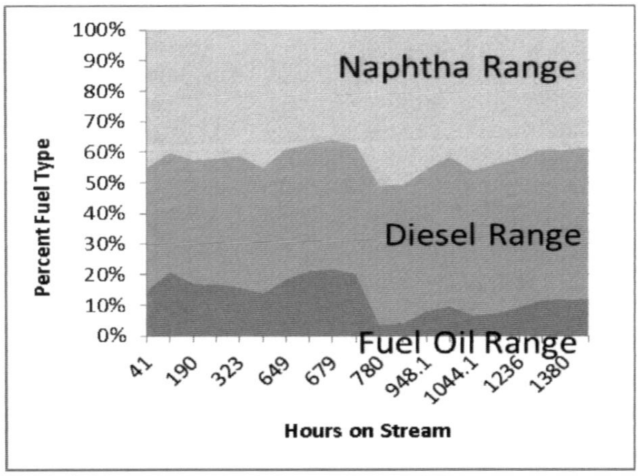

Upgraded FP oils typically contain naphtha range products. However, these can be distilled to specific cuts.

Bio-aromatics: 42.4-44.2 wt% jet fuel distillate by batch vacuum distillation

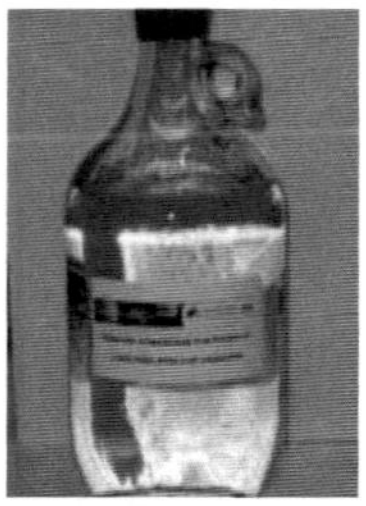

98% BioSPK and 2% renewable aromatics

PHOTO: Boeing U-787, Ed Clark
http://www.boeing.com/news/frontiers/archive/2009/september/sept09frontiers.pdf

GC-MS of upgraded oil at different TOS

Pacific Northwest NATIONAL LABORATORY
Proudly Operated by Battelle Since 1965

Oil density = 0.84; O wt% = 1.3%, HOS = 379h

Oil density = 0.87; O wt% = 2.5%, HOS = 600h

Process condition:
LHSV = 0.05-0.1 L/hr
T = 400-410°C

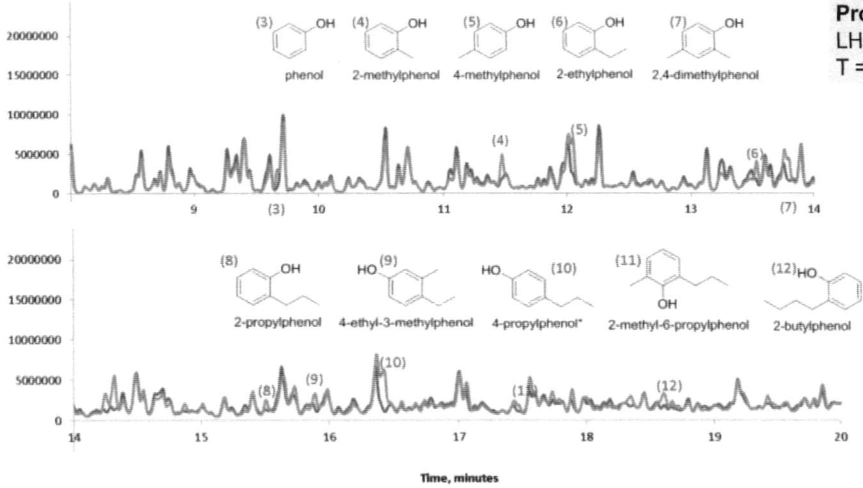

- Heavier (>C_3) recalcitrant phenols are formed/remain unreacted
- Possibly due catalyst activity loss (cracking and HDO)

Upgraded oil: GCxGC FID analysis

Pacific Northwest NATIONAL LABORATORY
Proudly Operated by Battelle Since 1965

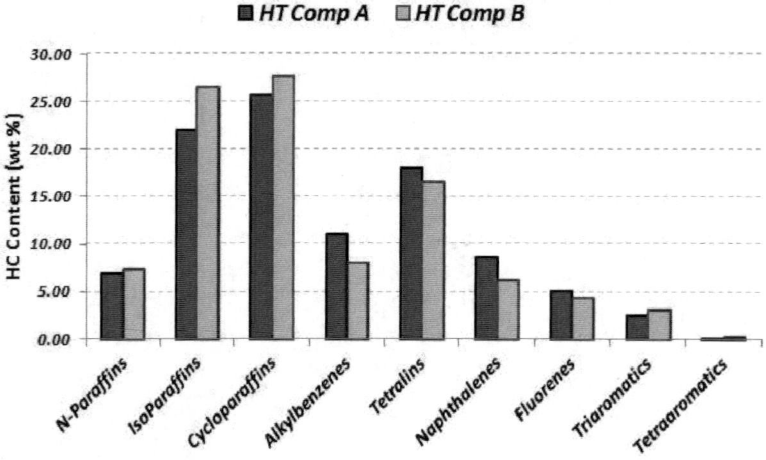

These two sets of upgraded oils showed similar HC contents.

27

Upgraded oil: GC-FIMS

Pacific Northwest NATIONAL LABORATORY
Proudly Operated by Battelle Since 1965

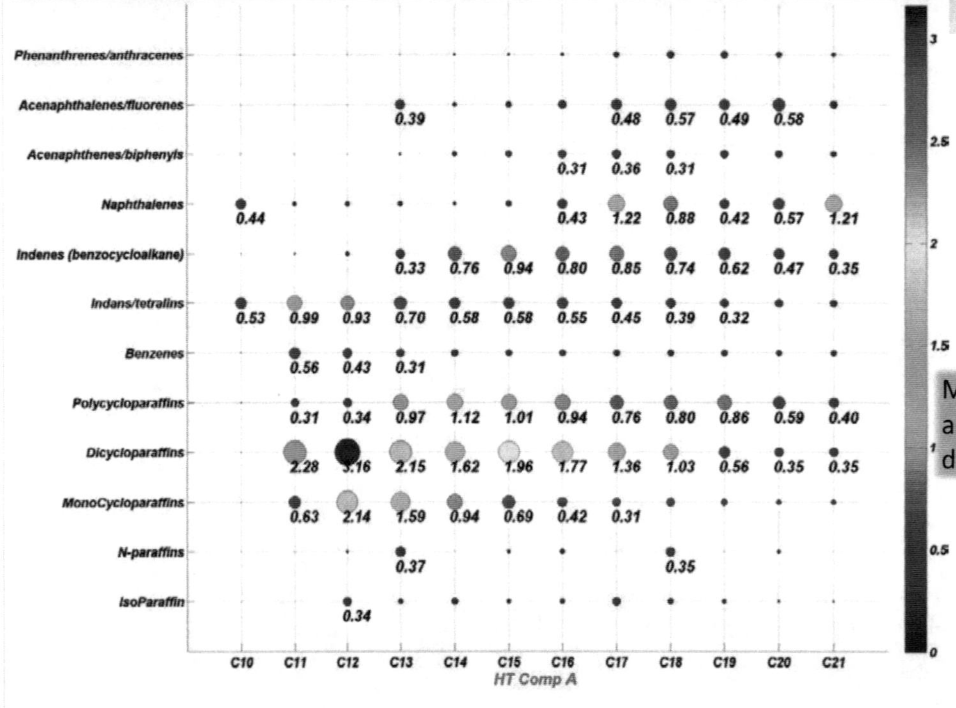

Most abundant are C_{12} dicycloparaffins

28

GCxGC TOF MS analysis of upgraded oils

Pacific Northwest NATIONAL LABORATORY
Proudly Operated by Battelle Since 1965

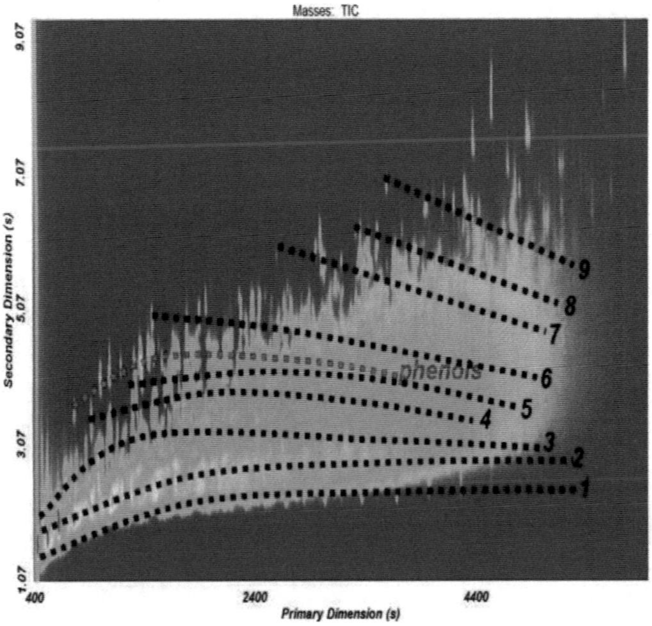

GC x GC – TOFMS total ion chromatogram (TIC) for upgraded oil composite with selected hydrocarbon classes where:

1: *n*-paraffins and isoparaffins
2: cycloparaffins/oleffins
3: alkylbenzenes
4: naphthenebenzenes
5: dinaphthenebenzenes
6: naphthalenes
7: acenaphthenes
8: fluorenes
9: phenanthrenes

Selected Ion Monitoring (SIM) was also implemented to identify specific peaks

29

Focus on phenols

Pacific Northwest NATIONAL LABORATORY
Proudly Operated by Battelle Since 1965

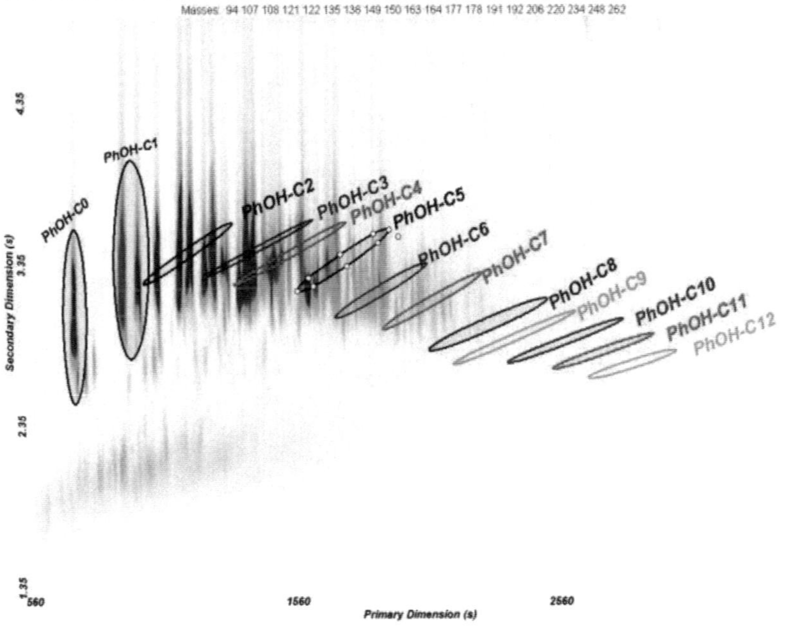

Observed phenols attached with up to 12 C chain

At O content of about 4.08%, phenols can account for about ~90% of oxygenates present.

30

What we learned

- Fast pyrolysis oil can be catalytically upgraded for more than 100 hours without plugging.
- Specific functional groups from bio-oil need to be removed/lessened during pre-treatment.
- More robust catalysts are needed.

Path forward

- Establish bio-oil feed and intermediate streams criteria.
- Advanced characterization methods for optimized upgrading performance and refinery integration
- Catalyst development.

2017 BETO Demonstration in pilot scale reactor

31

Acknowledgements

PNNL acknowledges the funding for this work from the Bioenergy Technologies Office (BETO) under the US DOE Energy Efficiency and Renewable Energy Office.

Thank you for listening.

Questions?

32

Conversion, selectivity and yield optimization of the simultaneous liquid-phase synthesis of isobutene and isoamylenes with ethanol over Amberlyst 35™

R. Soto, C. Fité, E. Ramírez, R. Bringué and F.Cunill*

Chemical Engineering Department. Faculty of Chemistry. University of Barcelona. Martí i Franquès 1-11, 08028-Barcelona.

*Corresponding author: email: fite@ub.edu, Tel.: +34934034769, Fax: +34934021291

A way to reduce the environmental impact derived from the gasoline usage is to limit the total emissions during the combustion process. It can be achieved by removing or substituting non-desired components. Examples of enhanced fuels are indeed the reduction of the C_5 olefins content in gasoline by its etherification with bioethanol to produce *tert-amyl ethyl ether* (TAEE) and the use of other oxygenates, such as bioethers and bioalcohols. Hereby high performance reformulated gasoline can be produced supplying a cleaner combustion, remarkable engine performances and whose scale-up and implementation at industrial scale could be easily carried out by means of revamping of actual *methyl tert-butyl ether* (MTBE) and *ethyl tert-butyl ether* (ETBE) plants. In this study the simultaneous liquid-phase etherification of C_4 and C_5 isoolefins with ethanol (EtOH) over Amberlyst 35™ is presented as a flexible and versatile process for refiners to produce simultaneously two important ethers (ETBE and TAEE) depending on the refinery requirements. The process itself implies the derived advantages from the process integration and intensification. Based on the fact that many refineries nowadays produce mixed C_4 and C_5 olefinic streams in Fluid Catalytic Cracking and Steam Cracking units [1], the process becomes attractive from an industrial point of view and deals with the endeavor of optimization of current fuels. In our previous work focusing on the same system [2], the thermodynamics of the simultaneous ETBE and TAEE production was deeply studied in the presence of water over commercial macroreticular acid ion-exchange resins. Interesting results were obtained, showing the feasibility to produce simultaneously both bioethers with a relatively high bioethanol usage as raw material. Since the involved etherification reactions are thermodynamically limited, the knowledge on the equilibrium conversions and selectivities under different experimental conditions, such as temperature and initial reactants ratio, is essential for chemical reactors design and to determine the industrial operating conditions.

This work studies the product distribution at chemical equilibrium for the simultaneous etherification of isobutene (IB) and isoamylenes (IA) with EtOH over the acidic ion-exchange resin Amberlyst 35™. Experiments were carried out in a 200 mL stainless steel jacketed batch reactor operating at 2.0 MPa and within the temperature range from 323 to 353 K. Samples were analyzed by GC/MS. The initial molar ratio alcohol to olefins ($R_{A/O}$) and the initial molar ratio of IB to IA ($R_{C4/C5}$) were both varied between 0.5 and 2. The catalyst was thermally pretreated to remove adsorbed water. Avoiding therefore possible unknown sources of water, a small amount (1 wt.% of the whole reaction mixture) of water was initially added as reactant in order to monitor the tertiary alcohols formation, tert-butyl alcohol (TBA) and tert-amyl alcohol (TAA).

Reactants equilibrium conversion (X_j) was found to decrease as temperature increased (Figure 1A), as expected for exothermic reactions [2]. As a whole, the obtained IB, IA and EtOH equilibrium conversions ranged between 69.8 and 97.9%, between 17 and 65.6% and between 32.3 and 97.0%, respectively, depending on the initial composition and reaction temperatures. The effect of the initial $R_{A/O}$ on reactants equilibrium conversion was noticeable

(Figure 1B). The higher olefin conversions were reached under initial stoichiometric excess of EtOH at the expense of lower alcohol conversion. On the other hand, the effect of initial $R_{C4/C5}$ on reactants equilibrium conversions was less pronounced.

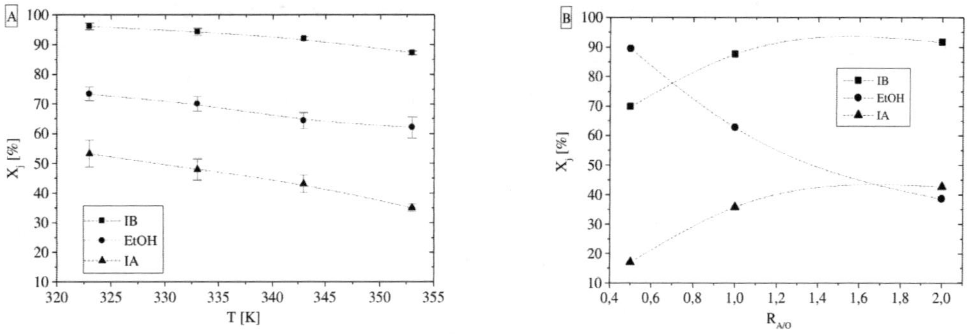

Figure 1— (A) Reactants equilibrium conversion versus temperature at $R_{A/O}$=$R_{C4/C5}$=1. Errors bars are referred to a 95% of confidence interval. (B) Reactants equilibrium conversion versus $R_{A/O}$ at 353K and $R_{C4/C5}$=1.

In all the cases, EtOH was found to react faster with IB than with IA, being more selective to ETBE than to TAEE except at $R_{C4/C5}$=0.5. Presumably, the catalyst showed high selectivity for the respective olefins toward ETBE and TAEE (>90%) within the assayed conditions and their value scarcely depended on temperature (Figure 2), which is desirable from an industrial point of view. Olefins selectivity toward ethers smoothly decreased as temperature increased. The effect of $R_{C4/C5}$ on the reactants selectivity was slightly more noticeable than that of $R_{A/O}$. The olefins (IB and IA) selectivity toward tertiary alcohols (TBA and TAA) was found to increase with temperature and it was maximum at $R_{A/O}$= 0.5 at each temperature and $R_{C4/C5}$. The selectivity of IA toward TAA was always higher than that of IB toward TBA. The amount of formed moles of TBA was always higher than that of TAA. Water affected drastically the olefins selectivity toward ethers formation since a 1% of initial water addition led to selectivity of olefins toward tertiary alcohols above 10 - 14%, for TBA and TAA respectively.

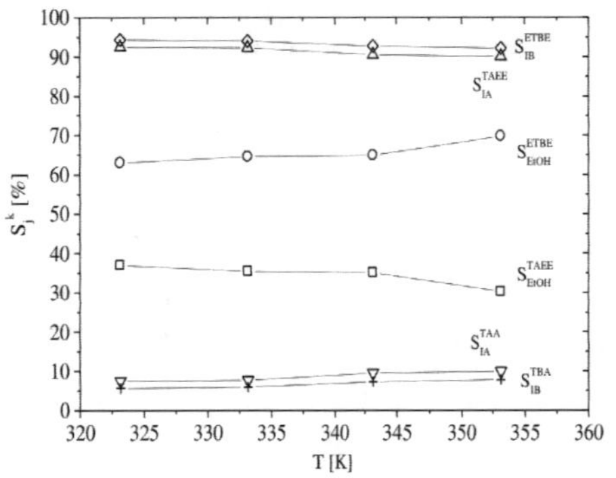

Figure 2 — Reactants selectivity toward products versus temperature at chemical equilibrium (after 360 min of reaction) and $R_{A/O}$=$R_{C4/C5}$=1.

Regarding the olefins selectivity toward dimerization products, diisobutene (C_8), IB and IA codimers (C_9) and diisoamylenes (C_{10}) were obtained as the main oligomerization byproducts and their formation was maximum at $R_{A/O}$= 0.5 and 353K. Maximum IB selectivity toward dimerization products was 13%, and 16% for the case of IA dimerization. IB yielded higher amounts of dimers at the aforementioned conditions in terms of number of moles. The dimers formation was found to increase linearly with reaction time. Neither trimers nor higher oligomers were detected under the conditions of this study.

Based on the fact that reaction yield is a suitable parameter to measure the industrial viability, which takes into account simultaneously the effect of reactants conversion and selectivity, experimental yield data were modeled by fitting the yield for both bioethers with respect to temperature and initial composition. A second order polynomial was fitted to data by means of the stepwise regression method. In order to center the variables of the experimental design, initial molar ratios were translated into initial molar fractions of alcohol in the initial reactant mixture (x_A) and of IB in the initial olefins mixture (x_{C4}). An example of the obtained response surfaces and the contour plot are represented in Figure 3 for the case of TAEE. The optimization of the experimental conditions to maximize ETBE and TAEE yields was carried out using the response surface methodology (RSM) and by means of the overlaid contours plot [3] for the responses maintaining constant x_{C4} at 0.5. This variable can be changed depending on the preferred ether (ETBE or TAEE) formation.

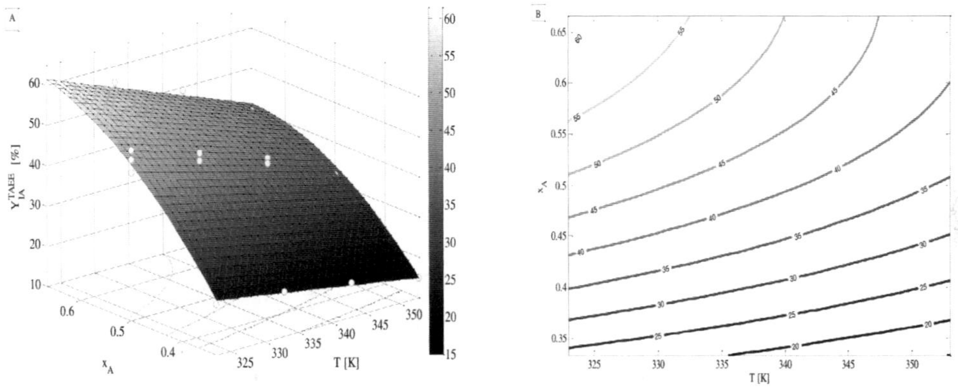

Figure 3 — (A) Response surface obtained considering IA yield toward TAEE as main response and X_A and temperature as variables at X_{C4}=0.5. Symbols (○) refer to the experimental yield data at chemical equilibrium (after 360 min of reaction). (B) Contours plot for the same conditions.

Within the experimental conditions, optimum range of temperature to produce simultaneously ETBE and TAEE was between 323 and 333 K and within $R_{A/O}$ values from 1.04 to 1.3. Thus, optimum conditions for maximizing both ethers formation are $R_{A/O}$ and $R_{C4/C5}$ close to the unity ($R_{C4/C5}$ can be varied depending on the preferential formation of one or another ether and depending on the refinery needs) and low temperatures that represent a compromise between chemical equilibria and kinetic limitations providing the lowest possible byproducts formation.

References

- [1] Fuel Ethers increase Refining Flexibility and Gasoline Production. The European Fuel Oxygenates Association. Accessed May 10/2014. <www.efoa.eu>

- [2] R. Soto, C. Fité, E. Ramirez, R. Bringué and F. Cunill. Equilibrium of the simultaneous etherification of isobutene and isoamylenes with ethanol in liquid-phase. Chem. Eng. Res. Des. 2014. 92(4), 644-656.

- [3] R. L. Mason, R. F. Gunst, J. L. Hess. Statistical Design and Analysis of Experiments (with Applications to Engineering and Science). Second Edition. Wiley- Interscience. 2003. New Jersey.

Hydrogen-alimented fuel cells (FC) have a strong potential to play a decisive role in the new energy system for the coming years. The production of H_2 pure enough to use it in fuel cells requires the development of very efficient catalysts for the WGS reaction. In our group several gold-ceria based catalysts have been developed presenting very promising results in this process [1,2]. The successful catalytic design makes mandatory an accurate knowledge about the reaction mechanism and the active species involved in the process. In order to address these issues a combination of several *in-situ/operando* characterization techniques is performed in this work using an optimized Au/CeO_2-FeO_x/Al_2O_3 catalyst. Synchrotron-based *in-situ* time-resolved X-ray absorption spectroscopy (TR-XAS) and *operando* DRIFTS during the WGS reaction are employed with the ultimate goal to establish structure-activity relations and to propose the most likely reaction pathways.

The obtained results may be summarized as follows:

1) Oxidation state from *in-situ* XANES: Au remains as metallic gold during the reaction. Ce evolves from Ce^{4+} to Ce^{3+} (actually complete reduction of ceria is not achieved). Fe is present as Fe^{3+} since the profile of the XANES matches to the hematite structure.

2) **Presence of oxygen vacancies on ceria**. Our Raman, XRD and UV-Vis data established the existence of oxygen vacancies in the ceria lattice, especially when iron is present.

3) Gold dynamics. *In- situ* EXAFS analysis indicates the competition between gold particles and water to nucleate in the oxygen vacancies of the support. This process is associated to the observed gold nanoparticles rupture and subsequent agglomeration depending on the reaction temperature.

4) Reaction intermediates. *Operando* DRIFTS data point out that the formation of either bicarbonate or carboxylic species play a key role during the WGS.

Considering these premises, an associative mechanism where the carboxyl intermediates domains the reaction can be proposed:

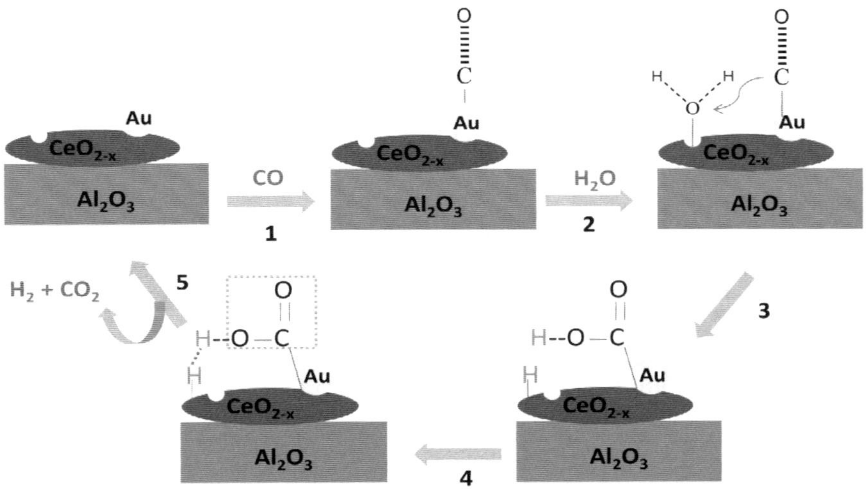

Figure 1: *Simplified WGS mechanism proposal for a Au/CeO_2-FeO_x/Al_2O_3 catalyst*

The main reactions pathways are:

1) Adsorption of CO on gold nanoparticles. 2) Activation of H_2O on the ceria oxygen vacancies. 3) The interaction between CO and activated H_2O on the metal-support interface. More specifically, this step represents the formation of the COOH intermediate due to the combination of CO and OH. This process may happen also in two steps. First the water rupture into OH and H and second the formation of the HO-CO bond. 4) Formation of the CO_2 and H_2. 5) Final release of the products.

Conclusions:

Our *in-situ/operando* study provides some relevant clues regarding the WGS mechanism and towards the catalyst design and future improvements. In particular, an associative route via carboxyl formation seems to control the process. The high catalytic performance is achieved when gold nanoparticles are in close contact with the structural defects of ceria and the oxygen mobility of the later is enhanced by the addition of promoters such as iron oxide.

References

[1] J. A. Odriozola, T. R. Reina, M. A. Centeno, S. Ivanova, V. Idakiev, T. Tabakova, Spanish Patent 201101163.

[2] M. González, T.R.Reina, S. Ivanova, M. A.Centeno, J.A. Odriozola J. Catal. 314 (2014) 1-9

382126 Fe-Promoted Au/Rutile for the Water-Gas Shift Reaction

Abstract:

The water-gas shift reaction (WGS) is an important reaction for hydrogen production and fuel cell applications[1], which are important for renewable energy. Au supported on oxides has been identified as an active catalyst for WGS. Supports play an important role in not only changing the WGS rates, but also modifying the types and chemical nature of the active sites[2,3]. Previous work has shown that Fe doping can promote the CO oxidation rate over Au/TiO_2 catalyst[4]. Here, a series of Fe-doped rutile (TiO_2) catalysts ranging from 0.25wt% to 5wt% Fe were loaded with ~1wt% Au and tested for WGS. WGS rates per mole of Au at 120 °C showed maximum promotion by a factor of 4 for 1wt% Fe content. Significant changes in kinetic parameters were observed with varying Fe content. Apparent orders with respect to CO decreased progressively and apparent activation energy increased progressively with Fe content increasing from 0 wt% to 5 wt%. As CO order decreased, WGS rates first increased, reached maximum at CO order about 0.3 and then decreased. From previous work, WGS rates over supported Au are influenced by the Au particle sizes[2,3], but for the catalysts reported here, similar average particle sizes (2.0±0.5 nm) obtained from Transmission Electron Microscopy (TEM) excluded the particle size effects as the source of the observed changes. Operando Fourier Transform Infrared Spectroscopy (FTIR) was performed on the 1wt% Fe-doped catalyst and the result was compared with Au supported on un-doped rutile. IR peaks at wavenumbers of 2097 cm^{-1}, 2044 cm^{-1}, 1967 cm^{-1} and 1888 cm^{-1} were observed. The peak at 2097 cm^{-1} can be assigned to CO adsorbed on metallic Au while the lower frequencies of 2044 cm^{-1} and 1967 cm^{-1} can be associated with CO adsorbed on partially negatively charged Au. The IR bands at lower frequencies were stronger than the corresponding peaks of Au supported on un-doped rutile and thus suggest a participation of Fe in the neighborhood of the site. Thus, the results showed that Fe-doping can modify the CO adsorption properties of Au/Rutile and that WGS rates can be correspondingly promoted with appropriate Fe content.

Elucidating the Fundamental Structure-Performance Relationships of H_2 Production by Cu/Cr/Fe Water-Gas Shift (WGS) Catalysts

Minghui Zhu,[a] Tulio C. R. Rocha,[b] Robert Schlogl[b] and Israel E. Wachs[a]

[a] *Operando Molecular Spectroscopy and Catalysis Research Laboratory, Chemical Engineering Department, Lehigh University, Bethlehem, PA 18015 USA.*

[b] *Fritz-Haber-Institut der Max-Planck-Gesellschaft, Department of Inorganic Chemistry, Faradayweg 4–6, 14195, Berlin, Germany.*[11]

Industrial production of hydrogen is mostly achieved by the water-gas shift (WGS) reaction ($H_2O + CO \leftrightarrow H_2 + CO_2$). To shift the equilibrium of the exothermic reaction to the right and maximize hydrogen production, the catalytic WGS reaction is performed at two temperature ranges. The high temperature WGS reaction stage was traditionally performed with the unsupported Cr_2O_3-Fe_2O_3 mixed oxide catalyst. In recent years, the addition of Cu to the Cr_2O_3-Fe_2O_3 mixed oxide catalyst was found to increase the performance of the HT WGS catalyst, but the promotion mechanism of Cu is still not well understood because of the lack of characterization studies under reaction conditions.

To advance the fundamental understanding of the promotion of Cu for this important industrial reaction for hydrogen production, a series of CuO-Cr_2O_3-Fe_2O_3 catalysts were synthesized and investigated with modern *in situ* (TPR, TP-DRIFTS and Raman) and *operando* (Raman-MS, XANES/EXAFS-MS and near ambient pressure XPS-MS) spectroscopy studies during the HT WGS reaction. The unsupported Fe_2O_3, Cr_2O_3-Fe_2O_3 and CuO-Cr_2O_3-Fe_2O_3 catalysts were synthesized by the co-precipitation method from the corresponding nitrates. The freshly calcined CuO-Cr_2O_3-Fe_2O_3 catalyst consists of surface Cr^{6+} species and a solid solution of Cr^{+3} and Cu^{+2} in the Fe_2O_3 bulk lattice. During the HT WGS reaction, surface Cr^{6+} reduces to Cr^{3+} and further dissolved in the iron oxide lattice, bulk $Fe_{2-x}Cr_xO_3$ transforms to bulk $Fe_{3-x}Cr_xO_4$ and Cu^{2+} reduces to large metallic Cu^0 particles. Redox experiments confirm that Fe and Cu are active sites during the HT WGS reaction, while the role of Cr is to stabilize the surface area of the $Fe_{3-x}Cr_xO_4$ component by retarding sintering. Addition of Cu promotes both the initial bulk transformation of $Fe_{2-x}Cr_xO_3$ to $Fe_{3-x}Cr_xO_4$ and, more importantly, the redox properties of the catalyst[12].

385402 Water-Gas Shift Reaction on Catalysts Prepared Using Supercritical Fluid Deposition

Abstract:

Catalysts composed of noble metals and reducible metal oxides are of great significance for modern energy applications. These applications include key reactions in hydrogen production such as water-gas shift (WGS) and preferential oxidation (PROX), as well as fuel cell electrodes. A major impediment to implementing hydrogen fuel cell technology is the need for safe, reliable, and efficient catalytic hydrogen generation for logistical small-scale operations. WGS is a crucial step in hydrogen enrichment, and effective catalysts that have been identified are platinum or gold in conjunction with ceria or other partially reducible oxides. For economic feasibility, minimizing the amount of platinum or gold required is vital. Recent research provides evidence that catalysts having extremely low loadings of these metals can still exhibit high activity, as the degree of reaction rate enhancement depends on interfacial area between active metal sites and oxides. Thus, methods of synthesizing supported catalysts with intimate contact between the two are essential.

This paper focuses on the synthesis and characterization of Pt/ceria WGS catalysts prepared using supercritical fluid deposition (SCFD), a technique by which organometallic precursors are adsorbed from a supercritical fluid solvent on to a support. Supercritical fluids (SCFs) offer unique advantages for interfacial deposition processes. Liquid-like density promotes dissolution of metal precursors; viscosity and diffusivities more akin to those of gases facilitate both transfer through the SCF and surface adsorption. These benefits make possible the reproducible, uniform distribution of organometallic precursors on the support. In our work, the support is high-surface area alumina to which ceria has been added by insipient wetness followed by calcination. The platinum is deposited using SCFD of platinum acetylacetonate from carbon dioxide, with subsequent calcination in air and reduction in hydrogen. The nominal loadings are 5% ceria and 1% platinum. Variations in preparation include reductive pretreatment of the ceria before Pt deposition, intended to disrupt the surface with oxygen vacancies that will ultimately enhance the deposition of Pt precursor on ceria-rich areas of the alumina surface.

Results of surface analyses, including temperature programmed reduction and carbon monoxide pulse titrations, indicate that the preparation results in both small nanoparticles (1-3 nm) and highly dispersed platinum. Further evidence for this is obtained using transmission electron microscopy. Catalyst activity testing takes place a flow reactor system using a simulated reformate stream containing CO, H_2, H_2O, and N_2. Turnover frequencies and activation energies for WGS on catalysts will be presented and correlated with variables in synthesis method.

389936 Alkali-Promoted Atomic Pt-Ohx Species on "Active" or "Inert" Supports Catalyze the Low-Temperature Water-Gas Shift Reaction

Abstract:

The water-gas shift (WGS) is an important reaction for hydrogen upgrading during fuel gas processing. Atomically dispersed Pt-O_x-support sites have been found to catalyze the low temperature WGS reaction on "active" supports, such as CeO_x [1]. However, on "inert" supports, such as alumina and silica, precious metal atoms are unstable and easily agglomerate into spectator nanoparticles. For such surfaces, we have recently shown that alkali metal oxides can be used as promoters of the platinum on silica [2], alumina [2], and even on carbon nanotubes [3], such that the high activity of the atomically dispersed Pt-OH_x- species is preserved. Here we have extended this work to other "active" and "inert" supports, including anatase, K type L-zeolites (KL), and mesoporous MCM-41. Through the use of alkali hydroxides, active and stable (up to 673 K) platinum catalysts were developed in this work. While the catalysts show large differences in activity without sodium promotion (Pt/Anatase> Pt/KL> Pt/MCM-41 ≈0 below 573 K), the sodium modified samples show the same, considerably improved activity in both product-free and full gas conditions for the same metal loading (0.5 wt.% Pt) on any of these supports. Modifying only the supports with alkali metals does not improve the catalytic activity if the Pt-O-alkali interaction is not established before calcination. The metal atom-centric Pt-OH_x-O_y-Na_zcluster is the active site, with the same intrinsic activity on any support. XPS analysis shows that platinum in these structures is close to Pt(II), same as for the alkali-promoted platinum on silica and alumina [2]. The addition of alkali metals does not change the apparent activation energy for this reaction. The presentation will include characterization results as well as stability evaluation under various operating conditions.

Acknowledgement

The financial support by the DOE/BES under Grant # DE-FG02-05ER15730 is gratefully acknowledged.

References

1. Fu Q, Saltsburg H, Flytzani-Stephanopoulos M, *Science* **2003**, 301, 935.
2. Zhai Y, Pierre D, Si R, Deng W, Ferrin P, Nilekar AU, Peng G, Herron JA, Bell DC, Saltsburg H, Mavrikakis M, Flytzani-Stephanopoulos M, *Science* **2010**, 329,1633.
3. Zugic B, Zhang S, Bell D, Tao F, Flytzani-Stephanopoulos M, *J. Am. Chem. Soc.* **2014**, 136, 3238.

388355 High Yield Production of Furfural from Hemicellulose Hydrolyzates Using a Facile Hybrid Enzyme-Chemo-Catalytic Technology

Abstract:

We propose an integrated process with low energy input for high-yield production of furfural from hemicellulose hydrolyzates. Due to the prevalent recognition that furfural is a versatile platform chemical for fuels and chemicals, there is a critical need for economically viable technologies for its production from the hemicellulose portion of biomass. Attempts to produce furfural in aqueous media (either from untreated biomass or from hydrolyzates) proved energy-intensive. Although hemicellulose hydrolyzates provide xylose, it is now broadly recognized that the best yields can be obtained by dehydration of the corresponding keto-sugar, xylulose, and by executing this reaction in non-aqueous reaction media (e.g. DMSO, gamma-valerolactone and ionic liquids (ILs)). The challenge then, for cost-effective and high-yield furfural production, is to devise a cohesive pathway to efficiently transfer xylose from biomass hydrolyzate into the non-aqueous reaction medium; produce furfural in high yield; and isolate it for downstream processing, all with low energy input and recycling of process streams.

Our technology addresses this challenge, and we propose a hybrid enzyme- and chemo-catalytic process that first converts xylose to more reactive xylulose and subsequently integrates with a downstream catalytic dehydration step that achieves high yields of furfural. The conversion of xylose to xylulose and its near-complete recovery is accomplished through a novel enzyme-based simultaneous-isomerization-and-reactive-extraction (SIRE) process. The recovered xylulose is then dehydrated to furfural in an acidic ionic liquid (IL) reaction medium and simultaneously extracted into a low-boiling tetrahydrofuran (THF) solvent to prevent side reactions, increase process yields and facilitate easy product recovery. Reactions of ketoses to furans can be conducted at milder operating conditions in IL-media compared to aqueous media (IL rxn T ~ 50 °C; aqueous rxn T > 150 °C) while achieving high product yields. Finally, a key feature of this technology is that it permits recycling and reuse of the catalysts, solvents and reaction media.

One Step Process for Production of Bio-BTEX and Bio-LPG from Biomass Derivative

Sungtak Kim, Michael Kai Mayeda, Erdem Sasmaz, and Jochen Lauterbach

We present a one-step process for producing petroleum gas (PG) and value-added organic chemicals such as benzene, toluene, ethylbenzene, and xylenes (BTEX) via deoxygenation of pentanediol (PeD) over HZSM-5 at a low temperature and atmospheric pressure, which can be offered as a promising method to displace petroleum-based PG and BTEX. Deoxygenation of PeD produced both gas product and organic liquid product (OLP). The gas product was analyzed by gas chromatography using a flame ionization detector and mass spectrometer (GC-MS). Fourier transform infrared spectroscopy was carried out to characterize acidic property change and dealumination of fresh and regenerated HZSM-5. The liquid product was analyzed by ^{1}H nuclear magnetic resonance spectroscopy and GC-MS. The results show that carbon yields to OLP and gas product were up to 63 % and 36 %. The OLP contains only aromatics, primarily BTEX; no oxygenated compounds were detected. The produced BTEX is usable for industrial solvent as well as primary chemicals for the production of polymers and transportation fuels. On the order hand, the gas product composes 92 % of C_2-C_5 range hydrocarbons, resembling that of commercial liquefied petroleum gas (LPG) based on american society of testing and materials method, and 8 % of CO, CO_2 and CH_4 mixture. Time on stream experiments indicate that HZSM-5 catalyst exhibits a stable BTEX and LPG production yield for 40 hours deoxygenation reaction. The high proportion of BTEX in OLP and LPG like fuel in gas product enables high carbon efficiency (up to 94 wt%), and the low coke formation allows easy catalyst regeneration process.

Key words: Pentanediol, deoxygenation, biofuel, bio-LPG, HZSM-5

375183 Production and Analysis of Pyrolysis Oils from the Fluidized-Bed Co-Pyrolysis of Agricultural Plastic Waste and Biomass Mixtures

Abstract:

Pyrolysis of biomass produces bio-oils that are highly oxygenated and acidic. Along with high water content and low higher heating values (HHV), these bio-oils are also unstable. Catalytic pyrolysis can alleviate these issues by producing bio-oils that consist of more desirable aromatic hydrocarbons. However, the low H/C nature of biomass leads to inefficient biomass carbon conversion and produces a great deal of CO, CO_2 and coke that eventually causes catalyst deactivation. The incorporation of a co-feedstock with a higher H/C ratio could address these issues. The catalytic co-pyrolysis of biomass and plastic has been shown to produce products that incorporate C from both feedstocks and increases the production of desirable aromatic hydrocarbons. The use of agricultural plastics as a co-feed for the pyrolysis of biomass has the added benefit of providing a method for disposal of this waste. Switchgrass and used hay bale covers composed of polyethylene (PE) were mixed and subjected to catalytic pyrolysis in the presence of HZSM5 in a bench-scale fluidized-bed system. The catalytic pyrolysis oils were analyzed and compared to oils produced via regular pyrolysis and pyrolysis using a tail-gas reactive pyrolysis (TGRP) method. Yields were tracked on a mass and carbon basis. Complete chemical and physical property analysis including CHNS/O composition, energy content, and acidity were conducted. Chemical compositions of the pyrolysis oils were determined and compared with those produced from biomass alone.

Abstract:

Conversion of Caffeine, Lipids, and Fatty Acids in Catalytic Fast-Pyrolysis of Spent Coffee Grounds.

Ari Fischer, Shoucheng Du, George M. Bollas

Department of Chemical & Biomolecular Engineering, University of Connecticut, Storrs, CT

Abstract

Fast pyrolysis of spent coffee grounds is investigated as a means of valorizing commercial coffee waste. Coffee is the second largest globally traded commodity with imports reaching 6.7 million metric tons in 2013, as reported by the international coffee organization.[1] Soluble coffee waste retains 65% of the raw coffee mass and is regularly disposed of in landfills where its environmental impact is experienced through oxygen intensive decomposition.[2] In this work, spent coffee grounds are explored as a feedstock to engineer alternative processes for the production of valuable chemicals, commodities, or fuels. In particular, coffee's residual fibers, proteins, lipids, and other organic species are chemically converted into fragmented products through catalytic fast pyrolysis. Products include a gas phase containing ammonia; hydrogen; methane; carbon monoxide; and carbon dioxide, liquid phase characterized by condensed vapors and aerosols, and a solid char mainly composed of fixed carbon with residual ash and inorganics. Pyrolysis conditions are tuned to improve liquid quality with reduced oxygen content and increased paraffins and olefins.[3] Introduction of shape selective ZSM-5 catalyst promotes decarbonylation and decarboxylation to form aromatics.

Table 1: Composition of spent coffee grounds by wt%[2]

Crude Protein	10-12
Crude fiber	35-44
Lignin	13-16
Cellulose	22-28
Lipids (ether extract)	22-27
Minerals	0.3-1

Like most biomass feedstocks, spent coffee grounds contain fibers in the form of lignin and cellulose. In addition, proteins and lipids are abundant in spent coffee grounds. Elemental analysis of spent coffee grounds is compared to that of Aspen wood in Table 2.[4] Coffee has a lower oxygen content and higher carbon and hydrogen fraction than the wood counterpart. Unlike wood, coffee has significant nitrogen content from caffeine, proteins, and other nitrogen containing compounds, which form ammonia gas as a pyrolysis product. Non-pyrolyzed nitrogen found in the char product was been shown to stimulate plant growth.[5] In this view, catalytic pyrolysis of spent coffee grounds is posing a challenging process engineering problem, in which the char product is best not to be burnt for the generation of the heat required for the endothermic pyrolysis reactions; rather it is a useful product. Nonetheless, as a first step towards advanced utilization of used coffee grounds, in-situ catalytic upgrading is studied in this work.

Table 2: Elemental Analysis of dried SCG and Aspen Wood

	Coffee Grounds	Aspen Wood (French et al.)
Carbon	53.6	49.5
Hydrogen	7.77	6.1
Nitrogen	2.17	0.1
Oxygen by difference	36.4	43.8

Conventional and catalytic pyrolysis of spent coffee grounds has been explored with fast and slow heating rates. Bok et al.[6] employed a fluidized bed reactor with sand to study char, liquid, and gas yields for 673K to 873K generating a maximum liquid product of 54.85 wt% at 823K. The liquid product contained a range of ketones, alcohols, alkanes, aromatics, caffeine, and acids with a water content between 23.96 to 32.93.[6] Kan et al.[7] performed slow pyrolysis in a TGA with NiCu-¦Ã-Al_2O_3 reporting significant fatty acids in the bio-oil product, which indicates resistance to pyrolysis.[7] Nonetheless, the nature of coffee derived bio-oil is not understood. This study employed fast pyrolysis experiments to trace product distribution and chemical yields with varying temperature and inclusion of a catalyst. Fast pyrolysis with heating rates of 500-1000 K/minute was executed in a drop bed reactor within a temperature range of 765 to 860 K. ZSM-5 in a 1:1 catalyst to biomass ratio was used to decompose the persistent carbonyls and carboxylic acids previously reported in waste coffee pyrolysis products. Caffeine, stearic acid, linoleic acid, and palmitic acid yields were quantified by GC-MS analysis. Gas yields of hydrogen, carbon dioxide, carbon monoxide, and methane were quantified with GC-TCD analysis. For pyrolysis to be a favorable approach over bio-diesel for upgrading spent coffee grounds, decarbonylation must be promoted. This study focuses on the unique lipid content of coffee grounds and follows trends in fatty acid decomposition and product yield with ZSM-5 and temperature.

Increasing the reactor temperature facilitates vapor cracking of fatty acids, as shown in Figure 1. Comparing 810K to 860K thermal experiments, acid yield decreases from 57.6mg to 34.7mg. Experiments with ZSM-5 catalyst exhibit lower fatty acid yields than thermal experiments at the same temperature. On the other hand, caffeine yields are fairly constant across each experiment. At 810K stearic, linoleic, and palmitic acids are abundant in liquid product while at elevated temperature with ZSM-5 the liquid retention of fatty acids is diminished. Acid cracking results in greater content of deoxygenated fragments such as aromatics and alkanes. Figure 2 shows an increase in the abundance of aromatics, alkenes, and alkanes with a decrease in acids content. To this end, conversion of lipids can be promoted by adjusting pyrolysis. In addition, ZSM-5 increases aromatic content in the liquid product yet decreases alkenes and alkanes. In conclusion, conversion is enhanced with elevated temperature while ZSM-5 catalyst effects liquid distribution.

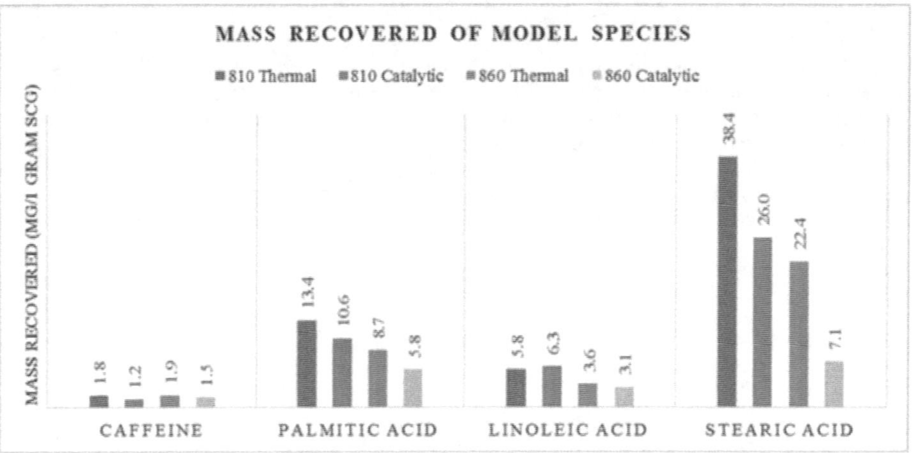

Figure 1: Mass of caffeine, palmitic acid, linoleic acid, and stearic acid recovered in thermal and catalytic pyrolysis in absolute temperature, quantified with GC-MS.

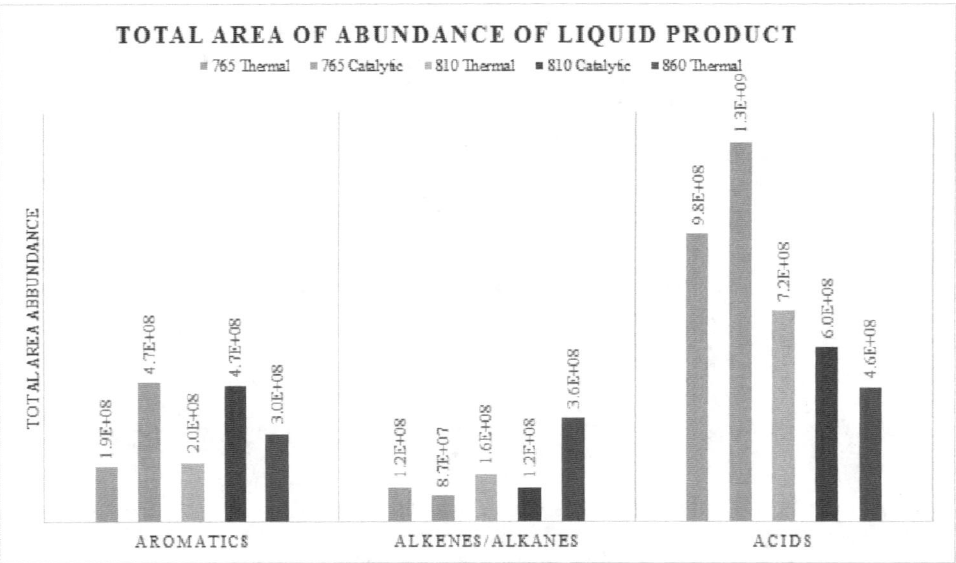

Figure 2: Recovery of aromatics, alkenes and alkanes, and acids from GC-MS abundance integrated areas for thermal and catalytic pyrolysis in absolute temperature.

Acknowledgments: This work was sponsored by the IDEA Grant of the University of Connecticut.

References

[1] *Imports of all forms of coffee by selected importing countries from all sources January 2014.* London: International Coffee Organization.

[2] Franca, A. S., & Oliveria, L. S. (2009). Coffee processing solid wastes: Current uses and future perspectives. In G. S. Ashworth, & P. Azevedo (Eds.), *Agricultural wastes* (pp. 155-180) Nova Science Publishers, Inc.

[3] Mohan, D., Pittman, C. U. Jr., & Steele, P. H. (2006). Pyrolysis of Wood/Biomass for bio-oil: A critical review. *Energy & Fuels, 20*, 848-889.

[4] French, R., & Czernik, S. (2010). Catalytic pyrolysis of biomass for biofuels production. *Fuel Processing Technology, 91*, 25-32.

[5] Vardon, D. R., Moser, B. R., Zheng, W., Witkin, K., Evangelista, R. L., Strathmann, T. J., et al. (2013). Complete utilization of spent coffee grounds to produce biodiesel, bio-oil, and biochar. *ACS Sustainable Chem. Eng., 1*, 1286-1294.

[6] Bok, J. P., Choi, H. S., Choi, Y. S., Park, H. C., & Kim, S. J. (2012). Fast pyrolysis of coffee grounds: Characteristics of product yields and biocrude oil quality. *Energy, 47*, 17-24.

[7] Kan, T., Strezov, V., & Evans, T. (2014). Catalytic pyrolysis of coffee grounds using NiCu-impregnated catalysts. *Energy & Fuels, 28*, 228-235.

APPLICATION OF THERMAL ANALYSIS FOR MEASURING TOTAL LIPID CONTENTS IN LARGE SCALE CULTIVATION OF FRESHWATER MICROALGAE

Dr. Bongmun Kang, Kyoto Institute of Technology, Kyoto-si, Kyoto-hu

Introduction

The biodiesl production rate and yield could be determined by monitoring the amount of lipid contained in microbial cells in large-scale cultivation. However, the characterization of biodiesel obtained from microalgae is affected by the laboratories with various analysis methods. Furthermore, the measurement variability in the large scale production is more higher than that in lab-scale ones. The large variability remains an important problem in measurement of algae productivity and compositional analysis. Especially, the total lipid contents measured by gravimetric methods based on the Folch and Bligh and Dyer showed surprising measurement variability among laboratories, inaccuracies in prevailing methods relative to reference procedures and have led to a desire to unify analysis methodologies. To overcome this problem, various high-throughput technologies were suggested for accurate measurement of total lipid contained in the microorganisms, especially oleaginous microalgae. However, these methods are difficult to measure total lipid contents in microalgae obtained from in-situ large-scale production without the pretreatment of wet sample. Because it is time-consuimg and tedius for quantifying total lipid contents in microalgal samples in large-scale production. The move from laboratory to large scale microalgal cultivation requires careful planning. It is imperative to do extensive pre-pilot demonstration trials and formulate a suitable trajectory for possible data extrapolation for large scale experimental designs. Considering newly advanced lipid quantification method, it is necessary for development of method for measuring total lipid contents without pretreatment of microbial cells obtained from large-scale cultivation.

In the current study, the thermal analysis performed with two-step linear temperature program was applied to the quantification of lipid contained in wet microbial cells produced in-situ large-scale production. If our method is proved to be a useful method, it may open the way to monitor the lipid induction in the wet samples obtained from large scale cultivation. Because it is important to know the time when the lipid induction is occurred for improving the lipid production during large-scale cultivation (raceway). The thermal analysis was performed with two-step linear temperature program in temperature range from 310 to 351 °C. The microorganism examined in this study was an freshwater microalgae, *Nostoc* sp.. We examined the relationship between the total lipid contents and heat evolveds in the temperature range from 310 to 351 °C obtained from wet microbial samples.

Title

Application of thermal analysis for measuring total lipid contents in large scale cultivation of freshwater microalgae.

Author/Byline Information

Bongmun Kang, Kyoto Institute of Technology, Kyoto, Kyoto-Fu, Japan
Ho-Sung Yoon, Kyeong-pook National University, Daegu, Kyeongsangbukdo, Republic of Korea.

Material and Methods

Nostoc sp. was grown in BG-11 agar medium (1.5 g $NaNO_3$, 0.04 g K_2HPO_4, 0.075 g $MgSO_4 \cdot 7H_2O$, 0.036 g $CaCl_2 \cdot 2H_2O$, 0.006 g $C_6H_6O_7$, 0.006 g $C_6H_{5+4y}Fe_xN_yO_7$, 0.001 g EDTA (disodium salt), 0.02 g $NaCO_3$, 1.0 mL Trace metal mix A5 (2.86 g $H_3 \cdot BO_3$, 1.81 g $MnCl_2 \cdot 4H_2O$, 0.222 g $ZnSO_4 \cdot 7H_2O$, 0.39 g $NaMoO_4 \cdot 2H_2O$, 0.079 g $CuSO_4 \cdot 5H_2O$, and 49.4 mg $Co(NO_3)_2 \cdot 6H_2O$ per liter) per liter and pH 7.1). The precultured *Nostoc* sp. cell was transferred to a 250-mL Erlenmeyer flask containing 100 mL of BG-11 medium and grown with shaking at 150 rpm and 28 °C for 21 days, respectively. Each seed culture of 100 mL was inserted in carboy containing 16 liter commercial liquid fertilizer, Bionex medium (5.1 % N, 10% P_2O_5, 5% K_2O; Biosanga, Busan, South Korea) and cultivated at air bubble condition and 28 °C for 21 days. The inoculation of 16 liter carboys were transferred to the raceway containg with 60 ton and cultivated for 15 days by using water-mill for water circulation. Cells were harvested by centrifugation at 3,600×*g* (= 9.80 m s^{-2}) for 1 min 30 sec. and then analyzed by the method mentioned in below. Consequently, the 60 ton raceway, containing 0.6 L of culture at 15 days were diluted to cause the induction of lipid synthesis through nitrogen deprivation. To do it, the cultures were immersed in a raceway thermoregulated at 28 °C using water-mill. After decreasing concentration of nitrogen in Bionex medium during adaptation of microalgal, N-deprivation was achieved automatically. A simultaneous TG-DTA instrument equppied with auto sampler (model DTG 60 A, Shimadzu Co. Ltd., Kyoto, Japan) was used to measure the heat evolved from microbial cells. The Shimadzu DTG 60-60H was capable of ramping a sample from 0.1 to 99.9 °C min^{-1}. The DTG furnace assembly was continuously flushed with air (21% oxygen and 79% nitrogen) at a flow rate of 250 mL min^{-1}. The wetted microbial sample of approximately 15 mg were placed in platinum pans of 5 mm in diameter. And 30 mg of α-alumina (α-Al_2O_3) powder (Shimadzu Co., Ltd., Kyoto, Japan) was used as an inert reference. This gave satisfactory reproducibility of TG-DTA curves. All chemicals used were of analytical grade. The two-step linear temperature program, which has been suggested by Kang [1], was applied to resolve the exothermic peak in specific temperature range in DTA curves. In addition, approximately 50 mg of centrifuged wetted microbial cells were was mixed with 450 μl chloroform–methanol (2:1 v/v) in a 2 ml Eppendorf tube. The mixture was vortexed for 20 min at room temperature and centrifuged at 20,000 × g for 5 min to recover the solvent phase. The solvent phase was transferred to a preweighed 15-ml Pyrex glass tube (Iwaki Co. Ltd., Japan). This extraction was repeated three times. After solvent evaporation, the weight of the crude lipid was determined by weighing the glass tube using an electronic scale. At least three measurements were performed with each microbial sample.

Results

1. The growth of Nostoc. sp in the large-scale cultivation

The growth curve of *Nostoc. sp* showed typical exponential increasement for 10 days in large scale cultivation (Fig. 1A). In this time, the concentration of phosphate and nitrogen were siginificantly decreased to approximately 8 and 14 mg mL^{-1} , respectively (Fig. 2A and 2B). Compared to the gradual decreasment of concentration of phosphate, that of nitrogen was reduced by approximately 50% only 6 days after large scale production. Simultaneously, the water quality was also determined by monitoring the DO, turbidity, and pH when *Nostoc. sp* was cultivated in large-scale reactors. The turbidity was increased about two-fold when incubating for 6 days (Fig. 1B). The increasment of turibidity is thought to result from exponential growth of *Nostoc* sp. At the same time, the concentration of oxygen in water is significantly reduced with cultivating *Nostoc* sp. (Fig. 2C). The recorded temperature of water remained approximately same from 27 to 30 °C (Fig. 2D). It was demonstrated that oxygen consumption was caused by the growth of *Nostoc* sp. than effect of temperature. When *Nostoc* sp. was inoculated, the pH was dramatically increased from 4 days and reached a maximum at around 6 days (Fig. 3A). The highest values (pH >8) at stationary phases were observed with *Nostoc* sp. from 6 days after algae cultivation. In the stationary phase of pH, the total lipid contents was exponentially increased and retained after 12 days inoculation (Fig. 3B). We found that optimal productivity of *Nostoc* sp. were obtained at pH 8.5 -8.6. Based on morphological observations, alkaline pH inhibited the growth of microalgae, thus diverting the energy to form TAG. In algal cultures, pH usually increases due to the photosynthetic CO_2 assimilation. The pH regulation could improve the nutrient removal and biomass productivity .

2. The relationship between heat evolved and total lipid contents obtained from Nostoc. sp

The heat evolved in the temperature range from 310 to 351 °C was gradually increased from 6 days and reached a maximum at around 12 dyas after inoculation of microbial cells (Fig. 3C). When the growth of *Nostoc* sp. was reached in the stationary phases, the energy accumulation in the cell was detected. There was no significant difference in the values (pH >8) from 6 and 12 days. Fig. 4 showed the relationship between the heat evolved in temperature range from 310 to 351 °C and total lipid contents. As a results, the linear relationship was detected between the heat evolved (H_E) and total lipid content of microbial cells. As like, there was a linear relationship of 98% between the H_E value and the total lipid content of the tested microorganisms. We find a statically significant correlation between the heat evolved from 310 to 351 °C of wetted microbial cell and it`s total lipid contents measured by gravimetry (Fig. 3B and 3C). However, there was no relationship between calorific value measured by bomb calorimetry and lipid contents. Therefore the linear relationship between the heat evolved from 310 to 351 °C and the amount of lipid determined by gravimetrically (Fig. 4) agrees with the thermal analysis using the two-step linear temperature program in specific temperature range. The higher correlation of total lipid contents could be explained why Fig. 4 shows a linear relationship with the heat evolved in the specific temperature range considered in this study. The temperature range from 310 to 351 °C used in this study , which differs from the specific temperature range previously found by the author before mentioned , from 280 to 360 °C for different type of oleaginous microorganisms. The specific temperature range is , as can be seen , a little difference may occurred by the type of lipid contained in oleaginous microorganisms. However, these results may prove the usefulness of thermal analysis performed with the two-step linear temperature program for the quantification of lipid of microalgae. Based on this method, the heat evolved of wet microbial sample was measured in the temeperature range from 310 to 351 °C. Because, the specific temperature range where reflect the total lipid contained in the microbial cell was previously investigated. The increase in the heat evolved in 310-351 °C was proportion to the total lipid contents of microbial cells (Fig. 4). The advances in present study is that, wet samples could be directed toward measurement of the heat evolved without the pretreatments including drying, washing and pulverizing. We could measure the heat evolved of wet samples directly obtained from the large-scale cultivation. If the cell lipid content information is available during the time course of the bioprocess, decisions on process control strategy can be made (for example, adjusting the carbon/nitrogen ratio of the medium, based on the output from the flow cytometry) so that lipid productivity can be increased.

3. The lipid induction of Nostoc. sp by N-starvation

It was showed in Figure. 5 that heat evolved of *Nostoc.* sp gradually increased to 3.41±0.50 KJ/g on 6 days decreasing to 2.24±0.90 KJ/g on 8 days before N-starvation. After N-starvation on 8 days, the heat evolved of *Nostoc.* sp significantly increased to 6.44±0.97 KJ/g on 12 days and was kept constant. Based on the equation shown in Fig. 4, the total lipid contents of *Nostoc.* sp. could be calculated from heat evolved of it. As a result, the total lipid contents of *Nostoc.* sp. on 6 days was estimated to be approximately 17 % which was 1.9-fold higher than total lipid content converted from the initial heat evolved (0 day). However, The total lipid contents calculated from heat evolved reached a maximum, approximately 31%, at around 12 days. The N-starvation on 8 days results in a 3.9-fold increase in the total lipid contents compared with initial value (0 day). Nitrogen is the single most critical nutrient affecting lipid metabolism in algae. A general trend towards accumulation of lipids, particularly TAG, in response to nitrogen deficiency has been observed in numerous species or strains of various microalgae. For example, lipid content of freshwater green alga *Chlorella vulgaris* could be significantly increased by 40% in low nitrogen-containing medium.

Conclusion.

In general, the quantifying cell lipid content was time consuming, it required the extraction of lipid from a large number of cells using organic solvents, evaporation of the solvent, and determination of the amount of lipid by weighing the dried extract. The thermal analysis performed with the two-step linear temperature program makes it possible to measure the amount of lipid present in wet microbial samples directly. Firstly, the wetted microalgal cell could be directly applied to thermal analyzer, as revealed by the fact that the cost of analysis is markedly reduced in oleaginous microganisms. Secondly, there is considerable interest in using this technology as the basis for lipid quantification. This method allows for the lipid quantification in wetted microalgal samples without the extraction using organic solvents, which caused high variability and inaccuracies in the total lipid contents among laboratories. This raises the possibility for the suggestions on standardization of the protocols for quantification of lipids produced from the large scale cultivation. More specifically, the total lipid contents was determined by monitoring the heat evolved of wetted microbial cells. The lipid quantification based on thermal analysis can provide invalulabe information about lipid induction. Finally, there is a conscious effort ongoing to suggest standard analytical methods so researchers and industry members in the field are able to compare and track oleaginous microalgae.

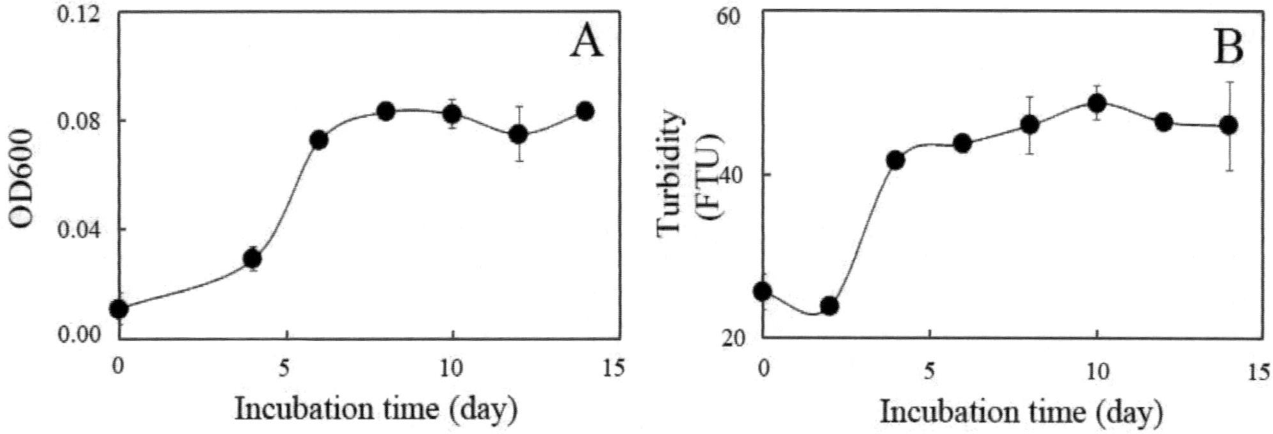

Fig. 1. The growth curve (A) and turbidity (B) of *Nostoc* sp.

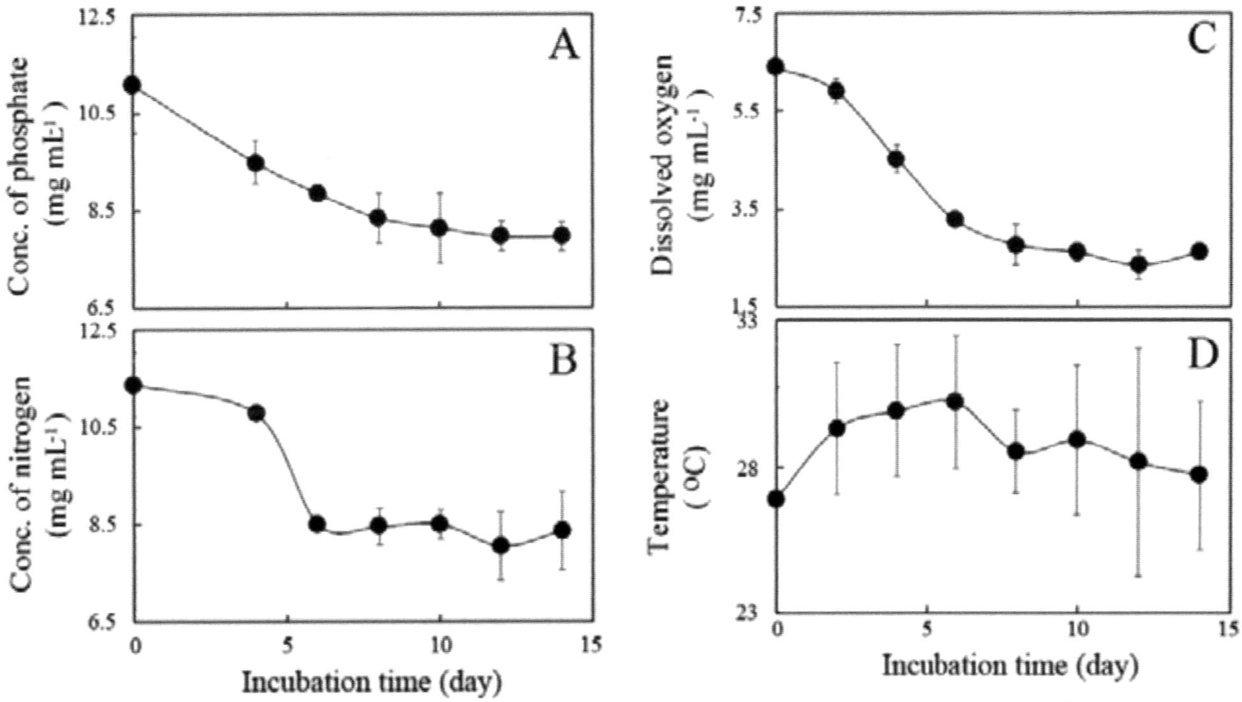

Fig. 2. The concentration of phosphate (A) and nitrogen (B), dissolved oxygen (DO) (C) and temperature (D) of water

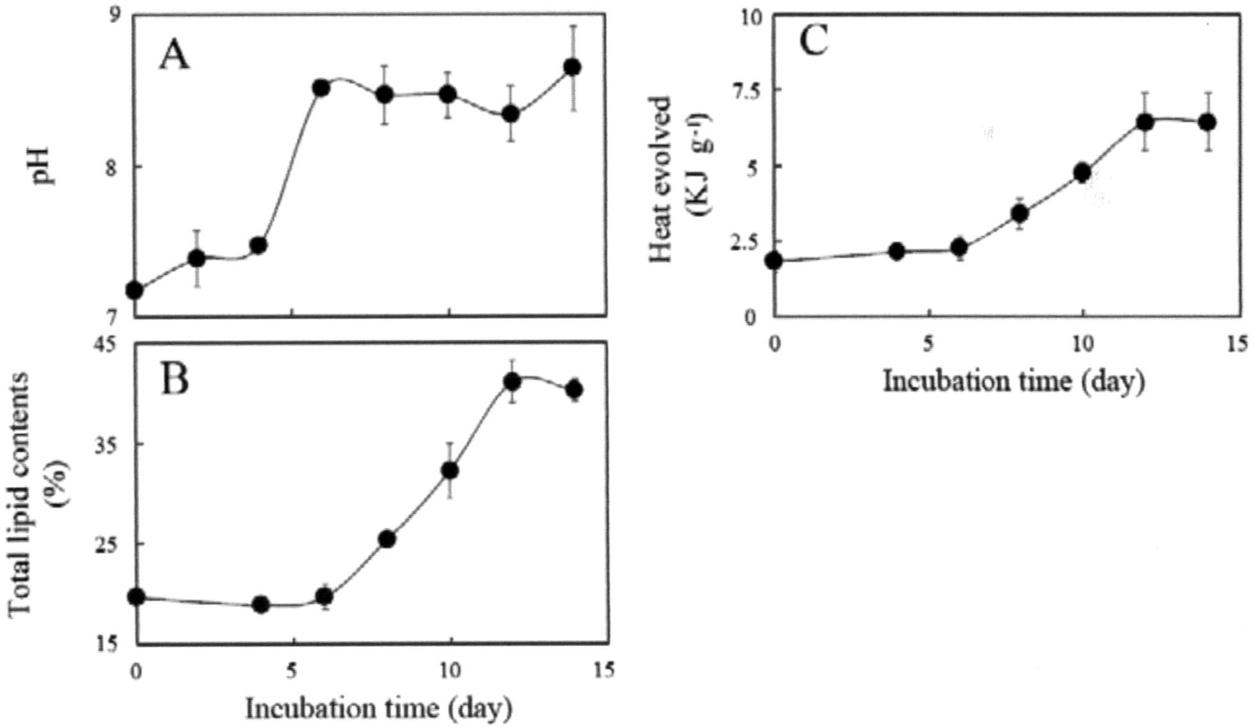

Fig. 3. The pH in the water (A), total lipid contents (B), and heat evolved (C) of microbial sample (*Nostoc* sp.)

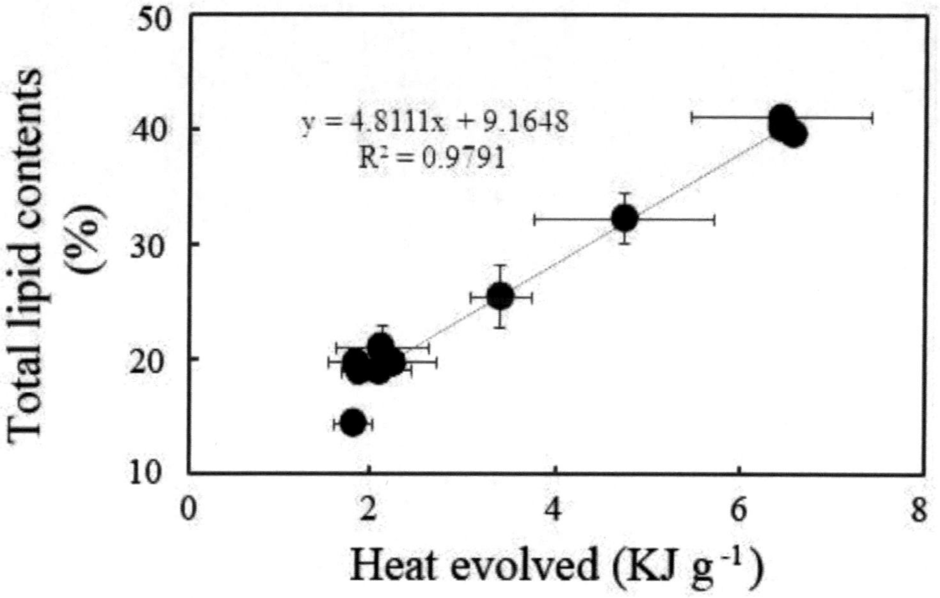

Fig. 4. The relationship between total lipid contents and heat evolved

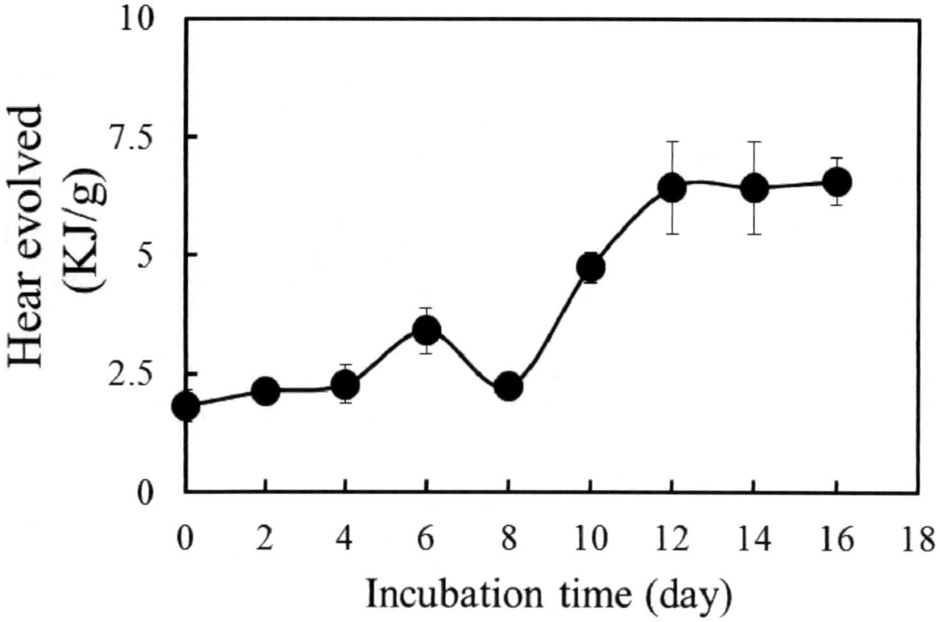

Fig. 5. The change of heat evolved by lipid induction.

References

1. Kang, B. M., Honda K., Okano K., Aki T., Omasa T. and Ohtake H. (2011) "Thermal analysis for differentiating between oleaginous and non-oleaginous microorganisms". *Biochemical Engineering Journal*, 57, pp. 23-29.

390058 The Conversion of Various Phenolic Compounds over Ru/TiO2: Active Site Requirements

Abstract:

One of the greatest challenges associated with the conversion of lignocellulosic biomass to fuels and chemical lies in the stable and selective conversion of lignin derived phenolics. Metals supported on reducible oxides such as Ru/TiO_2 have demonstrated the potential to preserve the carbon contained in methoxy groups and avoiding excessive hydrogen consumption of the ring upon deoxygenation of phenolic OH groups. Ru supported on TiO_2 also exhibits activity and stability enhancements for the conversion of phenolics when compared with Ru supported on other non-reducible supports. In this contribution, we have systematically varied the Ru/TiO_2 interfacial area, as well as the TiO_2 phase, and measured the initial rates of conversion of several model compounds containing hydroxyl and methoxy groups including, guaiacol, anisole, catechol and cresol. The implications of these results will be discussed in the context of kinetic models to describe the conversion phenolics such as guaiacol, where one must take into account several factors. These include the rate of conversion of each produced product (for example catechol and cresol) over the reduced sites, the interface, and the various metal sites. In-situ titration is also used to discriminate deoxygenation reactions that occur through the continuous production and healing of defects on the support vs. those facilitated by active sites located at the Ru/TiO_2 interface.

383593 First-Principles Investigation of the Mechanism and the Active Site for Hydrodeoxygenation over Ru/TiO_2

Abstract:

Using periodic density functional theory (DFT) we have explored the hydrodeoxygenation (HDO) pathways of acetaldehyde and phenol, two surrogate molecules for some of the over 400 different oxygenates from biomass fast pyrolysis [1], over supported Ru/TiO_2 catalysts. The main functional groups represented by acetaldehyde and phenol are aldehydes, ketones, and aromatic alcohols. These groups are abundant output products of fast pyrolysis, and catalytic conversion of these products to oxygen free compounds is of significant importance for bio-oil upgrade. There are three major HDO pathways: (1) hydrogenation (HYD), (2) direct deoxygenation (DDO), and (3) decarbonylation (DCN). Among these, DDO is the most desired pathway because it requires less hydrogen than HYD, and in contrast to DCN, the number of carbon atoms during HDO remains constant. In a systematic study of HDO pathways for acetaldehyde and phenol we have calculated the thermodynamic and kinetic parameters for a large number of possible elementary surface reactions with specific focus on product selectivity, i.e. the preference for C-H, C-O, and C-C bond scission reactions at each reaction intermediate.

In experimental studies Ru/TiO_2 has shown promising results in terms of selectivity and catalytic activity for HDO of different molecules (e.g. furfuran, acetone, phenol) [2-4]. To represent the supported Ru/TiO_2 catalysts we tested various active site models including metallic Ru(0001), pure rutile metal-oxide surfaces [RuO_2(110), TiO_2(110)], modified metal-oxide catalysts [1.0 ML RuO_2/TiO_2(110), Ru_1/TiO_2(110)] and a 10-atom cluster of Ru supported on TiO_2(110) [Ru_{10}/TiO_2(110)]. Our results show that the metallic Ru(0001) surface is not suitable for DDO. It preferentially catalyzes HYD for phenol, resulting in the fully hydrogenated products cyclohexane or cyclohexanone, or DCN for acetaldehyde. The dominant DCN pathway for acetaldehyde can be described as: adsorption

(CH_3CHO) → αC-H bond breaking (CH_3CO^*) → ßC-H bond breaking (CH_2CO^*) → C-C bond breaking (CH_2^*). At high hydrogen pressure CH_2^* may leave the catalyst surface as CH_4, while at low hydrogen pressure stepwise dehydrogenation leads to CH* and eventually surface carbon (C*), causing catalyst deactivation due to coke formation. HYD and DDO may occur on Ru(0001) and the formation of ethane or ethylene is feasible, but the removal of surface oxygen requires significant energy barriers ($E_a > 1.7$ eV) and limits the overall reaction. In turn, the accumulation of surface oxygen can lead to high oxygen coverages, partial oxidation or even the formation of a surface oxide.

Oxygen vacancy sites are expected to be important for HDO catalysis on metal-oxide surfaces. Hence, we initially derived a constrained thermodynamic phase diagram for the rutile $TiO_2(110)$, $Ru_1TiO_2(110)$, 1.0 ML $RuO_2/TiO_2(110)$, and $RuO_2(110)$ surfaces. Under typical HDO conditions (T > 500 K, $P(H_2) > 200$ bar) these surfaces are partially reduced and exhibit bridging hydroxyl groups (HO^{br}). Using this reduced state as reference surface we investigated vacancy formation pathways on each metal-oxide. Upon further hydrogen adsorption the surface hydroxyl groups can be removed as water along three different reaction pathways: (1) the common disproportionation pathway $HO^{br} + HO^{br} \rightarrow H_2O^{br} + O^{br}$, and two pathways involving molecular hydrogen adsorption at the coordinatively unsaturated Ti site (Ti^{cus}) (2) $Ti^{cus}\text{-}H_2 + O^{br} \rightarrow Ti^{cus}\text{-}H + HO^{br} \rightarrow Ti^{cus} + H_2O^{br}$, and (3) $Ti^{cus}\text{-}H_2 + HO^{br} \rightarrow Ti^{cus}\text{-}H + H_2O^{br}$. Among these pathways, reaction (3) occurs with the lowest energy barrier on Ru_1TiO_2, 1.0 ML RuO_2/TiO_2, and RuO_2. Due to weak interactions between molecular hydrogen and the Ti^{cus} site of TiO_2, reactions (2) and (3) are not availableand reaction (1) is favored instead [5]. In all cases, the desorption of the resulting H_2O^{br} forms the surface oxygen vacancy. This active vacancy site provides the preferred adsorption site for acetaldehyde and phenol, and plays an important role in the preferential activation of C-O scission over C-C scission reactions. Hence, the DDO pathway occurs most likely on the considered metal-oxides surfaces and follows the sequence of elementary steps given by: adsorption on surface vacancy (CH_3CHO^*-s) → isomerization (CH_3CHO^*-d) → αC-O bond breaking (CH_3CH^*) → ßC-H bond breaking (CH_2CH^*) → αC-H bond formation $(CH_2CH_{2(g)})$, leading to the desired product ethylene. In addition to the DDO pathway, metal-oxide surfaces can also catalyze the HYD pathway resulting in the intermediate formation of ethanol, which can subsequently be dehydrated to ethane.

The highest selectivity for ethylene is predicted on the $TiO_2(110)$ surface; however, the dissociation of hydrogen molecules and the formation of HO^{br} groups, the precursor to the formation of a required oxygen vacancy, are kinetically limited. The presence of Ru clusters on TiO_2 may facilitate hydrogen delivery from the gas-phase to $TiO_2(110)$ through hydrogen spillover, and we investigated this possibility using a Ru_{10} cluster model supported on $TiO_2(110)$. DFT results suggest that spillover of H atoms originating from the Ru cluster is kinetically hindered ($E_a \approx 1$ eV), but the Ru/TiO_2 interface is very active for hydrogen dissociation ($E_a \approx 0.1$ eV) and enables the formation of oxygen vacancy sites as required for C-O bond breaking reactions during HDO on TiO_2. The interface also provides sites with unique bifunctional properties: for example, the C-O bond in phenol adsorbed in a vacancy site at the interface can be selectively broken in a concerted step where hydrogen is provided from the Ru_{10} cluster and the eliminated OH group heals the bridging oxygen vacancy site on $TiO_2(110)$: $C_6H_5O^{br}H + H\text{-}Ru_{10} \rightarrow C_6H_{6(g)} + HO^{br} + Ru_{10}$.

In conclusion, detailed DFT calculations suggest that DCN is preferred over DDO and HYD on the metallic Ru(0001) surface, leading to surface carbon or methane, while on the investigated metal-oxide surfaces DDO and HYD are more likely than DCN. Oxygen vacancy formation on TiO_2 is limited by hydrogen delivery but can be facilitated at the Ru/TiO_2 interface. These vacancy sites selectively catalyze the DDO pathway of acetaldehyde and phenol, leading to the desired products, ethylene and benzene, respectively.

[1] Huber, G. W.; Iborra, S.; Corma, A. Synthesis of transportation fuels from biomass: chemistry, catalysts, and engineering. *Chem. Rev.* **2006**, *106*, 4044–98.

[2] Omotoso, T.; Boonyasuwat, S.; Crossley, S. P. Understanding the role of TiO_2 crystal structure on the enhanced activity and stability of Ru/TiO_2 catalysts for the conversion of lignin-derived oxygenates. *Green Chem.* **2014**, *16*, 645.
[3] Pham, T. N.; Shi, D.; Sooknoi, T.; Resasco, D. E. Aqueous-phase ketonization of acetic acid over Ru/TiO_2/carbon catalysts. *J. Catal.* **2012**, *295*, 169–178.
[4] Newman, C.; Zhou, X.; Goundie, B.; Ghampson, I. T.; Pollock, R. A.; Ross, Z.; Wheeler, M. C.; Meulenberg, R. W.; Austin, R. N.; Frederick, B. G. Effects of support identity and metal dispersion in supported ruthenium hydrodeoxygenation catalysts. *Appl. Catal. A Gen.* **2014**, *477*, 64–74.
[5] Rekoske, J.; Barteau, M. Isothermal reduction kinetics of titanium dioxide-based materials. *J. Phys. Chem. B* **1997**, *101*, 1113–1124.

389047 Acid/Metal Balance in Bifunctional Conversion of Cresol to Hydrodeoxygenated Compounds Via Alkylation, Dehydration, and Hydrodeoxygenation

Abstract:

The hydrodeoxygenation of phenolics has been a matter of great interest in recent studies as part of the biomass conversion efforts. We have recently proposed a mechanism that includes a fast enol-keto tautomerization step that precedes the ring hydrogenation. An unstable ketone intermediate forms over the catalyst surface and it can be hydrogenated via two possible paths. a) when the CO of this intermediate gets hydrogenated instead of the ring, a very reactive unsaturated alcohol (3-methyl-3,5-cyclohexadienol) is formed and it can be readily dehydrated to toluene, b) when the C=C bonds in the ring are hydrogenated, methylcyclohexanone is formed. The latter can be further hydrogenated to methylcyclohexanol. Over acidic supports, this alcohol dehydrates to methylcyclohexene; subsequent dehydrogenation produces toluene. A detailed kinetics study of the conversion of each of the possible intermediates yields strong support to the proposed mechanism.

Interestingly, when the reaction is conducted in the liquid phase on zeolite-supported metals (Pt or Pd), alkylation of the aromatic ring by the saturated alcohol becomes an important reaction pathway that produces C11 hydrocarbons of potential value as fuel components. In this contribution, we will discuss the importance of maintaining a suitable acid/metal balance and a favorable zeolite porous structure to maximize the yield of alkylated compounds.

384294 Catalytic Hydrodeoxygenation on Metal Carbides

Abstract:

Stoichiometric removal of oxygen is required for the chemical transformation of renewable biomass feedstock to energy carriers and other substitutes for petroleum derived products. Oxygen removal from biomass can be performed under catalytic hydrotreatment conditions (25-70 bar H_2, 300 – 400°C) using noble metal catalysts, but oxygen removal is often concurrent with excessive H_2 usage through hydrogenolysis and/or alkene/aromatic hydrogenation, producing low-value short-chain alkanes as opposed to light alkenes and aromatics.

We report vapor phase hydrodeoxygenation (HDO) of acetone, anisole, guaiacol, and m-cresol, representative of cellulose and lignin pyrolysis products over Mo_2C-based non-precious metal catalyst formulations that enable selective HDO at ambient pressure and low temperatures (98 – 200°C). The dominant HDO reaction products were unsaturated, deoxygenated hydrocarbons including propylene, benzene, phenol, and toluene with carbon selectivity ranging from 60-90%, and <1% selectivity to C-C bond scission products.

The HDO kinetics of these probe molecules were investigated as functions of contact time, reactant partial pressures, and temperature. This work proposes HDO mechanistic models to fit experimentally measured rates, and these kinetic models were further probed using transient isotopic experiments and the kinetic isotope effect. The HDO site requirements were probed via *in situ* titration experiments using pyridine, CO and CO_2 to assess the number and identity of active sites. The active catalytic phase for HDO was investigated as a function of catalyst oxygen content via *in situ* O_2 dosing experiments and systematic oxygen inclusion during catalyst synthesis to probe the effects of near-surface and bulk oxidation on catalytic HDO activity. The kinetics, mechanism, and site requirements of low pressure HDO on Mo_2C formulations will be presented.

388957 Surface Modification and Catalytic Activity of MoO_3 during Hydrodeoxygenation (HDO) of Lignin-Derived Model Compounds

Abstract:

Hydrodeoxygenation (HDO) is a major route of upgrading bio-oil. MoO_3 has been shown to be active for HDO of bio-oil model compounds. Specifically, the catalyst is selective towards cleavage of C-O bonds at temperatures of 320-350°C and H_2 pressures <= 1 bar. Importantly, this catalyst selectively cleaves stronger phenolic Ph-OMe over weaker PhO-Me bonds. Detailed reactivity studies were performed to gain insight into the nature of the active sites. An initial induction period was observed for the catalyst to achieve peak performance,which suggested a surface modification was required to render the catalyst active. The catalyst surface undergoes partial carburization as evidenced by the presence of oxycarbide- and oxycarbohydride-containing phases (i.e., $MoO_xC_yH_z$) at lower temperatures. Mo^{5+} species are hypothesized to be responsible for the HDO chemistry on the catalyst. The transformation of bulk phases and the surface modification of MoO_3 by carbon/H_2 are also investigated to understand the role of surface carbon in the stabilization and enhanced activity of the partially reduced MoO_3 surface. The partial surface carburization during reaction was found to play a role in stabilizing Mo^{5+} state on the catalyst surface and slowing down over-reduction of MoO_3 into inactive MoO_2 phase.

387495 Online Monitoring of Lignin Liquid Phase Conversion Processes By in-Situ ATR-IR Spectroscopy

Abstract:

Lignocellulosic biomass is becoming an increasingly important alternative resource for biofuels production. Lignocellulosic biorefinery operations must produce chemicals as well as energy vectors in order to be competitive, making use of *all*the major components of lignocellulose including lignin. Indeed, this aromatic polymer holds great potential for the sustainable production of renewable aromatics and phenolics. [1] Typical lignin valorization efforts entail the study of depolymerization and defunctionalization steps and almost inevitably involve complex reaction mixtures. In order to develop more efficient catalytic lignin valorization processes, more insight is clearly needed in the chemistry of lignin linkage cleavage and defunctionalization.

Our group has, for instance, reported on the solubilization and catalytic conversion of a wide variety of lignins in aqueous or ethanolic solvents. [2] A two-step route involving lignin depolymerization by liquid phase reforming step followed by hydrodeoxygenation of the resulting lignin-oil to yield, amongst others, lignin-derived BTX. [3] Improvements in the efficiency of this two-step conversion process requires further insight into the changes in molecular weight upon dissolution and depolymerization as well as insight in the changes in the chemical structure of lignin upon conversion. The application of in situ spectroscopic techniques for online monitoring of the liquid phase lignin conversion processes can prove very useful in this respect. Here, we report on our efforts to further develop the use of in situ spectroscopic techniques such as ATR-IR for on line monitoring of catalytic lignin conversion processes. Liquid phase reforming reactions are run in the presence of a heterogeneous Pt catalyst in ethanol/water at 225°C and 58 bar in a semi-batch autoclave equipped with an ATR-IR crystal. Changes in the fingerprint region provided insight into the rate, nature and extent of lignin depolymerization. The approach taken is expected to be of interest to the various liquid phase lignin depolymerization strategies that are currently being developed.

[1] Zakzeski, J., Jongerius, A. L., Bruijnincx, P. C. A. and Weckhuysen, B. M., Chem. Rev. 110 (2010) 3552.

[2] Zakzeski, J., Jongerius, A. L., Bruijnincx, P. C. A. and Weckhuysen, B. M., ChemSusChem, 5 (2012) 1602.

[3] Jongerius, A. L., Bruijnincx, P. C. A. and Weckhuysen, B. M., Green Chem., 15 (2013) 3049.

389285 Liquid Phase Hydrodeoxygenation of Phenolic Compounds on Platinum and Palladium Catalysts

Abstract:

The hydrodeoxygenation of phenolic compounds, which constitute a substantial portion of bio-oils, was investigated. Noble metals have been reported to be suitable catalysts for vapor phase or aqueous phase processing of phenolic compounds, but have not been tested in more complex reaction environments. The goal was to reveal the influence of other bio-oil components on catalyst stability and reaction kinetics. In situ attenuated total reflection IR spectroscopy was used to monitor the course of the reaction with high time resolution. Platinum and palladium supported on carbon served as the catalysts; typical reactions temperatures were 80 to 115 °C at a H_2 pressure of 250 psig.

The addition of small amounts (as little as 2 wt%) of alcohols to the aqueous reaction mixture suppressed conversion of phenol and of its hydrogenation product cyclohexanone. This effect can be ascribed to decomposition of the alcohol and poisoning of the catalyst and was more pronounced for primary than secondary alcohols. Conversion of phenol was more susceptible to this poisoning effect than conversion of cyclohexanone, which can be explained by different demands of these molecules on the reaction site. Another observation was that the performance of the noble metal catalysts varied strongly with the conditions of the reductive pretreatment, indicating that significant catalyst restructuring in water can occur.

The data demonstrate that competitive adsorption and structure-sensitivity are the dominating factors in the noble metal-catalyzed conversion of phenolic compounds and their hydrogenation products.

390343 Red Mud Catalyzed Pyrolysis of Olive Mill Wastewater Sludge

Abstract:

Olive mill wastewater sludge (OMWS) is one of the major environmental pollutants in olive oil industry using traditional or three-phase process. The major problem stems from the poor biodegradability of the OMWS because of its high phenolic compounds content. In most Mediterranean countries, olive mill wastewater are stored in ponds where the major part of water evaporates and the sludge dries up and is then landfilled for disposal. To address this serious environmental pollution problem, we developed fluidized bed catalytic pyrolysis of OMWS to produce pyrolysis liquids that are very stable, low viscosity (5-7 cP), neutral pH (6-7), and very high higher heating value (41 MJ/kg). The pyrolysis was conducted at 400-500 °C in a red mud catalyst bed. The yields of the organic fraction ranged from 28-35 mass% which is much higher than what obtains for other lignocellulosic biomass feedstocks. The char/coke yield ranged from 20-29 mass% while the gas yield ranged from 24-37 mass%. The ^{13}C NMR analysis of the liquid product showed predominance of aliphatic hydrocarbons and a small fraction of aromatic hydrocarbons. In the absence of the catalyst, the liquid product contained acidic groups and was more viscous.

372774 Oil Derived from Biomass: Qualitative and Quantitative Analysis of Oxygenates

Abstract:

Bio-oil is a complex liquid mixture of paraffins, aromatics, oxygenates and other biomass-derived components. Using current techniques we are able to characterize these oils, but are unable to account for as much as 35 wt% of bio-oil components. Main differences between bio-derived oils and petroleum based oils include the high water, acid and oxygen content of the former. Refining these bio-oils typically involves a hydrodeoxygenating step, and the type and amounts of oxygenates have an effect on this process. A better knowledge of oxygenates present in bio-oil have implication on properly evaluating bio-oil quality and may provide the key to solving processing issues encountered in the upgrading and refining. In this current climate of studying the possibility of integrating bio-oils as a co-feed in intermediate refinery streams an intimate knowledge of species present in bio-oils and their abundance is a key research need. The challenges include bio oil's propensity to form condensation products, reduce catalyst lifetimes, and its corrosiveness. We are working towards standardizing analytical methods to identify and quantify species such as carbonyls, carboxylates, phenolic and hydroxyl species in bio-oils. Our work in this regard will be discussed in this paper.

386280 Porous Graphitized Carbon Supported Catalysts for Lignin Depolymerization in Hydrothermal Conditions

Abstract:

Graphitized porous carbon supports was incorporated with dispersed catalytic sites, i.e. transition metal (Ni/Fe) through in-situ thermochemical synthesis or liquid-reduction deposition. The graphene structure and nano-scale active metal particles were characterized with TEM, SEM, XRD, Raman, TGA/DSC, physical and chemical adsorption. The advantages of in-situ thermochemical synthesis approach are in scalable chemical modification at the nanoscale and in tuning the surface active groups for selective lignin conversion. We assume that catalytically active metals distributed in porous grahitized carbon matrix will result in higher selective of lignin liquefaction and specific reaction pathways towards formation of aromatic derivatives in hydrothermal conditions. Lignin depolymerization products were analyzed and quantified with GC-MS. Mechanism of catalysts' deactivation was discussed based on characterization of fresh and used catalysts.

Biological Conversion of Methane: A Comprehensive Review

Kyle Stone[1], Q. Peter He[2], and Jin Wang[1]

[1] *Department of Chemical Engineering, Auburn University, Auburn, AL, 36849, USA*

[2] *Department of Chemical Engineering, Tuskegee University, Tuskegee, AL, 36088, USA*

Natural gas has emerged as a potential major feedstock and a more sustainable resource for the production of fuels and valuable chemical products due to two reasons: (1) the large abundance found in the U.S and (2) potential benefit of reducing greenhouse gas (GHG) emissions. The U.S. Geological Survey and the U.S. Energy Information Administration estimate that the U.S. has over 2,000 trillion ft^3 of technically recoverable natural gas (proved + unproved) that could last 50 years at the current consumption rate. Proven natural gas reserves have progressively increased since 2009. From natural gas, liquid fuel can be produced that decreases emissions by up to 50% compared to unconventional petroleum and decreases particulate matter by up to 40% compared to combustion of conventional diesel [1,2].

Methane is the prominent component of natural gas with smaller concentrations of other alkane gases. Chemically converting methane to liquid fuels through the Fischer-Tropsch process is a proven technology. However, the process is energy-intensive, requiring temperatures up to 800-1000°C and pressures of 50 bar in the conversion of methane. Additionally, the chemical route has high capital costs and low conversion efficiencies. According to the EPA, methane accounted for 9% of gas emissions in the United States in 2011 and a quarter of all emissions come from natural gas and petroleum industry. This is problematic because when compared pound for pound, methane's impact on climate change is over 20 times greater than CO_2 over a 100 year period. Therefore, there is a clear need to convert methane into liquid fuels through better routes in order to positively impact the energy sector while reducing GHG emissions.

An alternative method to the chemical conversion of methane is through the use of biological microorganisms called methanotrophs. This direct route of conversion has the potential to reduce capital expenses by more than 50% [3]. Additionally, the biological process achieves high selectivity at ambient temperature and pressure. Finally, literature through the past three decades has proven that these methanotrophs can convert methane into fuels or other chemicals highly desired by industry.

In this work, we provide a thorough review of available work on biological methane conversion. The existing results are organized according to different stages of methane conversion, which result in two distinct groups; methane activation and product formation. In methane activation, we discuss how this metabolic process occurs within these unique organisms. This includes an understanding of the enzymes that are key to this direct conversion and how they are regulated or inhibited through external environmental controls or intracellular mechanisms. Product formation is the second part of this work and focuses on how these cells can produce methanol and poly-β-hydroxybutyrate (a biodegradable plastic). In addition, possible productive pathways through co-culture systems with methylotrophs, a type of bacteria that are able to consume methanol, are considered. We developed a schematic flow chart to organize different reactions carried out by different strains of methanotrophs and methylotrophs. Finally, this review extensively covers the challenges of existing routes and possible solutions. This includes problems with the metabolism of the microbes and the external conditions imposed on them. Possible solutions discuss less known microorganisms and the outlook of metabolic engineering on the direct conversion of methane. It is the purpose of this work to provide a clear overview on the current status and challenges of biological methane conversion, therefore to provide the knowledge base for the design of novel, sustainable processes for fuels or chemical production from methane. More than 60 papers were reviewed in this work. Due to limited space all technical references are not listed.

1. "America's Energy Future: Technology and Transformation" (National Academy of Sciences, 2009)

2. U.S. Department of Transportation. OK-26-7015-00. Evaluation of Ultra-Clean Fischer-Tropsch Diesel Fuel in Transit Bus Applications, 2010.

3. ARPA_E Remote FOA. March 2013

367468 Modeling and Thermodynamic Properties Measurements of Levoglucosan for Process Design, Simulation, and Optimization in the Production of Pyrolitic Biofuel

Abstract:

Bio-oil produced from pyrolysis of cellulose biomass has attracted much attention in recent years as fuels substitution and its potential production of chemicals. Levoglucosan (1,6-anhydro-β-D-glucopyranose) is the major intermediate component in the pyrolysis of cellulose that controls char formation. Thermodynamic properties of levoglucosan are important in the simulation of process design and process development dealing with the pyrolytic reactions involving this compound that occur in the biorefinery.

In this work, we present a comprehensive analysis of the thermophysical and thermochemical properties of levoglucosan such as vapor pressure, enthalpy of combustion, and enthalpy of formation in the gaseous phase using static bomb combustion calorimetry, Knudsen effusion technique, and differential scanning calorimetry along with the computational calculations of the gaseous phase thermodynamic properties performed using the Gaussian 03 software package. Our measurement results were published in the Journal of Chemical Engineering Data (J. Chem. Eng. Data, 2013, 58, 1813–1821; DOI: 10.1021/je400207t).

Since levoglucosan is a hygroscopic compound, therefore, we will also discuss the detailed procedure of material handling and the challenges during the measurements to obtain accurate and reliable results.

The authors acknowledged Fundação para a Ciência e Tecnologia (FCT), Lisbon, Portugal and FEDER for financial support to Centro de Investigação em Química, University of Porto (strategic project PEst-C/QUI/UI0081/2013). I.M.R. and T.L.P.G. thank FCT and the European Social Fund (ESF) under the Community Support Framework (CSF) for the award of the research grants with references SFRH/BD/61915/2009 and SFRH/BD/62231/2009, respectively.

390593 Thermogravimetic Analysis and Fast Pyrolysis of Milkweed

Abstract:

Pyrolysis characteristics and fast pyrolysis of *Milkweed* was carried out in a thermogravimetric analyzer and a bubbling fluidized bed reactor. Total liquid yield of *Milkweed* was between 40.74% and 4.19 wt% between 425-550 °C. The gas yield increased from 27.90 wt% to 33.33 wt% with increasing reaction temperature. The higher heating value (HHV) of the *Milkweed* bio-oil (organic fraction) were relatively high (30.33-32.87 MJ/kg) and varied with reaction temperature, feeding rate and fluidization velocity. The selectivity for CO_2 was highest within non-condensable gases, and the molar ratio of CO_2/CO was about 3 at the different reaction conditions. The ^{13}C-NMR analysis, of the biooil showed that the relative concentration of carboxylic group and its derivatives was higher at 425 °C than 475 °C, which resulted in slightly higher oxygen content in bio-oil. Aliphatic hydrocarbons signals for the biooil were slightly higher at 475 °C compared to those at 425 °C. The pH of aqueous phase obtained at 475 °C was 7.37 which is the highest reported for any lignocellulosic biomass pyrolysis oils.

Alternate fuel synthesis in supercritical solvents

Anant Sarve , Mahesh N. Varma*, Shreeram S. Sonawane.

Department of Chemical Engineering, Visvesvaraya National Institute of Technology, Nagpur, 440010, India

Abstract: Mono-alkyl esters can be used as diesel fuel without any alteration in diesel engine. Synthesis of mono-alkyl ester namely by trans-esterification of edible and non-edible oil is one of the focus area for research due to high consumption of fuel all over world. This study mainly target synthesis of mono-alkyl ester using acid-base catalyst like sulfuric acid, sodium hydroxide and potassium hydroxide. This study also looks for use of heterogeneous acid-base catalyst and an enzyme for biodiesel synthesis in which the downstream separation of product is easier. Various aspect of non catalytic synthesis of biodiesel under supercritical condition is also considered for study in this work. Overall study includes study of effect of temperature, reactant concentration and catalyst concentration on conversion.

Keywords: Alternate fuel, Acid-base catalyst, heterogeneous catalyst, enzyme for alternate fuel, supercritical fluids.

*Correspondence address:
Dr. Mahesh N. Varma
Department of Chemical Engineering
Visvesvaraya National Institute of Technology
Nagpur: 440010
Email: maheshnvarma@gmail.com
Phone: +91-9273083142.

All Authors email
anant4u87@gmail.com
maheshnvarma@gmail.com
shriramsonawane@gmail.com

384246 Production of Biodiesel from Michigan Soybean Oil Using Supercritical Methanol with Acetic Acid

Abstract:

Biodiesel is an affordable and renewable transportation fuel. Within the United States it is commonly produced from soybean oil. In conventional production of biodiesel, the transesterfication reaction is typically catalyzed in batch reactors with reaction times greater than 1 hour. Supercritical methanol is an alternative to catalyzed biodiesel production resulting in significantly reduced reaction times at temperatures greater than 350 °C. The addition of acetic acid enables the transesterfication reaction to proceed in supercritical methanol at lower temperatures, nearing methanol's critical temperature of 240 °C. Lower temperatures could result in a decrease in the energy costs of supercritical methanol biodiesel production. The effects of temperature, concentration of acetic acid, methanol concentration, and reaction time on the production of biodiesel from Michigan soybean oil were evaluated in an agitated 500 mL Hastelloy C276 batch reactor. In addition, the effect of adding biodiesel as a recycle into the reactor to facilitate mixing was also evaluated. The product composition was characterized using ASTM D6584. The viscosity, acidity, and density of the biodiesel product were also evaluated.

BIODIESEL PRODUCTION FROM VEGETAL OIL AND ETHANOL VIA TRANSESTERIFICATION IN SUPERCRITICAL CONDITIONS

Victor Sidi Neto
IPT, São Paulo, SP - Brazil

João Guilherme Rocha Poço
FEI, São Bernardo do Campo, SP, Brazil
IPT, São Paulo, SP - Brazil

This work was done and presented to IPT by Victor Sidi Neto to get his Professional Master Degree in Industrial Processes (2012).

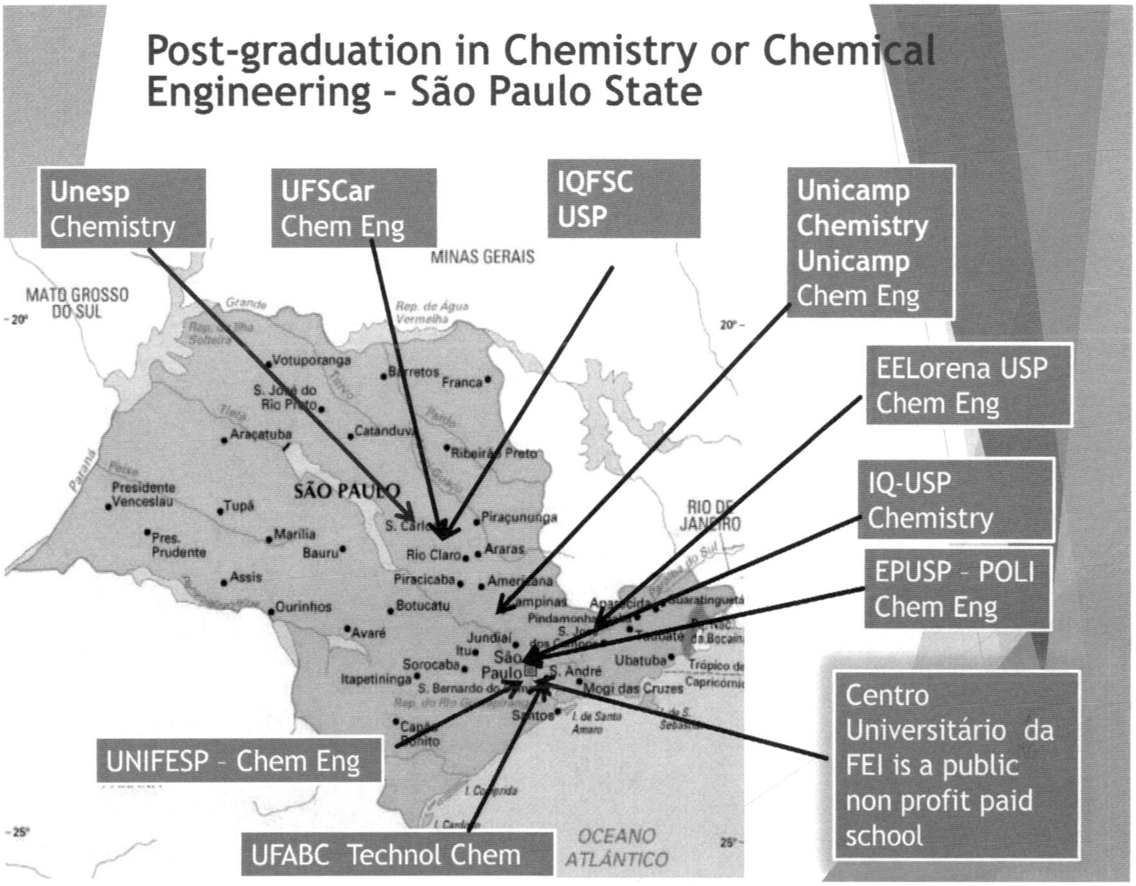
Post-graduation in Chemistry or Chemical Engineering - São Paulo State
Unesp Chemistry
UFSCar Chem Eng
IQFSC USP
Unicamp Chemistry
Unicamp Chem Eng
EELorena USP Chem Eng
IQ-USP Chemistry
EPUSP - POLI Chem Eng
Centro Universitário da FEI is a public non profit paid school
UNIFESP - Chem Eng
UFABC Technol Chem
MINAS GERAIS
MATO GROSSO DO SUL
SÃO PAULO
RIO DE JANEIRO
OCEANO ATLÂNTICO

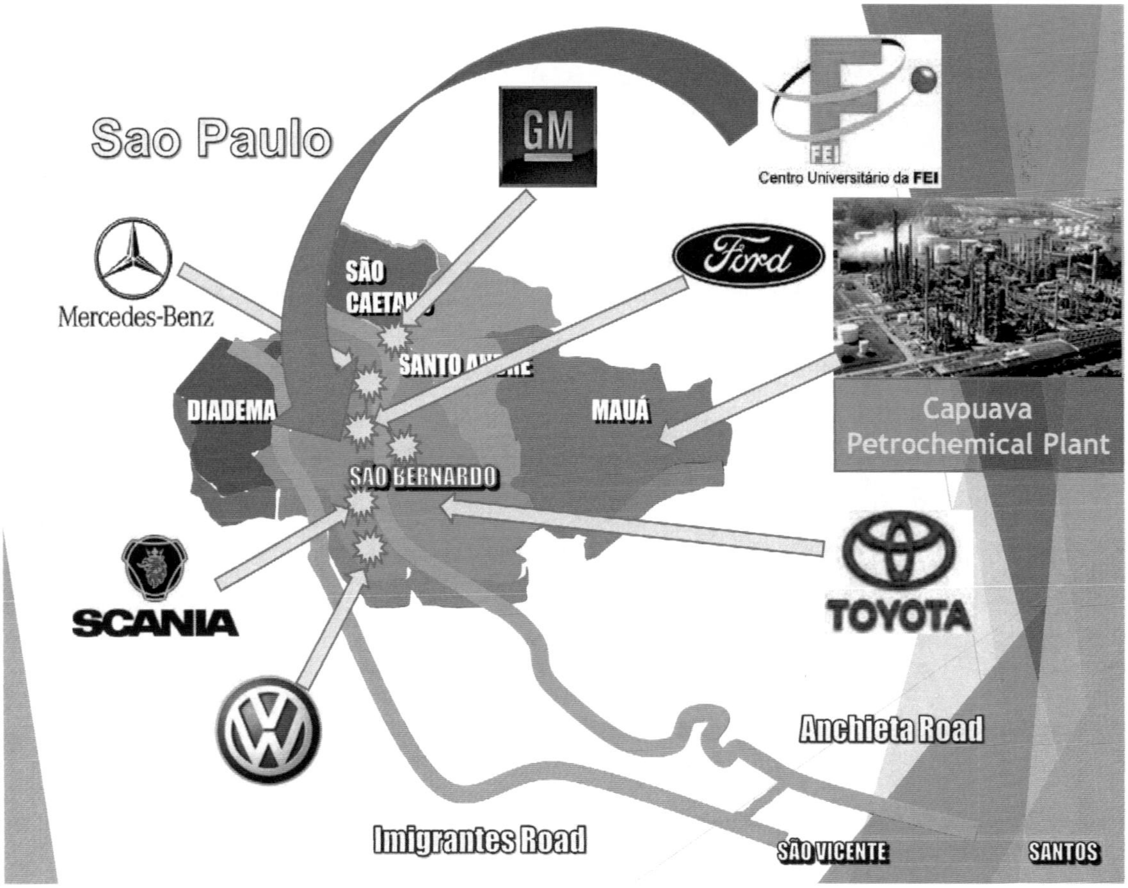
Sao Paulo
GM
Centro Universitário da FEI
Mercedes-Benz
Ford
SÃO CAETANO
SANTO ANDRE
DIADEMA
MAUÁ
SÃO BERNARDO
Capuava Petrochemical Plant
SCANIA
TOYOTA
Anchieta Road
Imigrantes Road
SÃO VICENTE
SANTOS

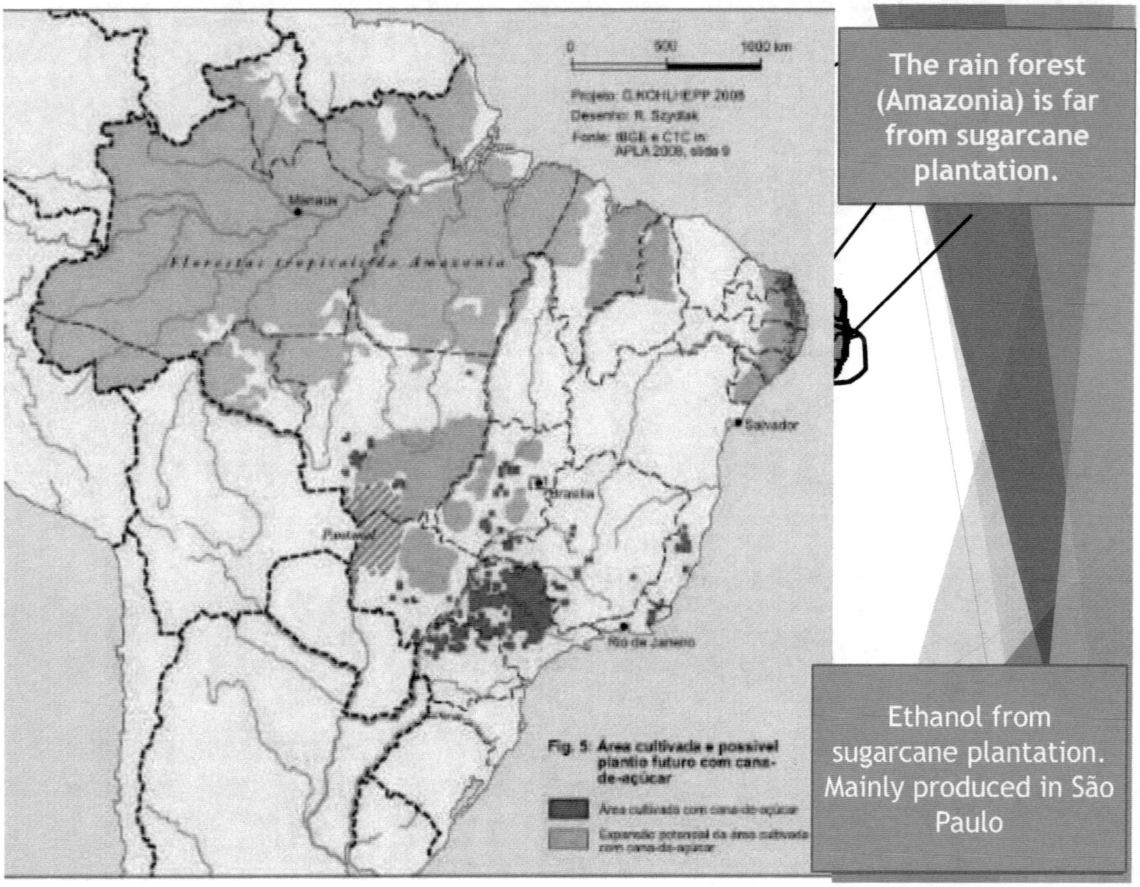

BIODIESEL = FATTY ACID ALCOHOL ESTER

- The word "Biodiesel" appeared first time in Chemical Abstracts in by Wang, R. (1988). Before, it´s was called by
  - "fatty acid methyl ester or fatty acid ethyl ester" that is obtained by the reaction of low molecular alcohols with triglicerides coming from the main part of vegetable or animal oils or fats.
  - The reactions of production of biodiesel was called interesterification (Feuge & Gross, 1949) or alcoholysis , but nowadays its is known as transesterification reaction

Wang, R. (1988) Development of biodiesel fuel. *Taiyangneng Xuebao* 9(4), p434-6 (apud in CAN 111:26233 - Source: Scifinder , CAS – American Chemical Society).

Feuge & Gros, (1949) *The J.A.O.C.S*, March p.97-102

STEPS IN BIODIESEL SYNTHESIS

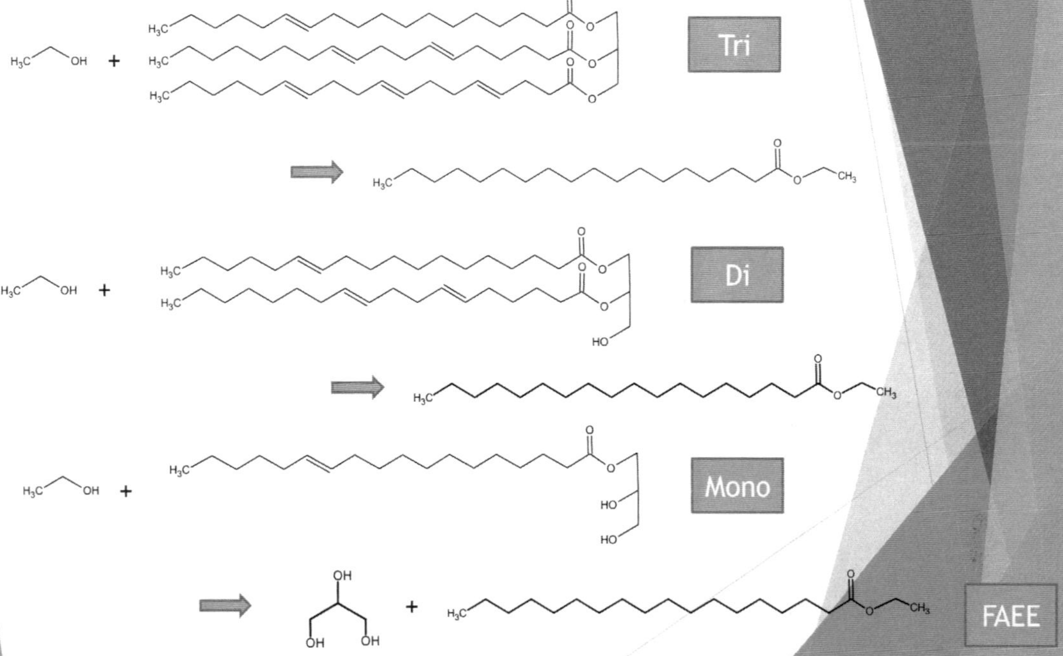

MECHANISM WITH BASE CATALYST

$$OH^- + CH_3OH \leftrightarrow CH_3O^- + H_2O$$

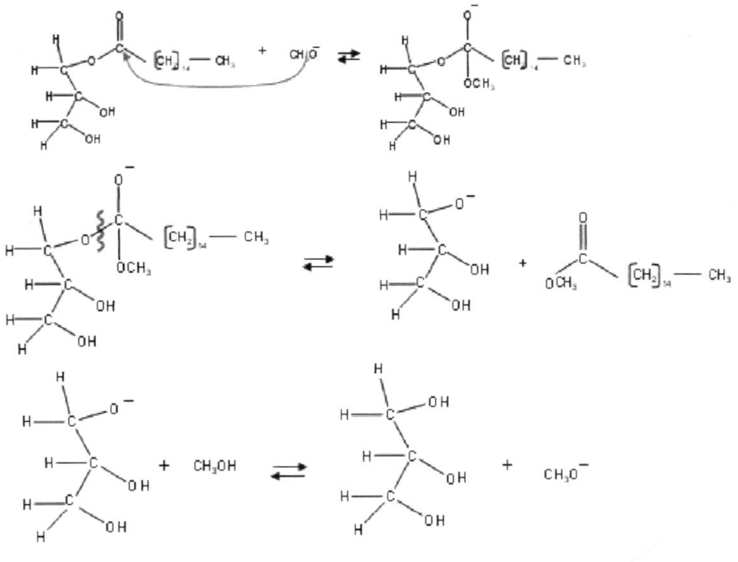

Source: Aranda, D. Seminário Biodiesel Shimadzu-SINC, 2007

NaOH 1% molar | $NaOCH_3$ 1% molar | Triazabiciclodeceno 3%

Make biodiesel seems to be very easy, but meeting specification is not so simple

This route raise some technical problems and imposes restrictions:

- Water in alcohol and FA source is a undesirable presence! Causes hydrolysis and soap formation.
- Free fatty acids waste the alcaline catalyst forming soaps.
- Ethylic biodiesel is not so easy to produce as methylic biodiesel.
- The reaction takes up to eight hour to complete.
- Multistep separation problems (emulsion and water consumption).

Motivation

Methods capable of overcoming such technical challenges, generating products with quality at least equal to the products currently obtained, would make the process more attractive in the economic, **technical and sustainable** points of view.

The **supercritical process** seems to be capable of generate high purity esters, without being influenced by the low quality of feedstocks and requiring separation processes that are simpler.

Motivation

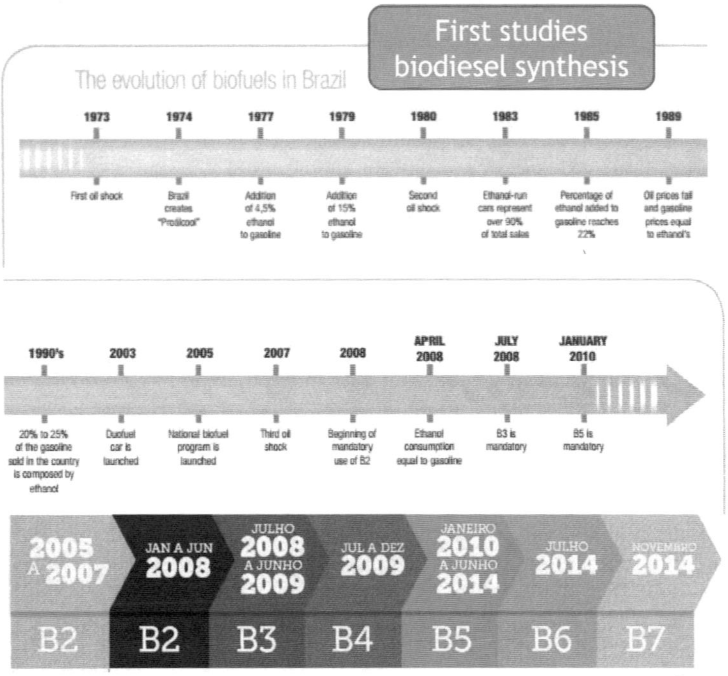

Source: ANP (2014)

Motivation

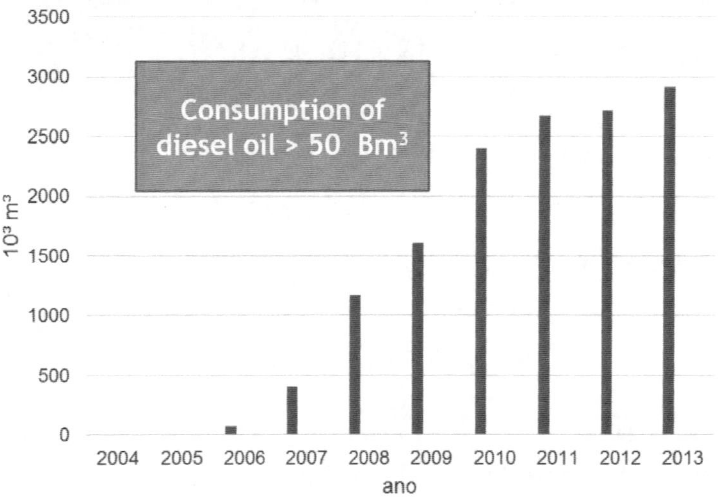

Biodiesel Production in BRAZIL
Source: ANP (2014)

Background

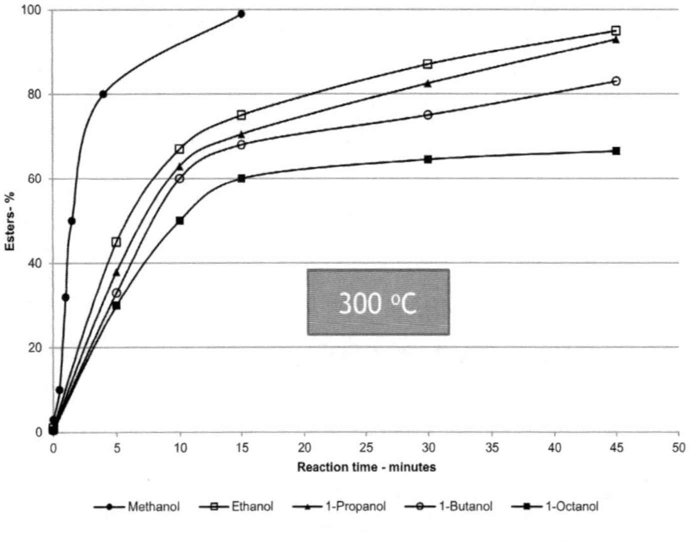

Supercritical ester formation from different alcohols presented by Warabi, Y. et al. (2004), *Bioresource Technology*, s.l., v. 91, p 283-287.

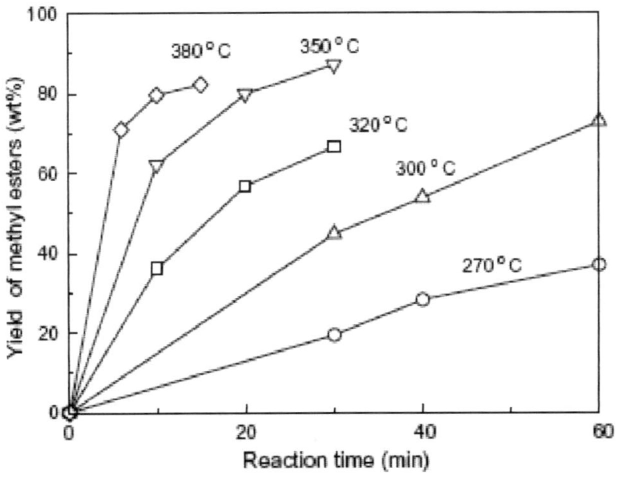

Supercritical ester formation from methanol.
Minami, E. Saka, S. (2006) Fuel v.85, p2479-2483

Objective

The present study aimed to produce biodiesel by supercritical method using soybean oil, oleic acid, and ethanol.

The reactions were performed with **anhydrous ethanol (99,5%)** or **hydrated ethanol** (92,8 %) in order to check how the presence of water in supercritical ethanol influences transesterification reactions and compare with the literature, which data were obtained with other experimental apparatus.

Experimental Procedure

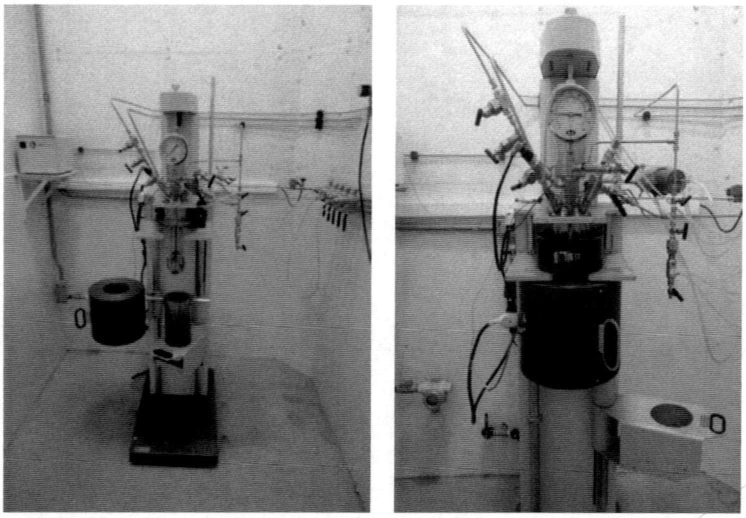

Batch Parr reactor. Max. Pressure 351.5 kgf/cm^2 a 500 °C
Source: Victor Sidi Neto, IPT (2012)

Raw materials and conditions

Molar ratio ethanol / triglyceride = 40
(Heating times varies from 77 to 110 minutes)

Alcohol	Fatty	Temperature (°C)	Sample time after temperature was reached (minutes)				
AEAC (anhydrous ethanol)	Soybean oil	280	0	20	40	60	120
		310	0	20	40	60	120
		340	0	-	-	60	120
		340	0	20	40	60	120
	Oleic acid	280	0	-	-	60	120
		280	0	20	40	60	120
AEHC (hydrated etanol)	Soybean oil	280	0	20	40	60	120
		310	0	20	40	60	120
		340	0	-	-	60	120
		340	0	20	40	60	120
	Oleic acid	280	0	-	-	60	120
		280	0	20	40	60	120
AEAC + Glycerol	**Ethylic ester**	**340 (inverse)**	**0**	**20**	**40**	**60**	**120**

Typical observed parameters

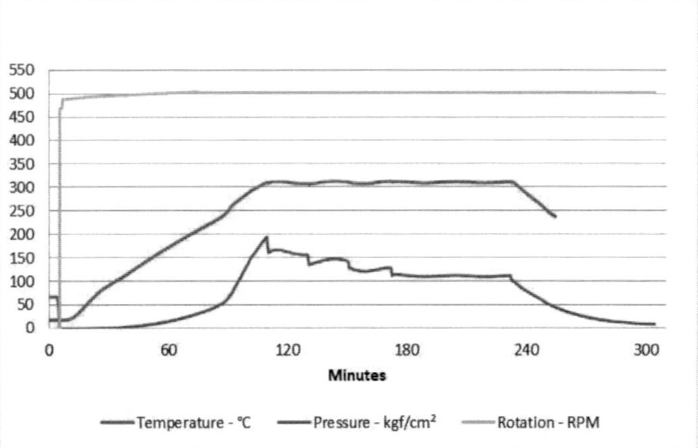

The following data were obtained with gas chromatography and Karl-Fischer measurements

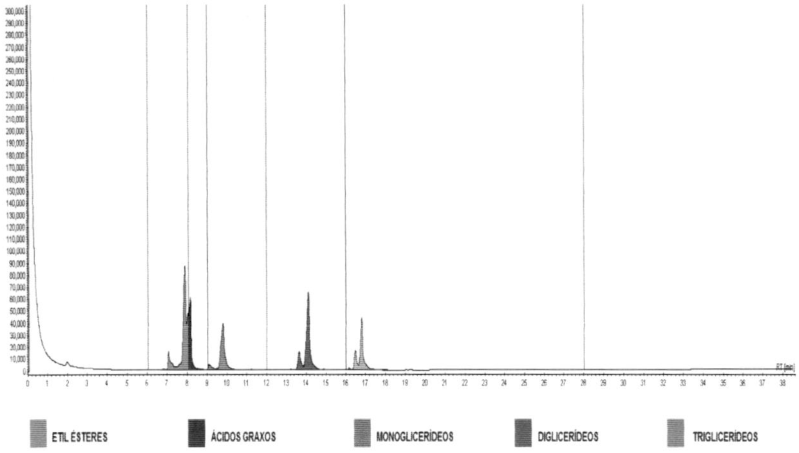

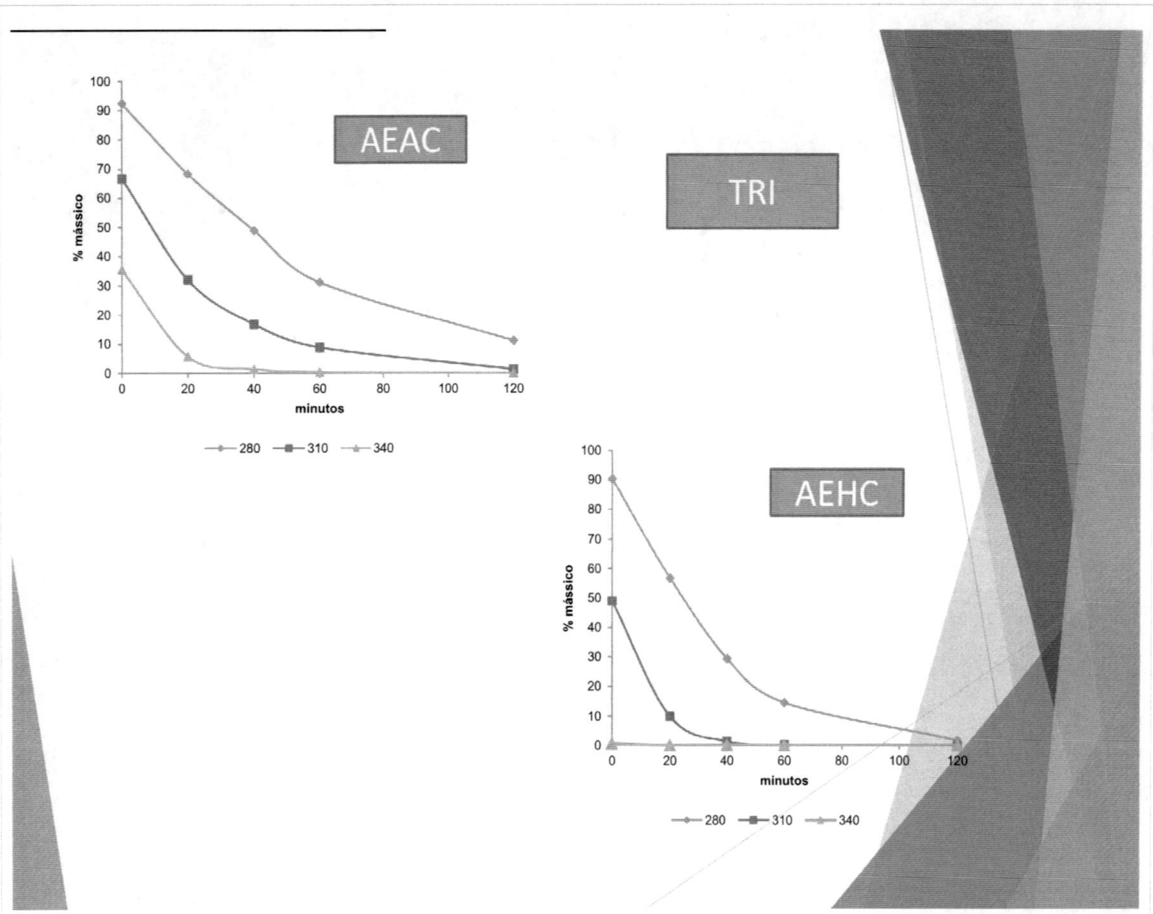
AEAC
TRI
AEHC
% mássico
minutos
280
310
340

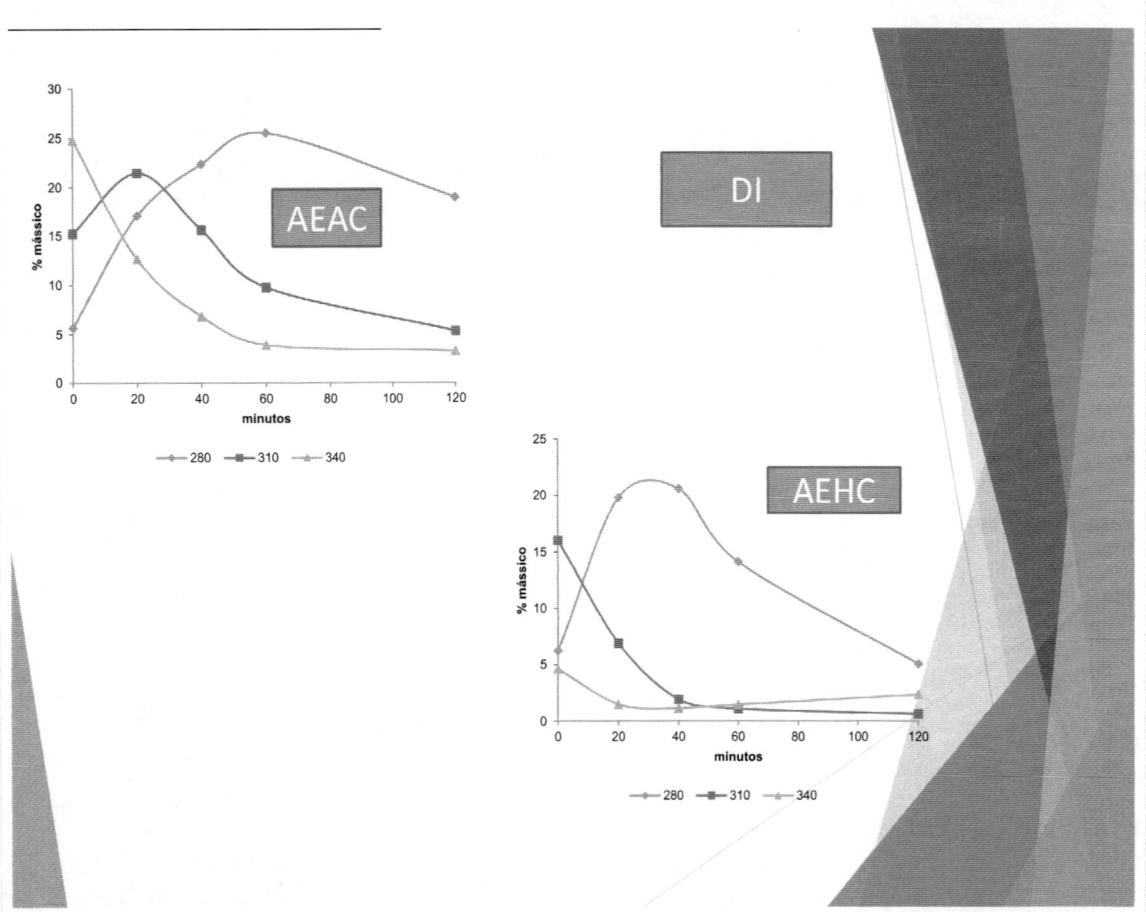
AEAC
DI
AEHC
% mássico
minutos
280
310
340

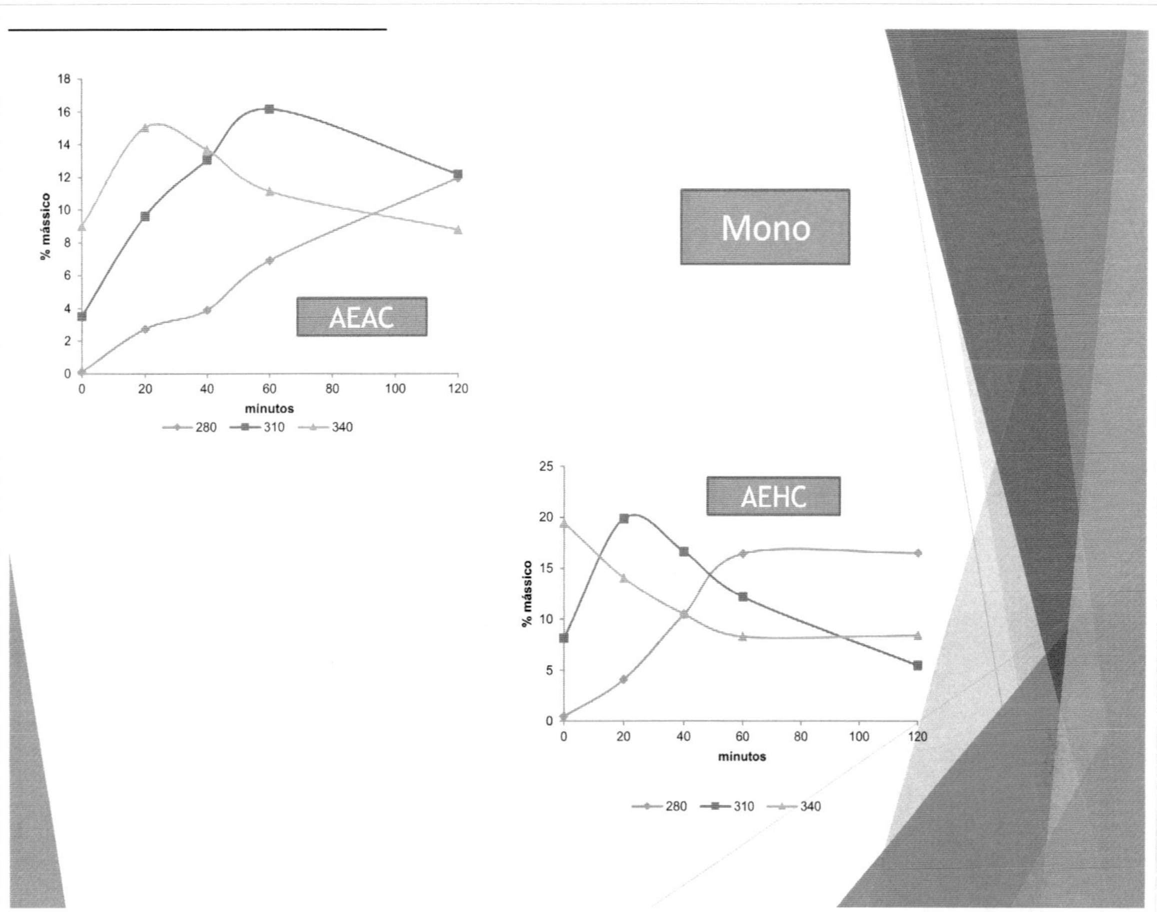
Mono
AEAC
% mássico
minutos
280
310
340
AEHC
% mássico
minutos
280
310
340

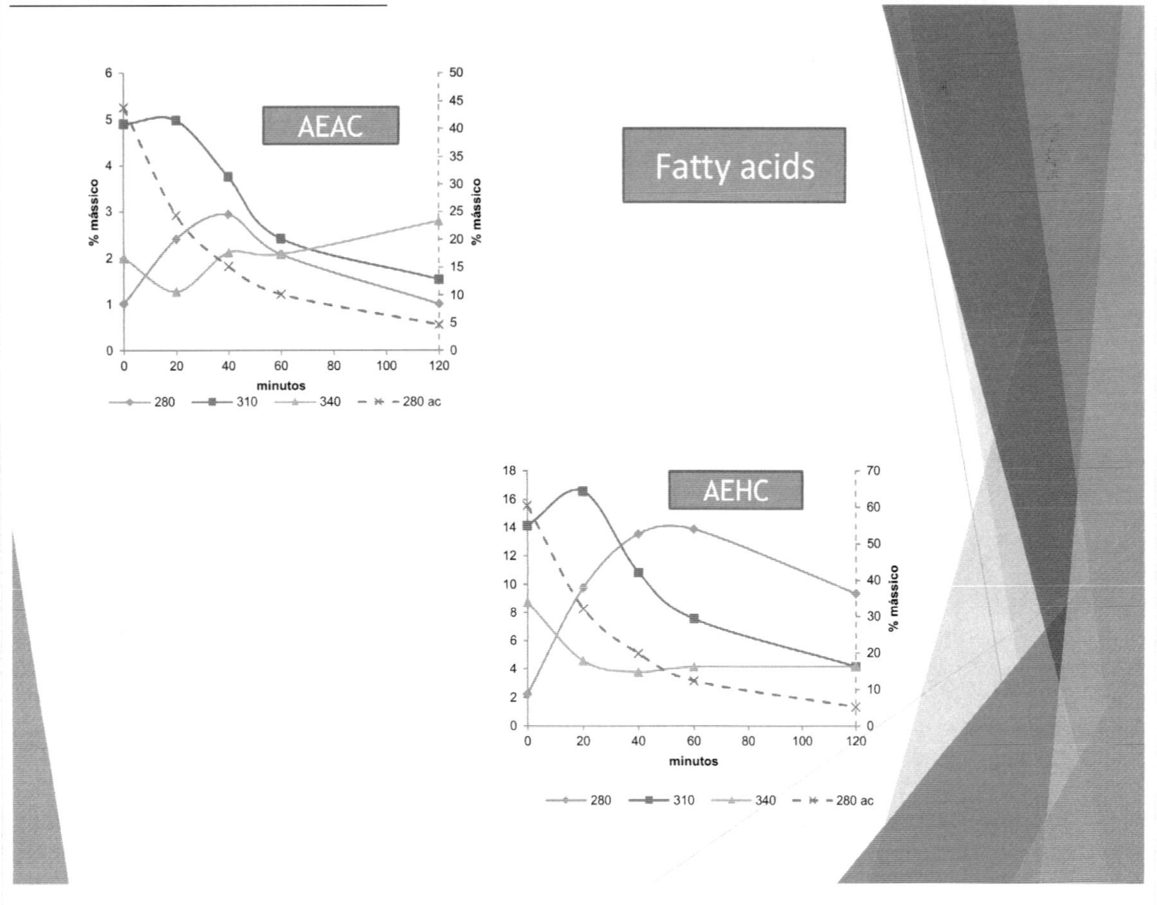
Fatty acids
AEAC
% mássico
% mássico
minutos
280
310
340
280 ac
AEHC
% mássico
minutos
280
310
340
280 ac

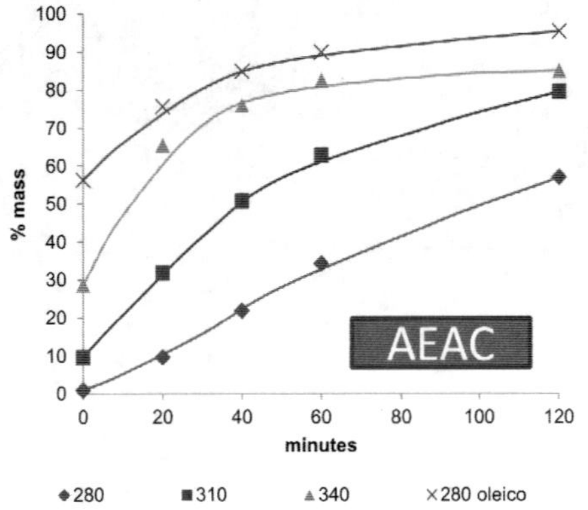

Esters - FAEE

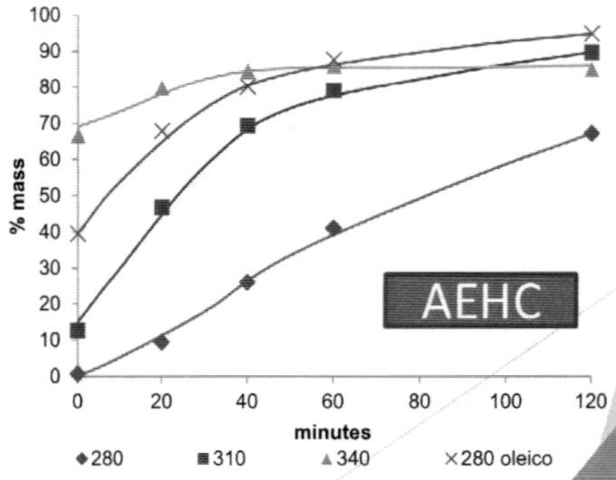

Kinectic modelling

triglyceride + ethanol ⇔ 3 ethyl ester + glycerol

$$[uEE] = 3 \cdot [triglyceryde] + 2 \cdot [diglyceride] + [monoglyceride] + [fatty\ acids]$$

$$\frac{d[uEE]}{dt} = -k \cdot [uEE]^{\alpha}$$

oleic acid + ethanol ⇔ ethyl ester + water

$$\frac{d[water]}{dt} = k \cdot [water]^{\alpha}$$

Model adjustment

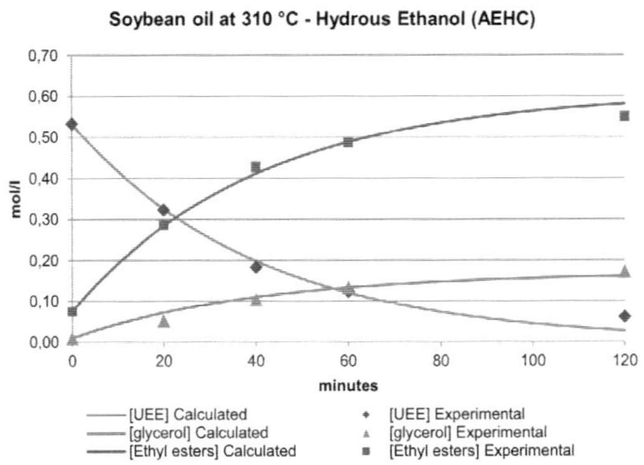

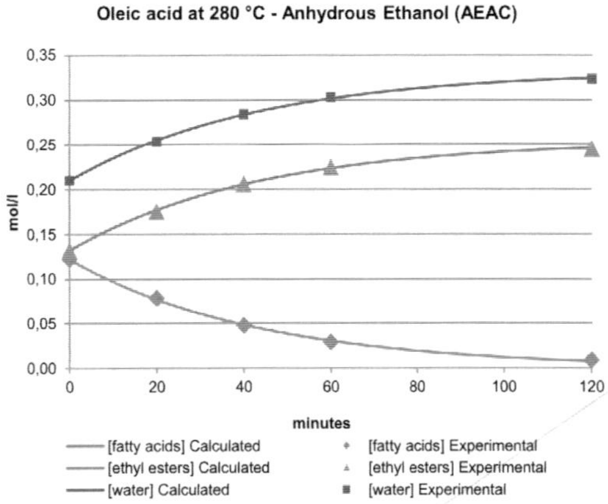

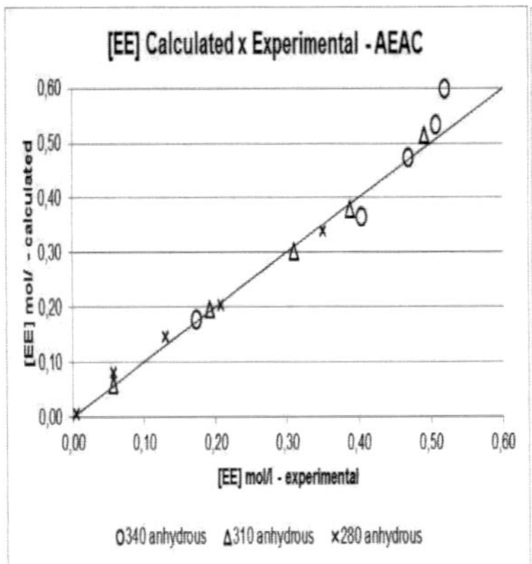

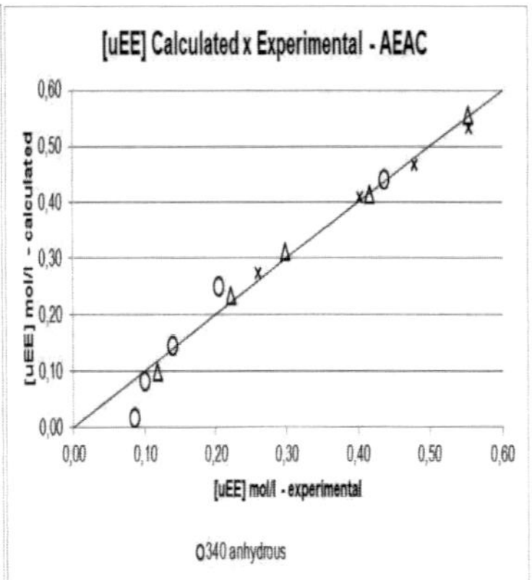

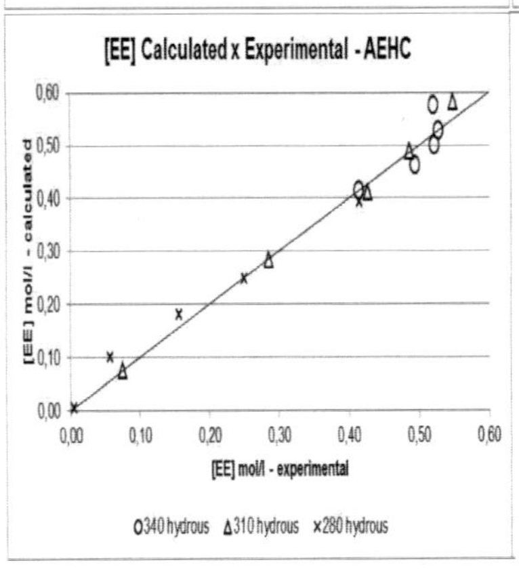

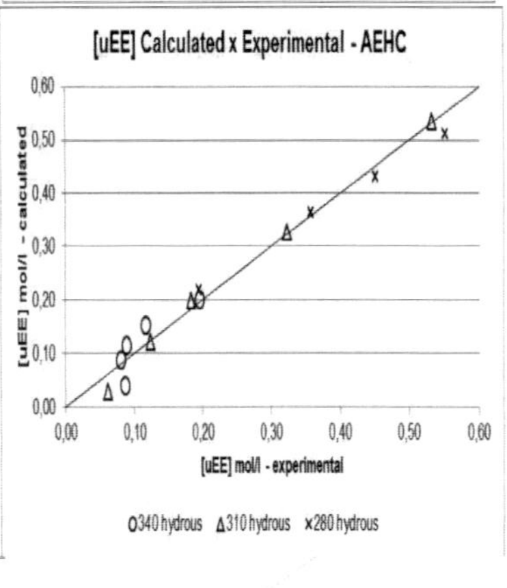

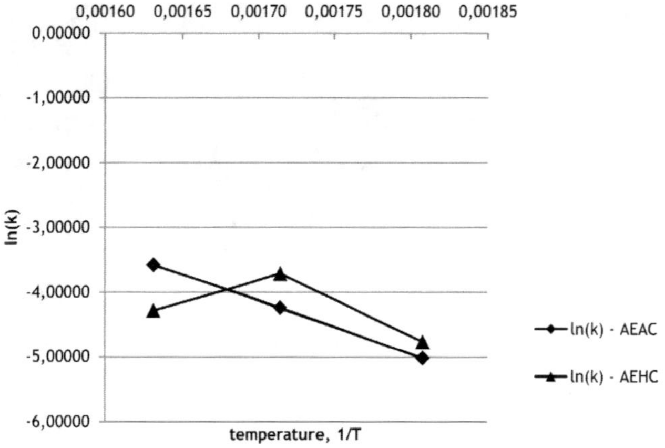

Activation energy (Ea)

- 67,65 kJ/mol for anhydrous ethanol
- 23,94 kJ/mol for hydrous ethanol.

Kusdiana and Saka (- 47 kJ/mol methyl ester)

Kusdiana, D; Saka, S. (2001) *Fuel*, s.l., v.80, p693-698.

Variation of concentration of the reactor componentes with time showing a reverse reaction at 340 °C.

Time (minutes)	Pressure (kgf/cm^2)	% Mass				
		EE	FA	MG	DG	TG
0	258,6	98,47	0,92	0,51	0,10	0,00
0 a 20	238,6	97,48	1,12	1,12	0,28	0,00
20 a 40	214,7	97,06	1,01	1,55	0,37	0,00
40 a 60	186,1	96,60	1,24	1,70	0,46	0,00
60 a 120	166,8	96,19	1,33	1,88	0,60	0,00

The maximum conversion and purity are limited

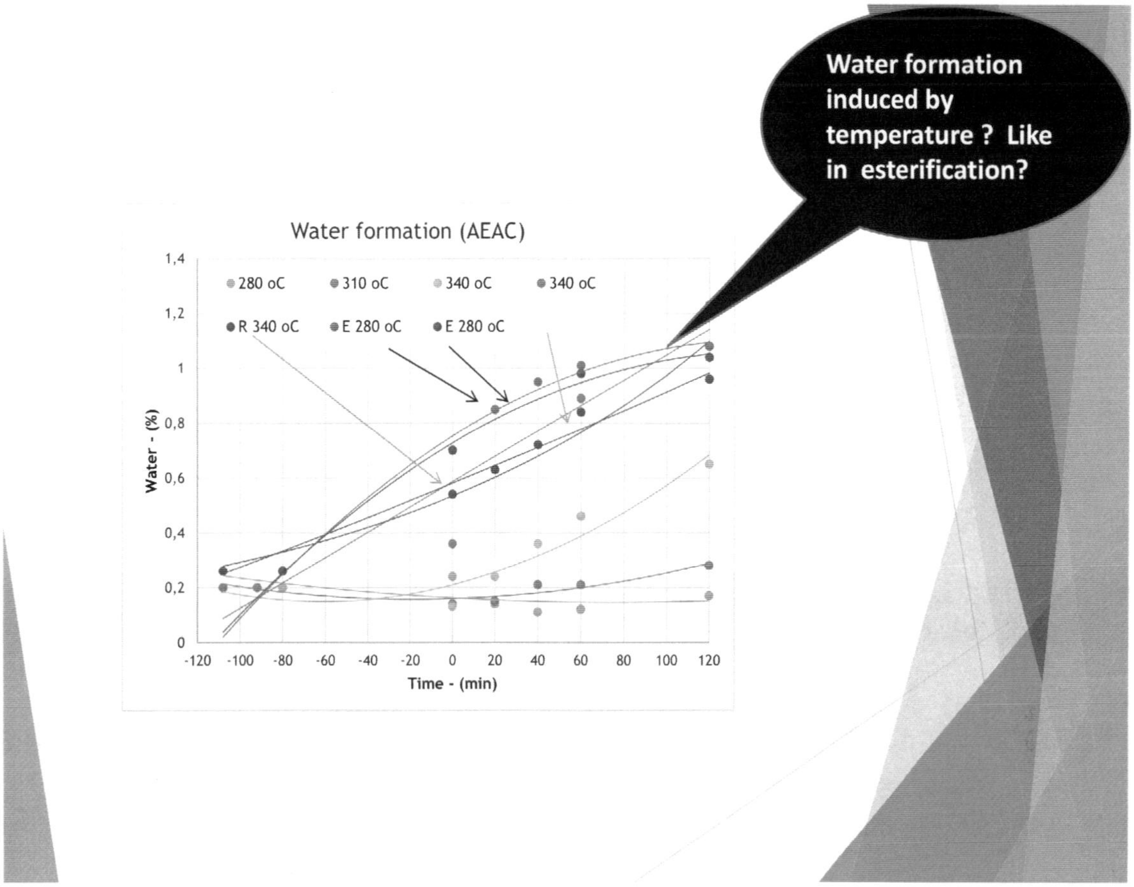
Water formation induced by temperature ? Like in esterification?
Water formation (AEAC)
280 oC
310 oC
340 oC
340 oC
R 340 oC
E 280 oC
E 280 oC
Water - (%)
Time - (min)

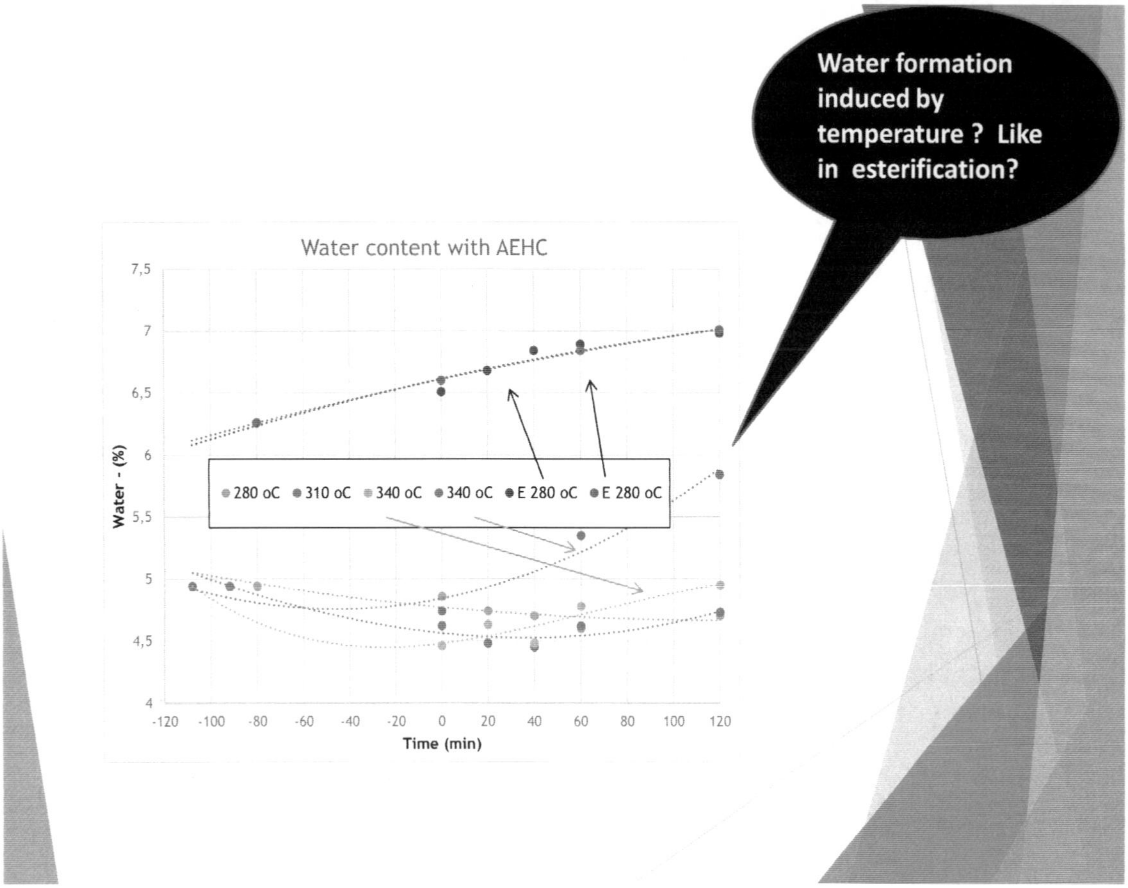
Water formation induced by temperature ? Like in esterification?
Water content with AEHC
280 oC
310 oC
340 oC
340 oC
E 280 oC
E 280 oC
Water - (%)
Time (min)

Conclusions

The experiments elucidated several critical points regarding transesterification processes and supercritical esterification with ethanol.

Conversion values and total time of reaction were evaluated in this case they are higher compared with the literature.

For the transesterification reactions with the refined soybean oil vegetable oil, the highest value of conversion to ethyl esters was around 90 %. Although this level of conversion was achieved, under the conditions tested, all conversions are lower than reported levels.

The presence of water in the reaction medium was decisive in the reactions with vegetable oil. High concentrations of ethyl esters were got at smaller intervals of time compared with anhydrous ethanol.

Conclusions

The use of hydrated ethanol could be positive for the productive process since it presence lowers cost which eventually would encourage its use in industrial production.

The data indicates that this method could be able to operate with lower quality raw materials and that pretreatment steps are not mandatory.

The reverse reaction occurs in this system with the possibility of causing limited ester yield.

It was also observed that water content increased in the most of experiments during the reactions. The origin of this water could not be elucidated in this work.

Acknowledgements

FEI CAPES FAPESP ipt INSTITUTO DE PESQUISAS TECNOLÓGICAS

Thank you for your attention

jgrpoco@fei.edu.br

The expanded abstract is acessible in AiChE material with references

390459 Promoting Efficient Adsorbent Bed Operation through Direct in-Situ Monitoring for Desulfurization of Distributed Energy Resources

Abstract:

Distributed energy resources such as biogas, landfill gas, and flare gas are becoming increasingly important alternative energy resources; however, many of these resources cannot be economically produced due to high gas cleanup costs with the majority of these costs coming from desulfurization. Due to the small size and distributed nature of these resources, adsorption remains the most common desulfurization technology for these applications, particularly where sulfur loads are less than 100 kg per day. Adsorption-based desulfurizers are typically designed and operated assuming they will continuously experience the maximum possible inlet sulfur concentration. During actual operation, the inlet sulfur concentration often fluctuates at levels lower than the design criteria. When this occurs and the adsorbent is replaced or regenerated at pre-determined intervals, a large portion of the bed is never utilized, resulting in significant waste and inefficient operation. While complicated systems that require analytical equipment (e.g. an online GC), continuous gas concentration monitoring, and a full understanding of the adsorbent's performance have been proposed to predict bed performance, it is not practical or economical to apply these solutions for small-scale applications. IntraMicron has addressed the issue of efficiently determining when to replace or regenerate an adsorbent bed with its bed life capacity sensor (BLS) which directly monitors adsorbent bed saturation. The BLS is based on spectroscopic techniques which are commonly used for adsorbent and catalyst characterization. These techniques that have been applied and extended to reliably determine bed saturation and to predict appropriate adsorbent replacement or regeneration timing. It has been demonstrated that this technology is effective for both IntraMicron-proprietary and other commercially available adsorbents. Furthermore, this technology is not significantly affected by the presence of common contaminants and/or competitive adsorbates. Since the BLS directly monitors adsorbent saturation, it is compatible with fluctuating inlet sulfur concentrations and does not require monitoring gas-phase sulfur levels with high-maintenance analytical equipment. The BLS allows maximum adsorbent utilization between adsorbent replacements or regenerations significantly reducing the waste-generation and OPEX of adsorption-based desulfurization systems. While the BLS is a stand-alone technology, it is commonly used within IntraMicron's Desulfurization Technology Suite (IM-DTS) for optimal desulfurization performance. IM-DTS is an economical, low-waste, environmentally-friendly sulfur removal and recovery system targeted at small-scale energy resources.

Thermophysical Surrogate Model Development of Transportation Kerosene Fuels

E. Khani and C. P. Chen

Department of Chemical and Materials Engineering
The University of Alabama in Huntsville
Huntsville, USA
chenc@uah.edu

Abstract

Aviation kerosene fuels such as Jet-A and JP-8 are traditionally used in gas turbines and jet engines. Recently, these aviation fuels are also considered for ground vehicles utilizing diesel engines. The spray combustion process in diesel engines requires detailed characterization of the ignition delay of the fuels used for the preparation of stratified fuel-air charge. Ignition delay phenomenon is dominated by the fuel heating and evaporating spray process. For computational modeling of the spray combustion processes, it is essential to have accurate values of fuel thermo-physical properties. Real kerosene fuels are typically composed of hundreds to thousands of components and are not feasible to be used in multicomponent spray combustion calculations. The object of this study is to develop a surrogate fuel, composed on a few components, to mimic the heating and evaporating behaviors of the real fuel. To this end, we have utilized a recently developed inversed batch distillation methodology to select a group hydrocarbon species that closely match the experimental distillation curve of a Jet-A blend. In addition, eight target properties including cetane number (CN), molecular weight, lower heating value (LHV), H/C ratio, density, viscosity, surface tension, and thermal conductivity are also accurately matched. From the hydrocarbon groups of typical kerosene fuels of linear paraffins, cyclo-paraffins and aromatics, we have identified the 4-componet surrogate fuel as: n-dodecane/iso-cetane/trans-decalin/toluene (mole fraction: 0.29/0.14/0.32/0.25). Detailed methodologies and data validations will be presented and discussed.

364259 Sulfur Removal Strategies for Biodiesel Production

Abstract:

High FFA fats, oils and grease are feedstock for ASTM 6751 biodiesel. These feedstocks contain higher levels of sulfur than traditional feedstocks. Common oil crops contain sulfur species that are water soluble and remain with the meal during oil processing. High FFA feedstocks from cooking and rendering operations contain oil soluble sulfur species that require additional processing steps to produce ASTM 6751 biodiesel that meets the 15 ppm sulfur specification. It all begins with good feedstock characterization.

376717 Molecular-Level Interactions in the Design of Reversible and Recyclable Flocculants

Abstract:

Previously we have developed a novel approach for the dewatering and harvesting of microalgae using polyamphoteric flocculants that can be recovered and recycled in response to their pH-dependent properties. Reversible and recyclable flocculants were shown to address the challenges of flocculant cost and downstream contamination that plague current algal biofuels processes. Here ampholytic polyacrylamide-based flocculants, with a pH-tunable charge character ranging from net positively-charged to net negatively-charged, will be compared for the dewatering of the microalgae system *Chlorella vulgaris* (UTEX 395) and the flocculation of a model colloidal suspension of homogeneously-charged (negatively-charged) silica microparticles. Direct comparison of the polyamphoteric flocculants with cationic, anionic, and nonionic commercial flocculants will also be provided. Molecular-level interactions are elucidated that enable the polyamphoteric flocculants to 1) initially adsorb to the particle surface and cause aggregation and 2) undergo facilitated desorption from the particle surface. By tuning these molecular interactions and operating conditions, the reversible and recyclable polyampholytes achieve similar initial flocculation efficiencies (>95%) as commercial cationic flocculants, however, can be recovered with >94% yields and recycled for over 5 flocculation processes.

Pour point calculation for fuels and lubricants in different model systems

M. Padaro, A. Göbel, M. A. Liauw, ITMC - RWTH Aachen University, Aachen, Germany

Introduction

Lubricants and fuels based on mineral oils are increasingly complemented by biobased alternatives. Beside viscosity or oxidation stability, a crucial parameter is the pour point, i.e. the temperature below which the fluid solidifies and becomes detrimental for the machines. Aim of this work is the investigation and comparison of a biodegradable biofluid with mineral oil based fluids and the forecast of the pour point based on the fluid composition. Only few equations for special tribological systems have been reported so far. A lubricant system has many and unknown compounds and is not a trivial system. To make calculations as easy as possible, different model systems should be reviewed. An alternative system should have few compounds of high purity and a well-defined molecular structure. Therefore the first model system are pure alkanes (C_7-C_{14}), pairs of which are mixed with each other and then a third alkane is added. Alkanes are used because they can also be found in fuels, are generally available and pour points for mixtures of them are still not reported. The second system are mixtures of a well known model diesel, the combination of hexadecane and methylnaphthalene, known for its analogy in the cetane number, and kerosene. Also investigations with methyl oleate as biodiesel compound can be done. Finally the pour point of the reported biobased lubricant with a mineral oil component should be calculated.

Experimental

All used compounds are commercially available except the biobased oils. The latter were made in a continuous plant in a two-step reaction where oleic acid methyl ester (OME) was converted with peracetic acid to epoxy stearic acid methyl ester (eSME, a solid above 283 K) and water. After separation of the water, the eSME and *i*-butanol reacted to yield the target molecule hydroxyl isobutoxy stearic acid methyl ester (HISM). Various side products are possible. Different oil mixtures are obtained upon variation of reaction time and temperature. Then the oil samples are mixed with a mineral hydraulic fluid (HLP 32). The pour point of all samples is determined with a

HCP 852 setup, according to ASTM standard D97. Melting points and enthalpy of fusion are measured via DSC (differential scanning calorimetry).

Method

The Clausius-Clapeyron equation has been integrated over temperature between the melting temperatures of the mixture (T_m) and of the pure compound (T^*_m) as well as over activity (a) under the assumption that both enthalpy of fusion (ΔH_m) and activity are temperature-independent.

$$\left(\frac{\partial T}{\partial \ln(a)}\right)_p = -\frac{R \cdot T^2}{\Delta H_m}$$

In addition, it was furthermore assumed that molecules can move independently from each other (alkanes have no strong interactions). Then, the activity of the added compound can be replaced by its molar fraction, yielding

$$T_m = \frac{1}{\frac{1}{T^*_{m1}} - \frac{R \cdot \ln(1 - x_2)}{\Delta H_{m1}}}$$

For the second compound the natural logarithm of x_2, ΔH_{m2} and T^*_{m2} are used. Now the temperatures of every experimental molar fraction can be calculated theoretical and plotted in a diagram. The experimental PPs can be compared with the theoretical calculated melting points due their affinity in alkanes.

Results

After the analysis of the experiments different propositions can be made. The properties of the biodegradable lubricant strongly depend on the composition. On one hand, certain proportions of HISM and HISI are necessary to decrease the pour point; on the other hand, the solid compounds increase the pour point immensely. In the tests with HLP 32 the addition of biodegradable lubricant (with high pour point) depressed the pour point of the mixture even below that of the pure hydraulic fluid. This is discussed in the framework of colligative properties. Also in the experiments with the alkanes, similar observations for binary mixtures above C_{10} were made. Alkane combinations below C_{10} showed linear trends. In the ternary mixture of C_{10}, C_{12} and C_{14} some colligative areas were found.

In the diagram, three systems containing tetradecane are plotted. Each point describes experimental data points of a binary mixture and each of the graphs the theoretical calculated trend of the function for each compound. The intersection of the curves for two components is the lowest PP of that binary system.

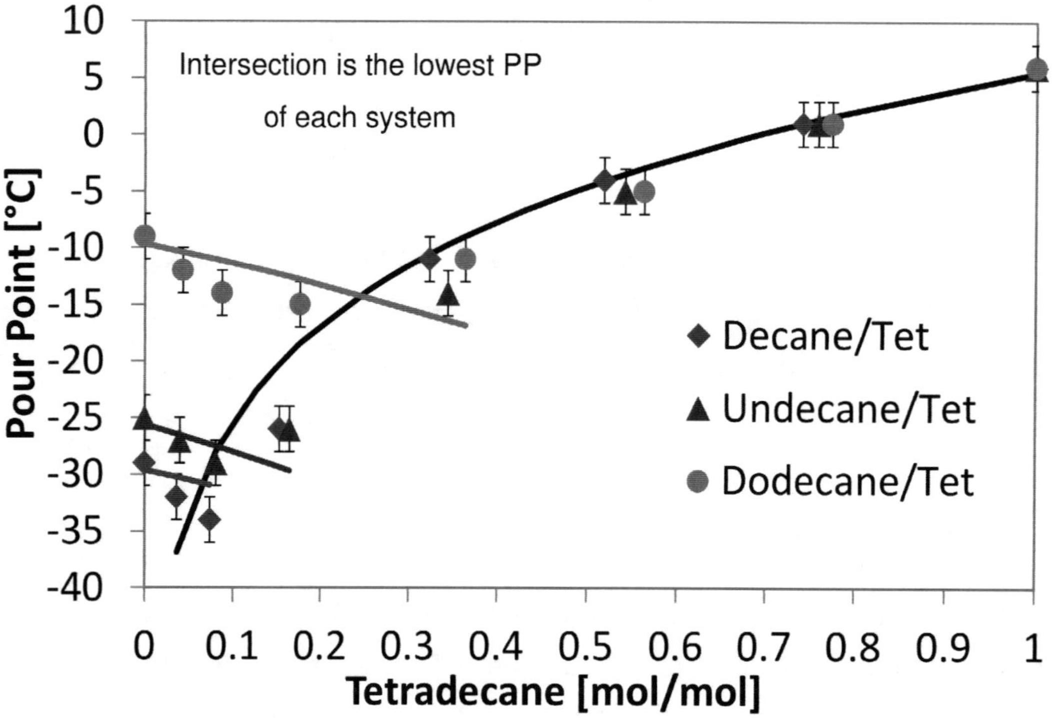

Figure 1: Calculated fits of: decane (green), undecane (blue), dodecane (red), tetradecane (black)

The results show that the calculated ideal compositions fit to the experimental data and also the calculated PPs (lines) are close to the measured ones (dots). Hence only through determining the melting temperature respectively the PP and the enthalpy of fusion of each pure compound in a multicomponent mixture the ideal composition for the lowest PP and the PP itself can be calculated. This could be applied also for this sample of a ternary system with decane, dodecane and tetradecane (see table). Experimental measurements gave two minima *exp1* and *exp2* (although of these points is a binary mixture), but the calculation (*theo1)* gave an even lower theoretical PP. The experimental measurement of this composition (*exp3*) confirmed that in this composition the PP is really lower than before. Also the calculations with two compounds combined as one new pseudo-single compound, with newly determined melting temperature and enthalpy of fusion of the mixture, worked (*theo2)* and gave the same result for the composition.

Alkane (mp.)	n_{exp1}	n_{exp2}	n_{theo}	n_{exp3}	n_{theo2}
Decane (-30 °C)	**82.7**	**77.0**	**77.8**	**77.8**	**77.3**
Dodecane (-10 °C)	**17.3**	**16.1**	**18.2**	**18.2**	**22.7**
Tetradecane (6 °C)	**0.0**	**6.9**	**4.0**	**4.0**	
minimum PP	**-37 °C**	**-37 °C**	**-34 °C**	**-38 °C**	**-34 °C**

If the method is applied to a system with kerosene and the model diesel, again the theory fits the experiment. Only the minimum for ideal composition does not fit, the reason for this is that no depression of the pour point of the blend is experimentally observed below that of pure kerosene.

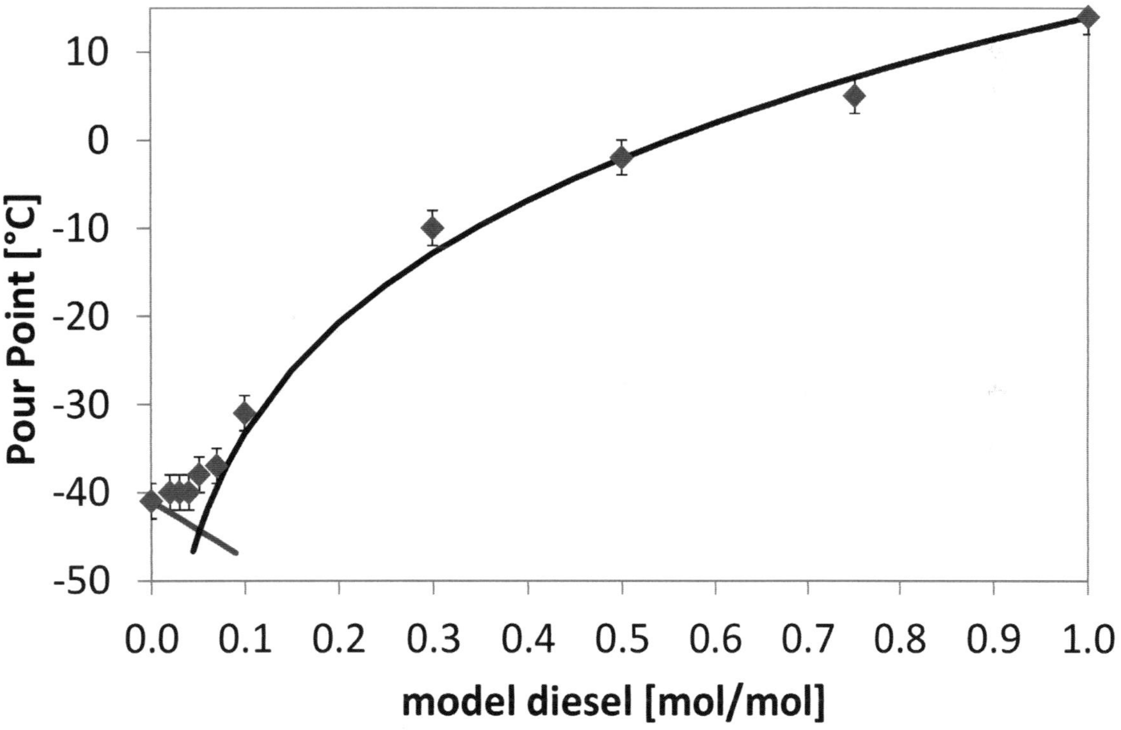

Figure 2: Calculated fits of kerosene (green) and model diesel (black), experimental data (red)

This was observed with other experiments/mixtures of model diesel and kerosene. Nevertheless the depressant of the PP for the model diesel could be described perfectly for compound ratios up to 90 %.

Conclusion

The aim to calculate pour points and the depression for different model systems was reached. So far alkane systems and model diesel system could be described independently from the amount of compounds with this method. With this idea the possible compounds for a successful pour point depression could be located. This method has the further advantage that the ratio is directly known in which the compounds have to be mixed, and which temperature they probably will reach, making only few (if any) control experiments.

Perspective

In the future the method will be tested under consideration of the made simplifications and with adding suitable factors also for the introduced lubricant system.

390475 Oxidative Sulfur Removal for Distributed Energy Resources

Abstract:

Small-scale distributed energy resources such as biogas, landfill gas, and flare gas are alternative fuel sources that can contribute significant greenhouse gas emissions if they are not utilized. The major challenge associated with utilizing these small-scale resources is the high costs associated with gas clean-up, especially sulfur removal. IntraMicron's Oxidative Sulfur Removal (OSR) is an innovative sulfur removal and recovery system developed to address the current need for scalable desulfurization technology. It contains a selective catalytic process that transforms up to 99% of sulfur-containing species into elemental sulfur using air as oxidant. The elemental sulfur is recovered through sulfur condensation as a useful byproduct. The catalyst is chemically inert to most contaminants in these energy resources and it can maintain its high catalytic activity during the longer term operations. The OSR catalyst maintains near 100% selectivity to elemental sulfur even in the presence of 10 times the stoichiometric amount of air. As a result, the strict oxidizer control based on inlet sulfur concentration is not required, which significantly simplifies OSR operation. This is particularly beneficial for applications with varying sulfur levels. For applications where relatively high sulfur concentrations are acceptable (from a few ppm to 25 ppm), the OSR reactor(s) and sulfur condenser can operate stand-alone, which can completely eliminate the use of adsorbents leading to a low-cost and environmentally-friendly desulfurization process. In cases where adsorbents are needed for low sulfur applications, an adsorbent bed equipped with IntraMicron's Bed Life Sensor (BLS) can be utilized to minimize adsorbent consumption. Due to its process robustness and simplicity, OSR provides up to a 65% reduction in desulfurization cost, up to a 31-fold reduction in reactor volume, and up to an 85-fold reduction in solid waste generation. With these advantages, OSR enables the economical and environmentally-friendly utilization of numerous stranded, sulfur-laden resources that currently cannot be produced due to technical and economic limitations.

Furfural has been shown to be an attractive platform chemical in biomass upgrading routes [1]. Its combination with aliphatic and aromatic alcohols and diols has the potential to yield useful precursors for gasoline, jet and diesel blendstocks, as well as chemicals. In this work, we show the utility of transfer hydrogenation and aldol-type coupling for this combination and we investigate the mechanism of the transfer hydrogenation and that of the coupling over homogeneous and heterogeneous catalysts. Initial optimization studies show that mixtures of Ni(dppe)Cl_2 and K_2CO_3 are efficient catalysts for the dehydrogenative cross-coupling, giving aldol adduct yields higher than 90% for a variety of alcohols. Furthermore, the presence of both Ni(dppe)Cl_2 and K_2CO_3 is required for alcohol and furfural cross-condensation, the role of the Ni^{2+} complex being the dehydrogenation of the alcohol to the reactive aldehyde and that of K_2CO_3 being the aldol condensation between furfural and the aldehyde. This is shown by the results of reactions in the presence and the absence of either of the two components of the catalytic system and also by a reaction progress study that shows the aldol adduct is a secondary product of the reaction. Also, further reaction studies over this system suggest that dehydrogenation proceeds via a Ni(II)-mediated transfer of a hydride from a butoxide to a furfural.

In order to make this process attractive and relevant from an industrial standpoint, we propose carrying it out in heterogeneous conditions. Heterogenizing this reaction requires the replacement of the base and the metal with heterogeneous versions thereof. For this reason, we chose a nickel-substituted hydrotalcite material and evaluated its reactivity in the gas phase and the liquid phase. Surprisingly, increasing the Ni content results in a decrease of both dehydrogenation and aldol condensation reaction rates. This suggests that the presence of Ni^{2+} sites is not required for the chemistry to take place. Instead, we show that the dehydrogenation of butanol takes place via a transfer of hydrogen from the butanol to the furfural, in a Meerwein-Ponndorf-Verley-type reaction [2]. Both this reaction, as well as the aldol condensation reaction are known to be base-catalyzed [2-4]; our results suggest that the presence of Ni in the catalyst decreases its basicity and hence the reaction rates.

For the dehydrogenation step, kinetic isotopic effect (KIE) studies with 1,1-d_2-butanol and butanol-OD excluded the possibility of O-H bond cleavage as the kinetically relevant step as opposed to C-H bond scission in the dehydrogenation step. These observations also precluded the possibility of a concerted transfer of the surface proton to the carbonyl of the furfural and the hydride transfer from butoxide to the carbonyl functionality of furfural. The first-order dependence of the dehydrogenation rate on butanol and furfural at low pressures and the zero-order dependence at higher pressures of butanol indicate that both furfural and butanol participate in the rate-determining step of the dehydrogenation and that butoxide is the most abundant surface intermediate on the active sites of hydrotalcite.

On the other hand, for the aldol condensation step, enolization of the *in situ* formed butyraldehyde or a C-C bond forming step between the butyraldehyde enolate and furfural have been reported to be kinetically relevant steps in the aldol condensation reactions over basic oxides [3,4]. During gas-phase aldol condensation reactions using butyraldehyde and furfural as reactants, we observed a first-order dependence on furfural and a zero-order dependence on butyraldehyde. While the observed order of the reaction with respect to butyraldehyde is zero, the established aldol condensation mechanism requires a first-order intrinsic dependence on butyraldehyde, since it participates both during the enolization and the C-C bond forming events of aldol condensation. As such, the independence of the rate of aldol condensation from the butyraldehyde pressure indicates the saturation of butyraldehyde enolate on the hydrotalcite surface. On the other hand, the first-order dependence on the furfural pressure indicates that the C-C bond formation is the kinetically relevant step for the aldol condensation under the experimental conditions. Since β-hydroxy aldehyde is not detected in the product stream during entire course of the reaction, dehydration step is believed to proceed rapidly under the reaction conditions.

1. Binder, et al. *ChemSusChem* 3, 1268 – 1272

2. Ivanov, et al. *J. Mol. Catal.* 91, 45 – 59

3. Rekoske and Barteau *Ind. Eng. Chem. Res.* 50, 41 – 50

4. Zhang, et al. *Appl. Catal.* 40, 183-190

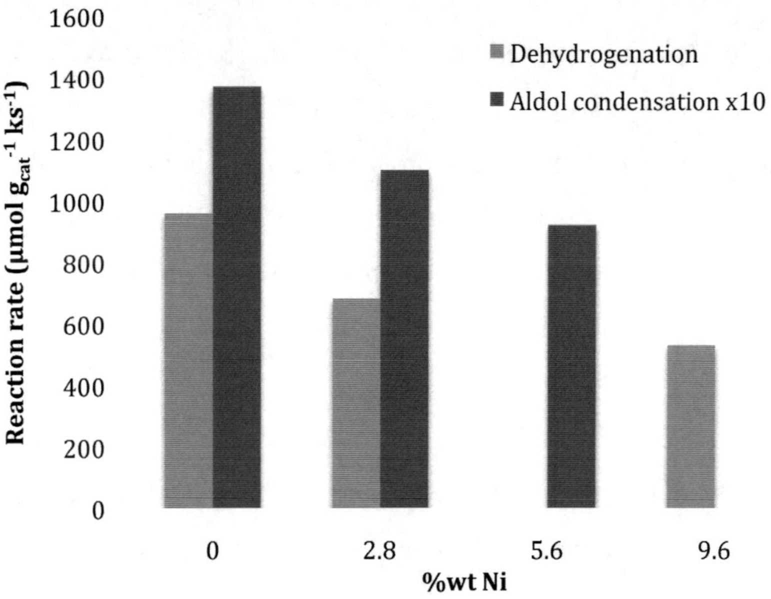

Figure 1: Effect of Ni content on dehydrogenation and aldol condensation rates. 413 K, 150 mg catalyst, 0.95 kPa furfural, 2 kPa butanol, balance He

Hydrodeoxygenation (HDO) of Guaiacol over Ceria-Zirconia Catalysts

Chukwuemeka Okolie, Sarah M. Schimming, Onaje D. LaMont, Michael König, Andrew D. D'Amico, Carsten Sievers

AIChE Annual Meeting
November 20, 2014.

Georgia Institute of Technology
School of Chemical & Biomolecular Engineering

Background - Pyrolysis

Thermochemical decomposition of organic material at elevated temperatures in the absence of oxygen

1000 kg wood

Flash pyrolysis →

Combustible gas (150 kg)

Pyrolysis oil (700 kg)

Char (150 kg)

Georgia Institute of Technology
School of Chemical & Biomolecular Engineering

BTG – http://www.btgworld.com/2005/pdf/technologies/folder-pyrolysis.pdf

2

Background – Pyrolysis oils

- Higher energy density than biomass
- Complex mixture (Volatile and non-volatile compounds)
- High oxygen content (~ 40 wt.%)
- Water content (~ 20 wt.%)
- Typically less than 40 ppm sulfur
- Immiscible in mineral oil
- Not as stable as mineral oil

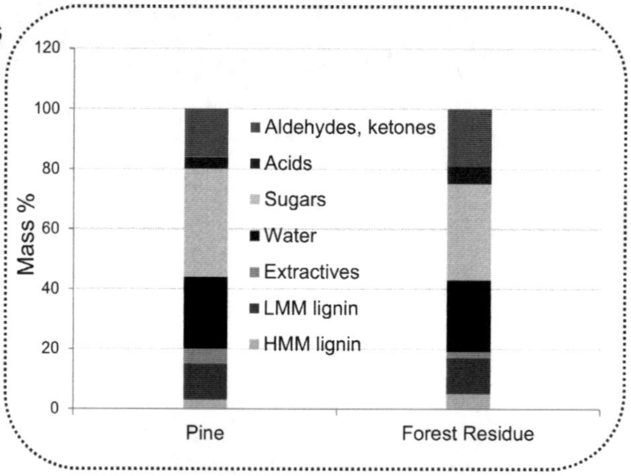

Poor fuel properties

A. Oasmaa, *Biomass & Bioenergy*, 14, 103-113 **1998**.

Georgia Institute of Technology
School of Chemical & Biomolecular Engineering

3

Background – Upgrading Pyrolysis-oils

Pyrolysis-oils are considered as a source for traffic fuels and chemicals

The upgrading of pyrolysis-oils is required to eliminate oxygenated groups, this improves
→ Thermal stability
→ Chemical stability
→ Heating value
→ Volatility

Hydrodeoxygenation (HDO) is an evolving technology needing
→ Moderate temperatures
→ Moderate pressures
→ catalysts

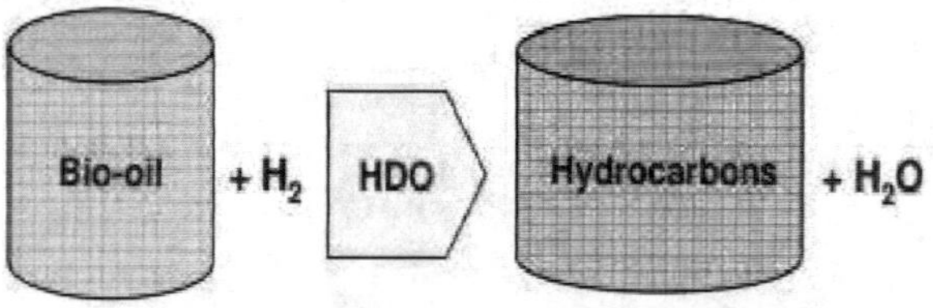

HDO is a reaction in the excess of hydrogen to produce hydrocarbons and water

Georgia Institute of Technology
School of Chemical & Biomolecular Engineering

4

Background – Guaiacol as a Model Bio-oil

Wood based pyrolysis oils are complex mixtures

(Carboxylic acids, alcohols, ketones, carbohydrates, degraded lignin, water etc.)

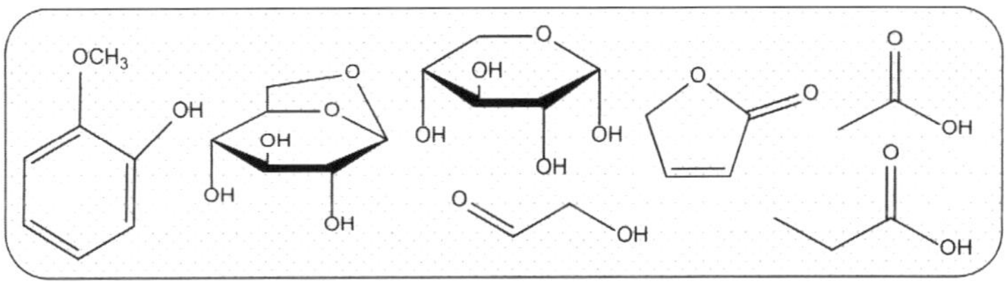

Difficult system to study

- Viscosity problems, large amount of side reactions, solubility problems

Guaiacol is a good model compound that simulates the behavior of oxygenated pyrolysis oil

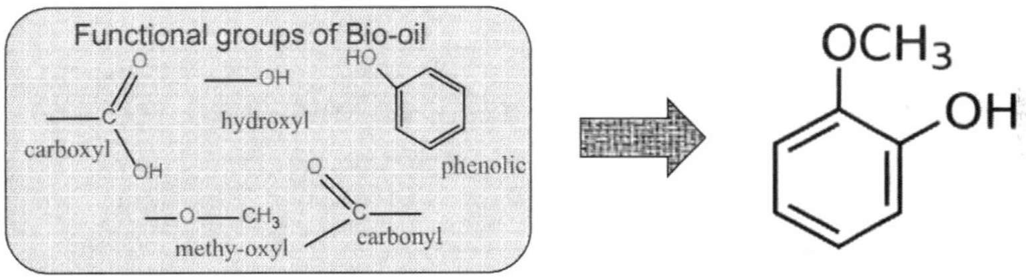

Georgia Institute of Technology
School of Chemical & Biomolecular Engineering

5

HDO Catalysts

Most studies on HDO of pyrolysis oils use supported noble metals or sulfided NiMo or CoMo catalysts.

- Sulfide catalysts can bind heteroatoms (S, N, O) of organic compounds in defect sites and subsequently remove them from the parent molecule.
- During HDO, sulfide are converted into inactive oxides. Toxic H_2S needs to be supplied to keep them in their active form.

Goal: Development of sulfur-free HDO catalysts with comparable properties.

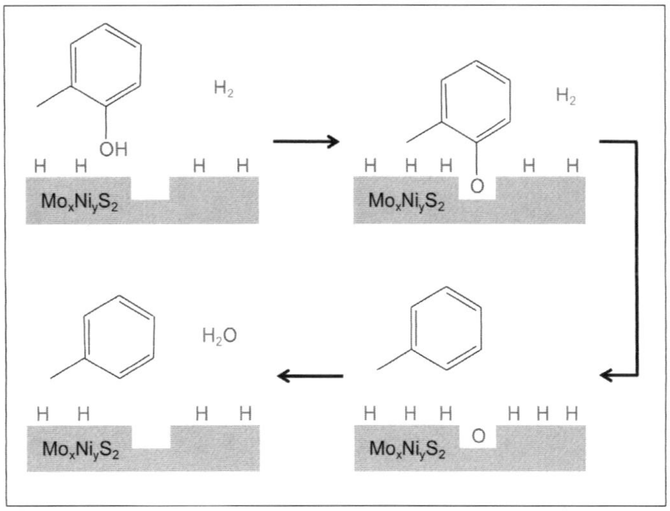

Mechanism: Oxygen atoms are bound to defect sites and subsequently replaced by hydrogen.

E. Furimsky, Appl. Catal. A-Gen. 199 (2000) 147-190.

Georgia Institute of Technology
School of Chemical & Biomolecular Engineering

6

Ceria-Based HDO Catalysts

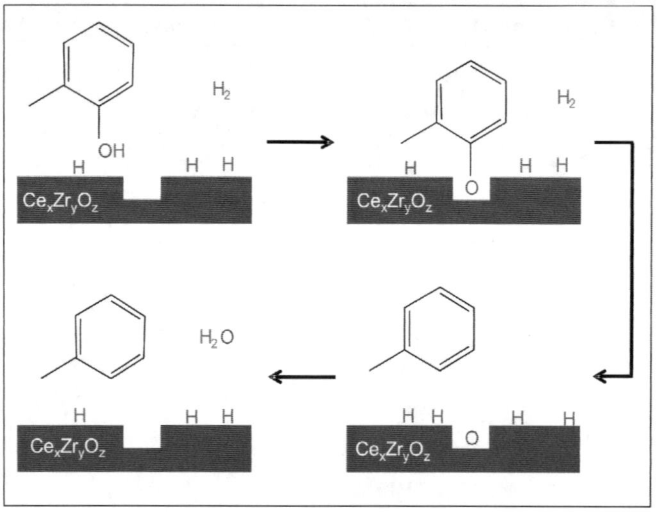

Requirements for oxide-based HDO catalysts:

- Formation of oxygen vacancies at realistic reaction temperatures.
- Ability to activate hydrogen.

Georgia Institute of Technology
School of Chemical & Biomolecular Engineering

7

Catalysts Synthesis

Synthesis by precipitation / coprecipitation:

- CeO_2 – Ce100
- ZrO_2 – Zr100
- $Ce_{0.82}Zr_{0.18}O_2$ – Ce82
- $Ce_{0.60}Zr_{0.40}O_2$ – Ce60
- $Ce_{0.46}Zr_{0.54}O_2$ – Ce46

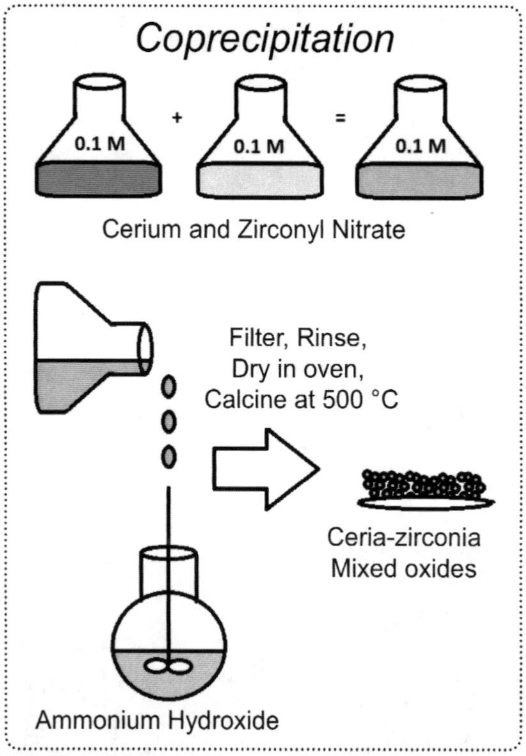

Georgia Institute of Technology
School of Chemical & Biomolecular Engineering

8

X-Ray Diffraction

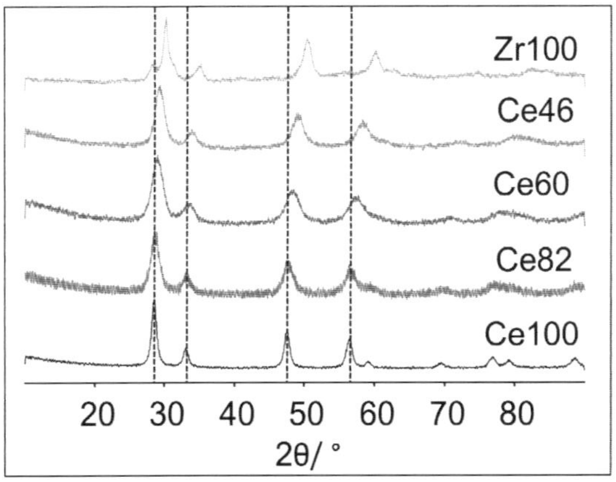

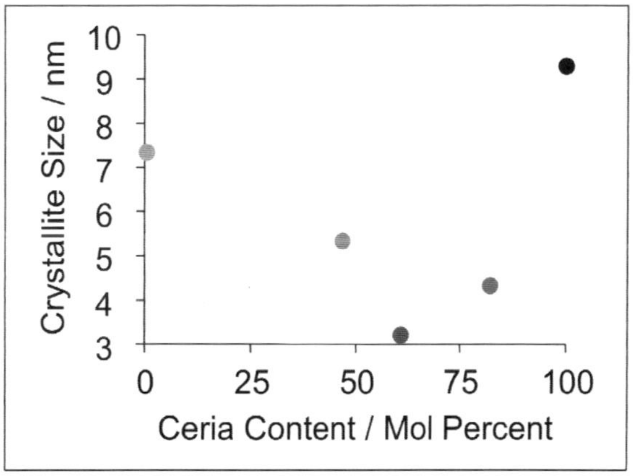

- Ceria-zirconia retains the CaF_2 structure of ceria.
- Distance between crystal planes decreases, with increasing zirconium content.
- All mixed oxides have smaller crystallite sizes than ceria.
- Crystallite size reaches a minimum at Ce60.

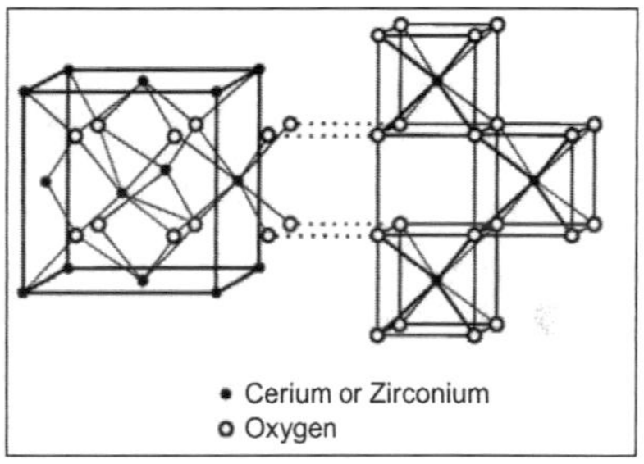

Georgia Institute of Technology
School of Chemical & Biomolecular Engineering

Mamontov, E.; Egami, T.; Brezny, R.; Koranne, M.; Tyagi, S., *J. Phys. Chem. B* **2000**, *104* (47), 11110-11116.

9

N_2 Physisorption and OSC

BET Surface Area:

- All mixed oxides have a higher surface area than ceria
- Ce60 has the highest surface area

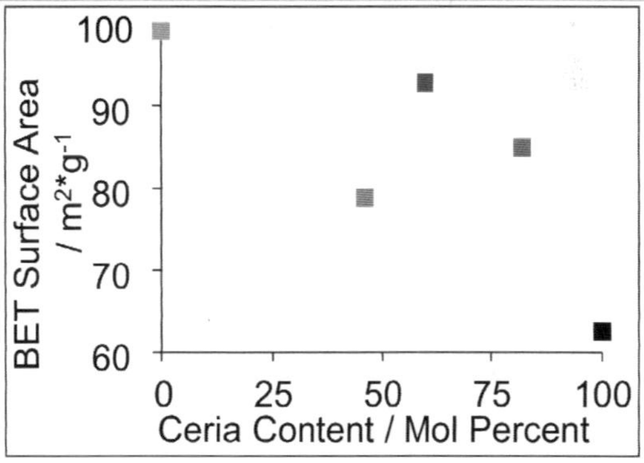

Oxygen Storage Capacity:

- No formation of oxygen vacancies at 200 °C
- At 300 °C, oxygen defects form
- Largest concentrations of vacancies on Ce60 and Ce82

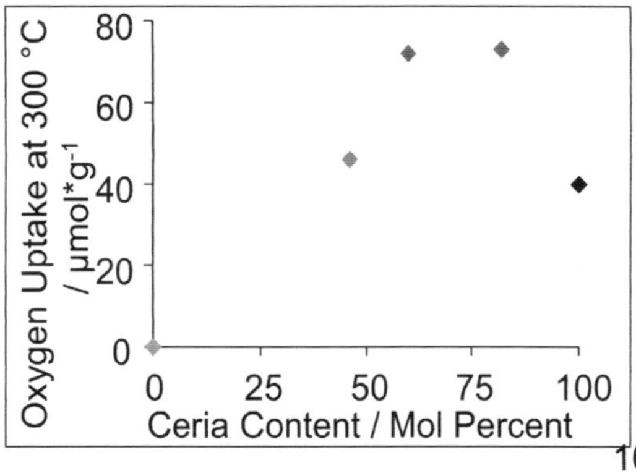

Georgia Institute of Technology
School of Chemical & Biomolecular Engineering

10

Temperature Programmed Reduction

- Surface reduction starts around 200 °C.
- Addition of zirconia reduces the temperature for bulk reduction.
- Ceria-zirconia samples also show increased surface reduction below 300 °C.
- Ce60 has highest H_2 consumption below 300 °C.
- Reducibility of surface sites is essential for HDO activity.

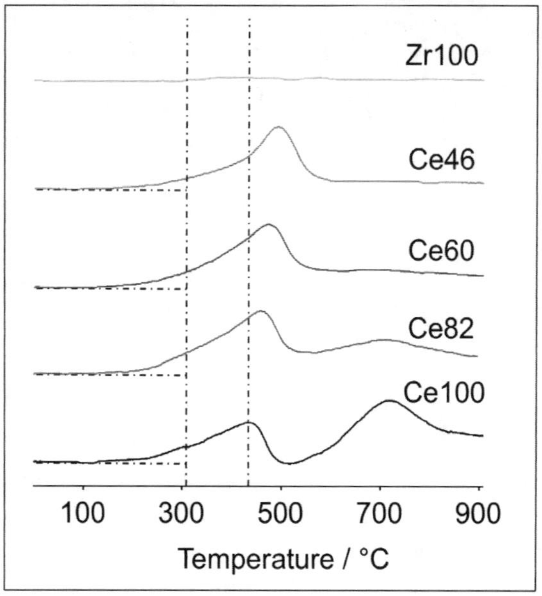

Georgia Institute of Technology
School of Chemical & Biomolecular Engineering

11

H_2-D_2 Exchange on Ceria-Zirconia

$$H_2 + D_2 \leftrightarrow 2HD$$

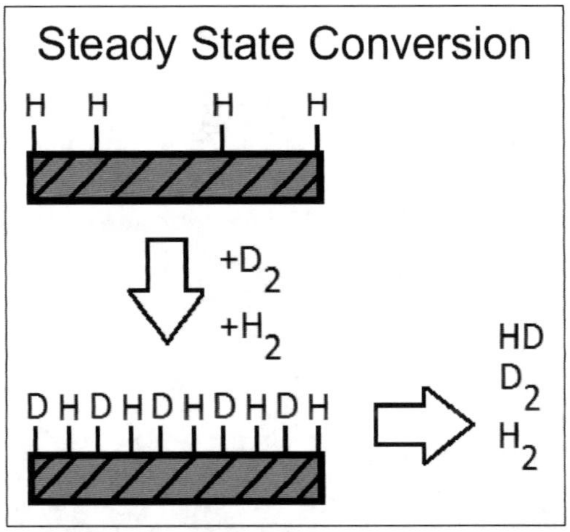

H_2-D_2 exchange:

- Test reaction for hydrogen dissociation
- Hydrogen is readily activated over ceria-zirconia catalysts
- Up to 10.5 hydrogen atoms per nm^2 participated in isotopic scrambling at 300 ^oC
- HDO reactions over catalysts are not limited by the amount of reactive hydrogen on the surface

Georgia Institute of Technology
School of Chemical & Biomolecular Engineering

12

Experimental

Simulation of wood based bio-oils
→ 4.5 wt.% of guaiacol (GUA) in decane and octane

Trickle bed reactor
→ 1 bar
→ H_2 flow 40 mL/min
→ 300 - 400°C
→ 72 hours

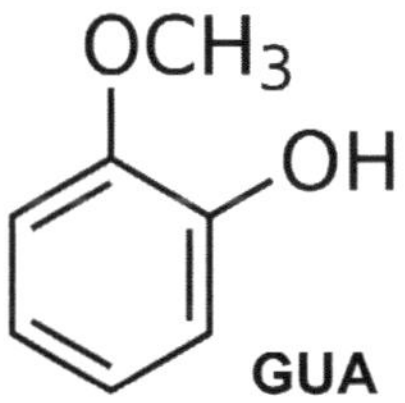

GUA

Catalysts
→ 500 mg in reactor on quartz wool

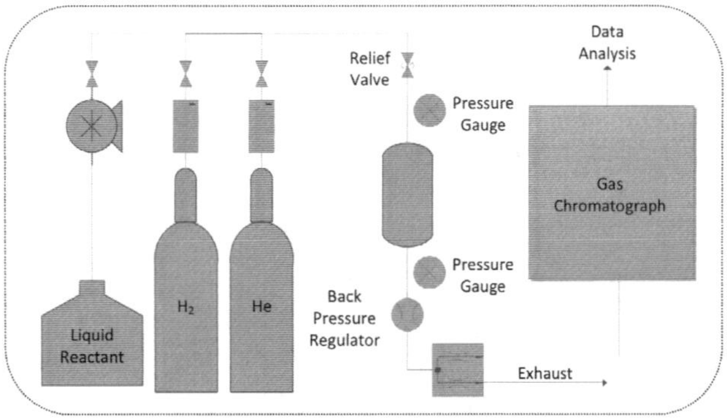

Georgia Institute of Technology
School of Chemical & Biomolecular Engineering

Reaction Products - Experimental

❖ Catalysts

Pure ceria catalyst: Ce100
Mixed ceria & zirconia catalysts:
Ce82
Ce60
Ce46

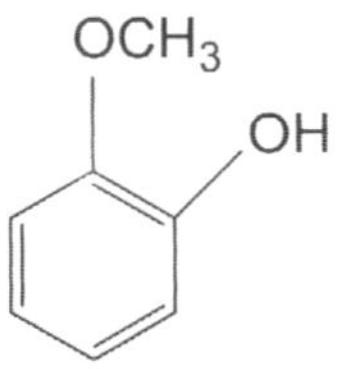

Guaiacol

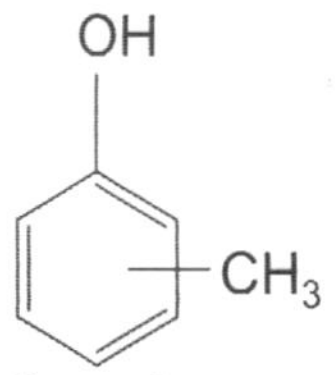

Cresols

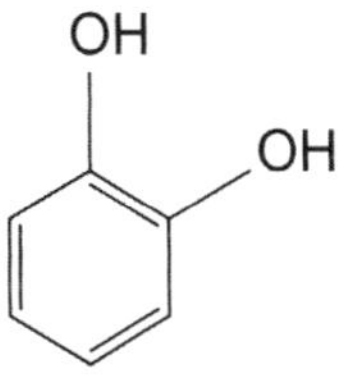

Catechol

Conversion: $X[\%] = \frac{F_{g,in} - F_{g,out}}{F_{g,in}}$ x 100

Product yield: $Y_p\ [\%] = \frac{F_{p,out}}{F_{g,in} - F_{g,out}}$ x 100

Product Selectivity: $S_p[\%] = \frac{Y_p}{X}$ x 100

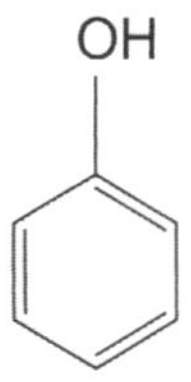

Phenol

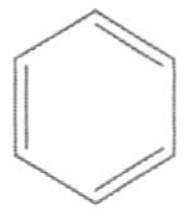

Benzene

Georgia Institute of Technology
School of Chemical & Biomolecular Engineering

HDO: Variation of Contact Time

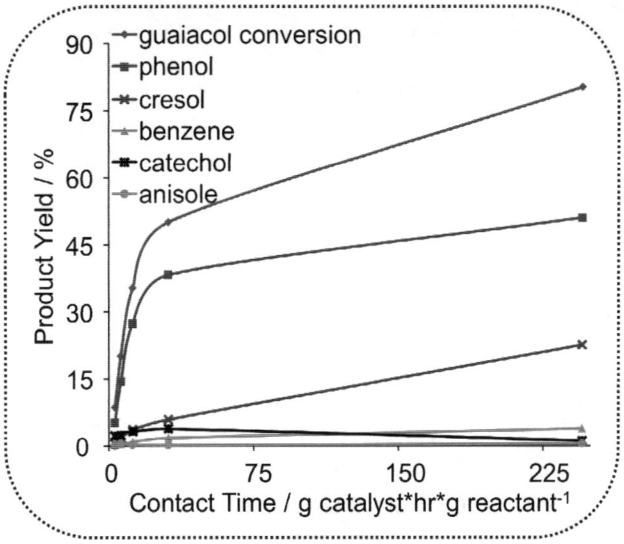

Catalyst: 0.5 g of Ce82
Temperature: 400 °C
Pressure: 1 bar
Hydrogen flow: 40 mL/min
Liquid flow: 0.001 to 0.08 mL/min
0.001 mL/min repeated
Total run time: ~ 72 hours

With increasing contact time:
- Total HDO product yield increases
- Phenol, benzene, and cresol yields increases
- Catechol yield passes through a maximum

Ceria-zirconia has limited activity for conversion of phenol.

School of Chemical & Biomolecular Engineering

15

Suggested HDO reaction scheme

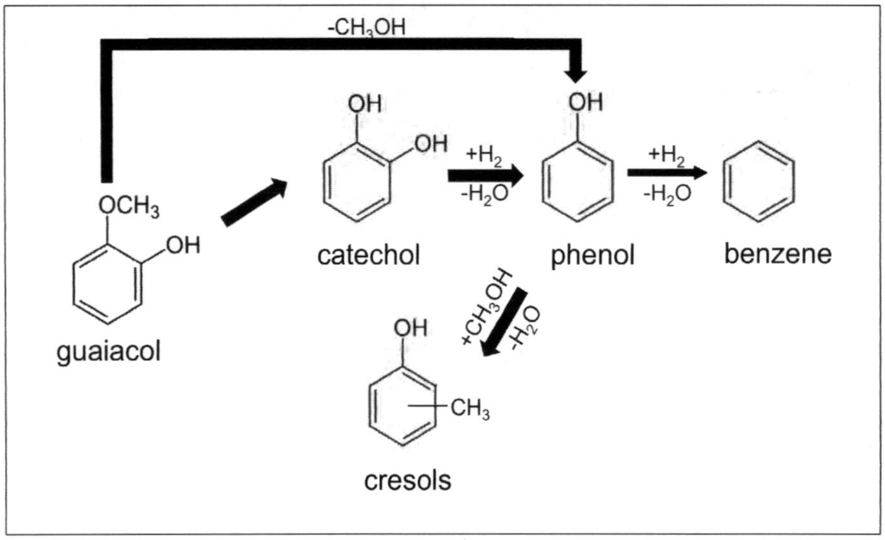

- Catechol is an intermediate.
- Cresol is formed by transalkylation reaction between guaiacol and phenol.
- Ceria-zirconia has limited activity for conversion of phenol.

Georgia Institute of Technology
School of Chemical & Biomolecular Engineering

16

Variation of temperature

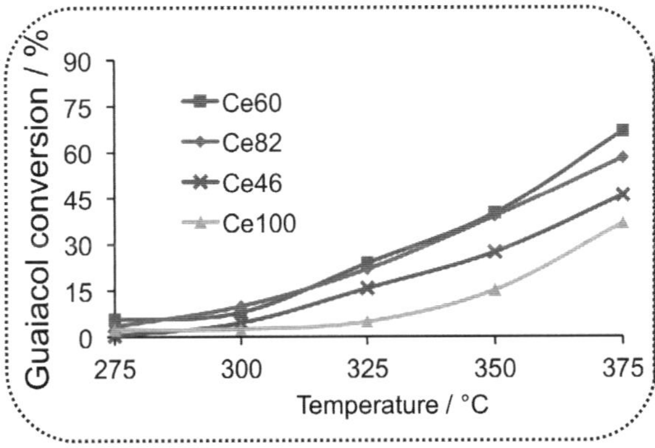

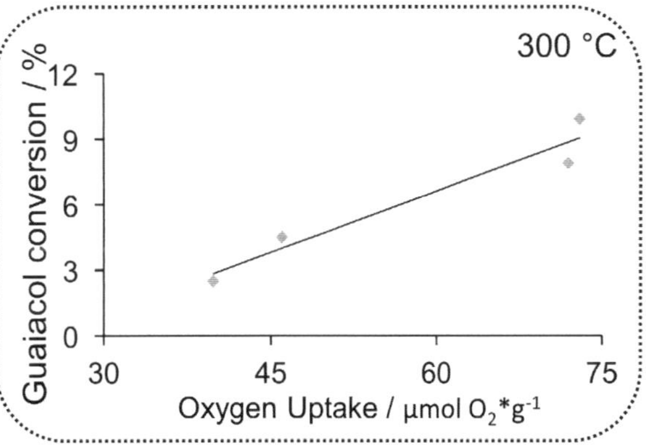

- All mixed oxide catalysts have a higher activity than pure ceria.
- Ce60 and Ce82 have the highest activity.
- Guaiacol conversion at 300 °C scales linearly with the oxygen storage capacity at 300 °C.

Temperature: 275 °C – 375 °C
Pressure: 1 bar
Hydrogen flow: 40 mL/min
Liquid flow: 0.008 mL/min
Total run time: ~ 72 hours

Activation Energies for Guaiacol Conversion

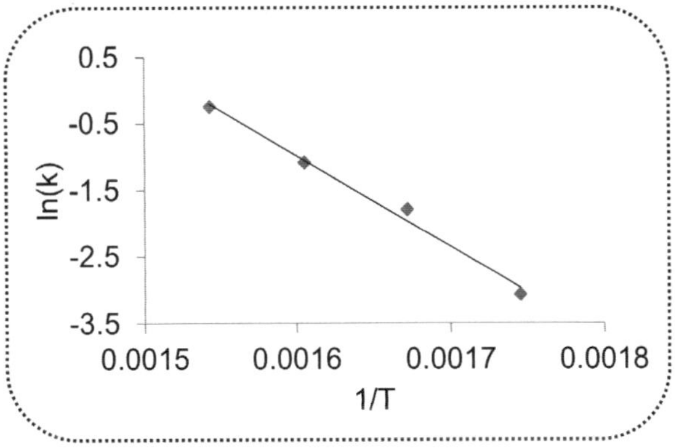

Catalyst	E_A (kJ/mol)
Ce100	131
Ce82	97
Ce60	111
Ce46	114

- Approximation: Guaiacol conversion is a single, irreversible, pseudo-first order reaction.
- All mixed oxides have lower activation energies than pure ceria.
- Ce82 has the lowest activation energy.
- The activation energies compare to values of 50-124 kJ/mol in studies over CoMo or NiMo on Al_2O_3.[1]

[1]E. Furimsky, Appl. Catal., A 199 (2000) 147.

Conclusions

- Addition of zirconia to ceria improves reducibility at below 300 °C.
- The crystallite size is the most important descriptor for the performance of ceria-zirconia catalysts.
- Ceria-zirconia catalysts are active for HDO of guaiacol with phenol as the major product.
- Ceria-zirconia catalysts did not deactivate over the course of the reaction due to coking or water in the feed as other common HDO catalysts do

Georgia Institute of Technology
School of Chemical & Biomolecular Engineering

19

Acknowledgements

People

- Allyson Rogers
- Tiorra Ross
- Dr. Sievers' group members at Georgia Tech

Funding

- Renewable Bioproducts Institute
- International Paper
- NewPage Corporation
- Georgia Tech start-up funds

Georgia Institute of Technology
School of Chemical & Biomolecular Engineering

20

THANK YOU!!!

Georgia Institute of Technology
School of Chemical & Biomolecular Engineering

385639 Novel Polymeric Solid Acid Catalysts for Cellulose Hydrolysis

Abstract:

Biofuel derived from lignocellulosic biomass is one of the leading renewable energy to replace fossil-base transportation fuels. The main drawback of the current biochemical processing of biomass is the cost of pretreatment, cellulase enzymes as well as the slow enzymatic hydrolysis process. A synthetic enzyme mimic polymeric solid acid catalyst overcomes all these limitations. Dual functional polymeric solid acid catalysts are synthesized in order not only to catalyze cellulose hydrolysis but also to facilitate its dissolution. The acidic polymeric chain, poly (styrene sulfonic acid) (PSSA) immobilized on a glass/membrane substrate via atom-transfer radical polymerization (ATRP) catalyzes biomass hydrolysis. Its neighboring poly (vinyl imidazolium chloride) (PIL) chain grafted via UV-initiated radical polymerization helps solubilize lignocellulosic biomass and enhance the catalytic activity. UV-initiator 4-ethoxy-5-oxo-4,5-diphenylpentanoic acid (BEE-COOH) was successfully synthesized and immobilized prior to UV polymerization. Cellulose hydrolysis was conducted in 1-methyl-3-ethylimidazolium chloride [EMIM]Cl as well as in aqueous solution. An optimal ratio of the two polymer chain densities as well as optimal polymer chain lengths were found for achieving highest catalytic conversion of cellulose substrates. These polymeric catalysts are stable in ionic liquid solvents and maintain high catalytic activity after more than 12 repeated runs. Our novel polymeric solid acid catalysts demonstrate total reducing sugar (TRS) yield over 94% after 12 runs in [EMIM]Cl. Methods for ionic liquid regeneration and reuse were also developed. More importantly the catalytic activity of these polymeric acids can be further tuned by ring substitution and copolymerization. These polymeric solid acids are reusable, environmental friendly and potentially superior catalysts for many other organic reactions requiring acid catalysts. Theoretical investigations are conducted to obtain fundamental insight into their high catalytic activity.

383348 Continuous Transesterification at Near-Critical and Supercritical Ethanol Conditions

Abstract:

Biodiesel is a renewable petroleum diesel replacement produced predominately through the transesterification of triglycerides using an alcohol, most often methanol, with an alkaline catalyst. Since methanol is typically produced from natural gas, the majority of methanol comes from non-renewable sources. One way to make biodiesel truly renewable is to instead use ethanol in place of methanol; however, reaction times with ethanol are greatly increased over those needed for methanol. It is possible to reduce the reaction time to 30 minutes or less by carrying out transesterification at near-critical and supercritical conditions of ethanol. Since no catalyst is used, saponification is suppressed thus increasing tolerance of water and free fatty acids. For this presentation, a constant molar ratio of 41:1 ethanol to canola oil was used. Temperatures ranging from 250°C to 350°C and residence times of between 2 and 20 minutes were examined to optimize the production of ethyl esters and determine yield and selectivity. Reactions were carried out in a continuous, custom made 101 mL Haynes® 282 Alloy tubular flow reactor.

Hydrocracking of DDGS Corn Oil over Transition Metal Carbides Supported on Al-SBA-15: Effect of Metal Ratio, Preparation Method, and Fractional Sum of Metal Electronegativities

Basem Al Alwan[a], Steven O. Salley[a,b], and K. Y. Simon Ng[a,b,c]

[a] *Department of Chemical Engineering and Materials Science, Wayne State University, Detroit, MI 48202, USA*

[b] *National Biofuels Energy Laboratory, NextEnergy, Detroit, MI 48202, USA*

[c] *Alternative Energy Technology, Wayne State University, Detroit, MI 48202, USA*

Abstract

Economic, ecological, and environmental issues associated with fuels derived from fossil sources have spurred research to develop alternative liquid fuels that are sustainable and renewable. Triglycerides from vegetable oils may be a suitable source for liquid biofuel production (green diesel) via catalytic hydrocracking. Low quality vegetable oils such as distillers dried grains with solubles (DDGS) corn oil has been considered as an economical alternative to using fresh vegetable oil as a biofuel feedstock in order to avoid competition with food markets. The hydrocracking of DDGS corn oil over bimetallic carbide catalysts was explored for green diesel production. A catalyst composed of nickel-tungsten (Ni-W) carbide supported on Al-SBA-15 was designed based on the ability of nickel to adsorb and activate hydrogen and the potential of tungsten for hydrogenation reactions. Four different Ni-W ratios (1:9, 1:1, 2:1, and 9:1) were prepared by impregnation method in order to study the effect of metal ratio on the catalyst structure, activity and selectivity. The catalysts showed significant differences in activity and selectivity, with the catalyst having a Ni-W ratio of 9:1 achieving 100% conversion of corn oil and 100% selectivity to diesel for two days. Results indicate that minimizing metal alloy formation and enhancement of the metal dispersion leads to higher activity, selectivity and durability of the catalysts.

A Dendrimer-Encapsulated-Nanoparticles (DENP) method was employed in order to minimize alloy formation and increase the metal dispersion on the support. The catalysts prepared by the DENP method showed greater activity than the catalyst prepared by impregnation method for the hydrocracking of DDGS corn oil.

The effects of the fractional sums of the electronegativities of the transition metals on the catalyst activities and selectivities were investigated. The closer the fractional sum of the transition metal electronegativities was to the electronegativity range of noble metal catalysts (2.0–2.2), the better was the catalyst performance. The effects of doping a promoter (Ce) on the catalyst electronegativity and activity were also studied.

377104 Ni-Co/r-GO Catalysts for Hydrogen Generation from Chemical Hydride System Containing Ammonia Borane

Abstract:

Renewable energy attracts more and more attention in recent years. Among various green energies, hydrogen is one of the most promising energy due to its higher efficiency and power density, as well as its environmental benignity. Commonly, the hydrogen energy could be effectively extracted through devices such as fuel cells. Nowadays, the researches related to proton exchange membrane fuel cells (PEMFCs) become one of popular topic because PEMFCs have high energy efficiency and low operation temperature. Hence, the transportation of hydrogen to PEMFCs is the main topic in the future. In general, there are four methods of hydrogen storage: (1) high-pressure tanks; (2) liquefied hydrogen; (3) metal hydrides; and (4) chemical hydrides. Chemical hydrides not only have higher hydrogen storage capacity but also generate hydrogen quickly in the presence of adequate catalysts. Ammonia borane (NH_3BH_3) is a prospective compound in chemical hydrides. The hydrogen energy density of NH_3BH_3 is 19.6 wt%. Furthermore, NH_3BH_3 and its spent products are rarely toxic and flammable.

In this work, NH_3BH_3 and Ni-Co/r-GO catalysts were pulverized by using high energy ball milling machine (denoted as solid-state AB composite). Reduced graphene oxide (r-GO) was adopted as a light-weighted catalyst support. The Ni-Co/r-GO catalysts were characterized with SEM, EDS, TEM, VSM and ICP-AES. Hydrogen was generated by injecting adequate water from a syringe to initiate the hydrolysis reaction of solid-state AB composite. The amount of water dosage and loading of as-prepared catalysts will be discussed in this study. Hydrogen profiles were recorded by a mass flow meter (MFM) installed on a computer and temperature profiles were observed by a thermocouple at the same time.

AIChE
The Global Home of Chemical Engineers

天津大学
Tianjin University

The Nature of Active Copper Species in Cu/SiO_2 Catalysts for Hydrogenation of Methyl Acetate to Ethanol

Yue Wang, Yongli Shen, Yujun Zhao, Xinbin Ma*

School of Chemical Engineering and Technology
Tianjin University

AIChE Annual Meeting
Nov 20 2014, Atalanta

Background

- **High-octane rating; great antiknock performance**
- **High oxygen content; reduce the CO, NO_x emissions**

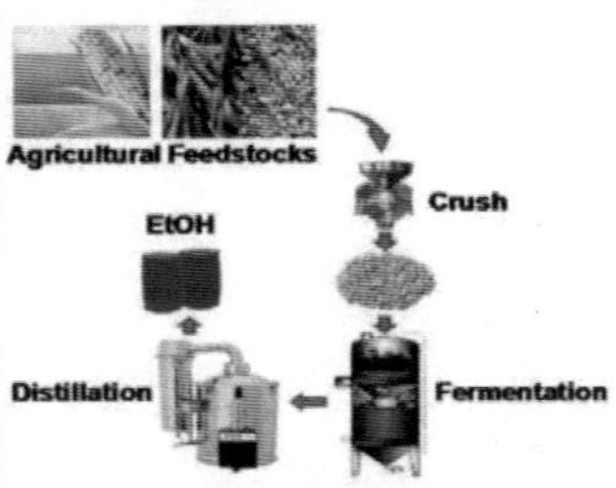

Fermentation
Restriction of cereal grains as raw materials

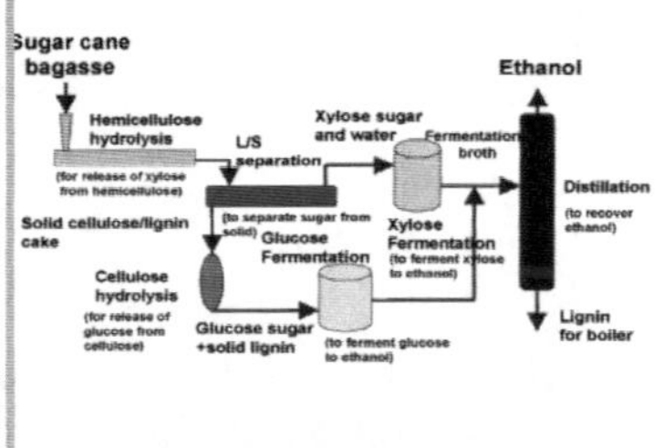

Cellulose hydrolyzation
Low energy - density and poor conversions

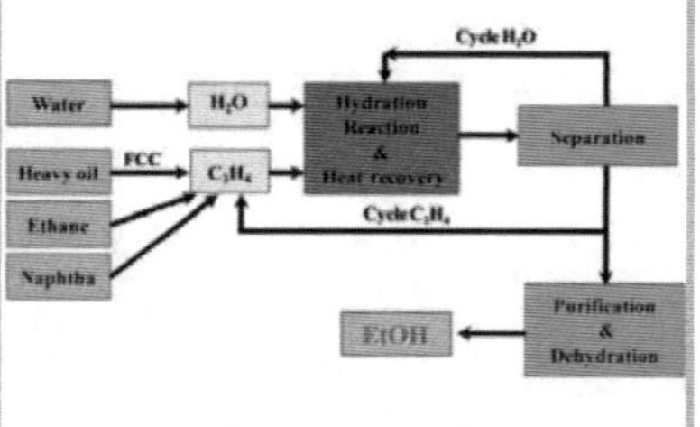

Ethylene direct hydration
Low selectivity, high energy and water consumption

Background

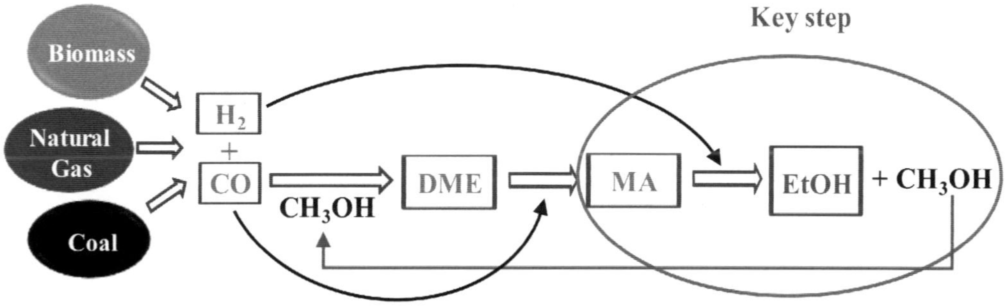

The new route for ethanol synthesis from syngas via dimethyl ether

Process is short, low-cost, low discharge, low energy and water consumption, suitable for characteristics of Chinese energy framework (rich in coal, lack of petroleum and natural gas)

Reviews of literatures related

Cu^0-Cu^I:

- **Cu^0 sites could dissociate H_2;**
- **Cu^+ sites could function as electrophilic or Lewis acidic sites to polarize the C=O bond via the electron lonepair in oxygen**

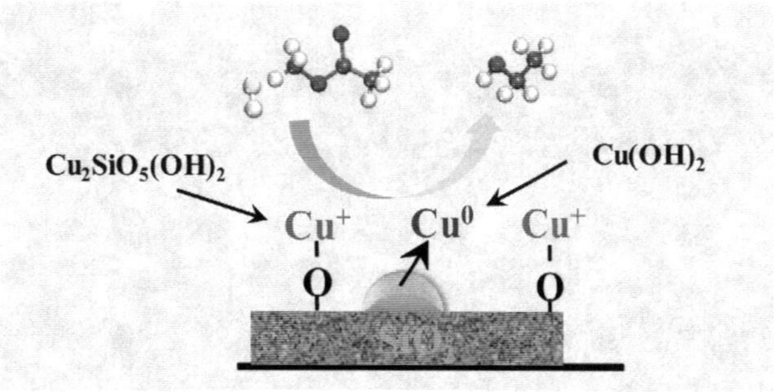

J. Catal., *2000, 193, 16-28*
J. Catal., *2008, 257, 172-180*
J. Phys. Chem. C, *2009, 113, 11003-11013*
Catal. Commun., *2011, 12, 412-416*

Cu^0:

- **Metallic Cu was the active site for hydrogenation** — *Energy & Fuels*, *2009, 23, 2843–2844*
- **Highly dispersed metallic Cu is in favor of improvement the catalytic performance** — *ACS Catal.*, *2012, 2, 1958–1966*

The strategy——Manipulation of Cu^0/Cu^I

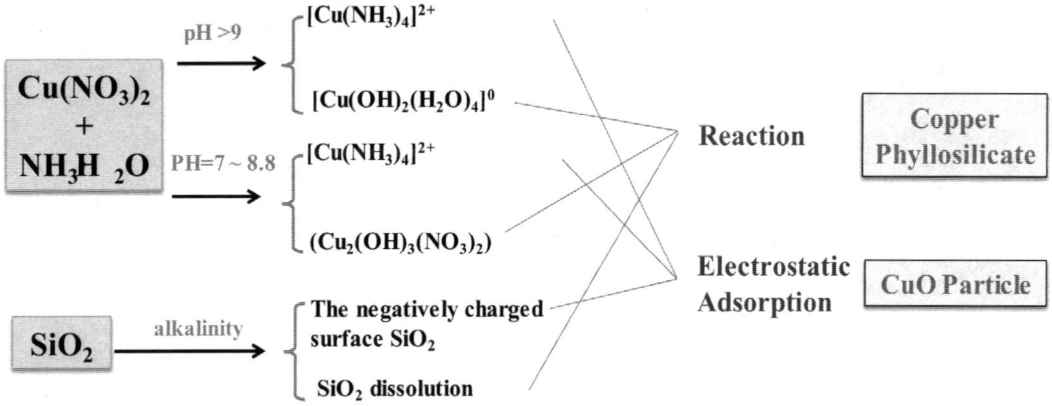

Target: relationship between active species and catalytic activity

Challenges: to exclude influences of other factors (geometry, channel structure, surface area……)

- the amount of Cu loading
- interaction time between cuprammonia and silica sol
- the SiO_2 surface groups

Toupance et al., J Phys Chem B, 2002, 106, 2277-2286; *Chen et al., J Catal, 2008, 257, 172-180.*

The structure of catalysts

Catalyst	Metal loading (wt%)[a]	S_{BET} (m^2/g)	D_{pore} (nm)	Crystallite size (nm)	
				by XRD[b]	by TEM
$5Cu/SiO_2$-303-4h	4.9	254	10.4	ND[c]	2.1
$5Cu/SiO_2$-353-4h	5.0	250	11.7	ND	1.8
$5Cu/SiO_2$-4h	5.0	263	13.6	ND	2.2
$10Cu/SiO_2$-303-4h	9.8	309	7.7	ND	2.9
$10Cu/SiO_2$-353-4h	9.9	302	12.4	ND	2.0
$10Cu/SiO_2$-4h	9.9	294	15.7	ND	2.4
$20Cu/SiO_2$-303-4h	19.7	381	6.8	4.5	4.0
$20Cu/SiO_2$-353-4h	19.6	382	10.7	4.8	3.9
$20Cu/SiO_2$-4h	19.5	409	10.8	ND	2.9
$20Cu/SiO_2$-16h	19.6	414	11.0	4.5	3.5
$20Cu/SiO_2$-24h	19.5	406	9.9	4.7	3.6
$20Cu/SiO_2$-40h	19.6	421	10.5	ND	3.1
$30Cu/SiO_2$-4h	29.0	480	8.6	3.6	4.1
$40Cu/SiO_2$-4h	37.2	558	6.2	3.8	4.7

a Metal loading was determined by ICP-OES.
b Diameter of Cu particles calculated from the XRD data by Scherrer equation.
c ND: not detected.

XRD

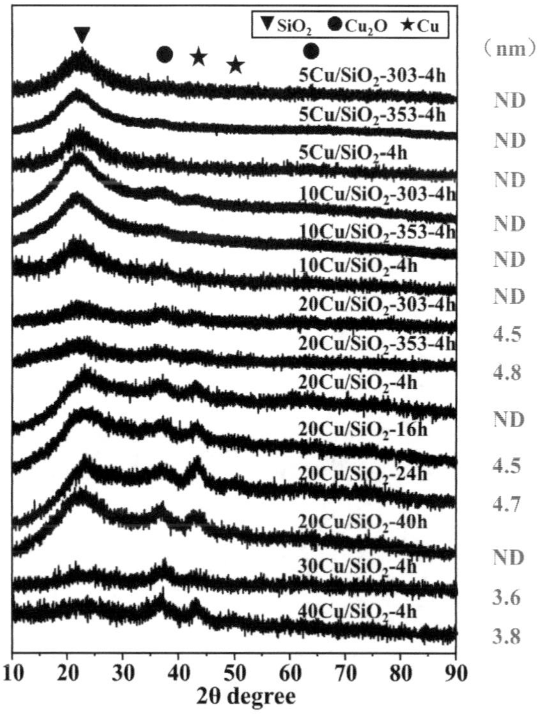

- A diffraction peak at 2θ of 43.3° and 50.4° characteristic of fcc Cu.
- The peaks at 2θ of 36.4° attributed to the Cu_2O (111) plane, indicating that a portion of copper exists as Cu (I) after reduction.
- The broad peaks at around 22° are assigned to amorphous SiO_2.
- The catalysts were successfully prepared with highly dispersed sub-5 nm metallic copper particles

TEM

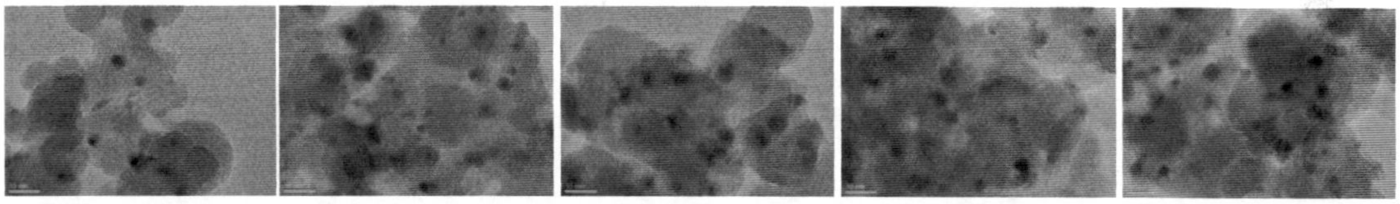

5Cu/SiO_2-303-4h 5Cu/SiO_2-353-4h 5Cu/SiO_2-4h 10Cu/SiO_2-303-4h 10Cu/SiO_2-353-4h

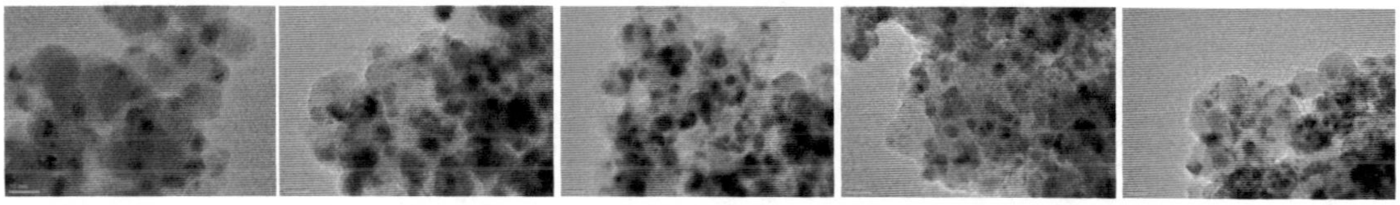

10Cu/SiO_2-4h 20Cu/SiO_2-303-4h 20Cu/SiO_2-353-4h 20Cu/SiO_2-4h 20Cu/SiO_2-16h

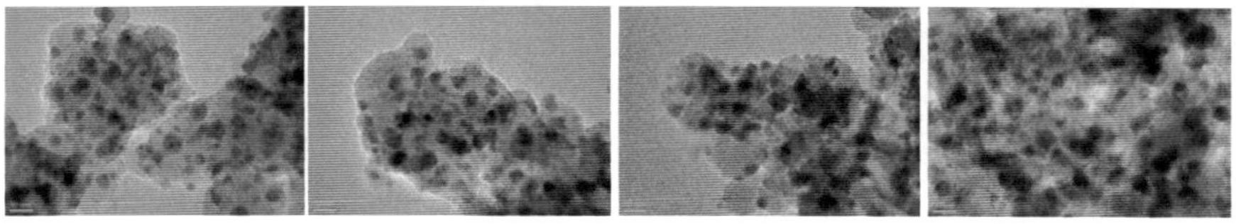

20Cu/SiO_2-24h 20Cu/SiO_2-40h 30Cu/SiO_2-4h 40Cu/SiO_2-4h

Catalytic performance

Catalyst	MA conversion (%)	Selectivity (%)[b] EtOH	EtOAc	Others[c]	$S_{Cu(0)}$ (m²/g)	$S_{Cu(I)}$ (m²/g)
5Cu/SiO$_2$-303-4h	12.3	44.3				
5Cu/SiO$_2$-353-4h	22.1	43.0				
5Cu/SiO$_2$-4h	19.8	44.4				
10Cu/SiO$_2$-303-4h	26.4	43.3				
10Cu/SiO$_2$-353-4h	51.2	52.4				
10Cu/SiO$_2$-4h	42.3	50.0				
20Cu/SiO$_2$-303-4h	79.7	76.4	23.5	0.1	[illegible]	
20Cu/SiO$_2$-353-4h	59.9	59.4	40.6	0.0	39.6	9.4
20Cu/SiO$_2$-4h	61.2	62.2	37.7	0.1	24.4	10.3
20Cu/SiO$_2$-16h	73.4	70.9	29.0	0.1	23.9	13.8
20Cu/SiO$_2$-24h	81.7	77.8	22.2	0.0	22.1	20.5
20Cu/SiO$_2$-40h	73.8	71.5	28.4	0.1	23.5	14.8
30Cu/SiO$_2$-4h	62.7	65.3	34.6	0.1	37.6	12.2
40Cu/SiO$_2$-4h	66.3	67.3	32.6	0.1	44.2	13.9

face area, and concentration of surface Cu$^+$ sites, which essentially enhanced catalytic performance. CO-chemisorption FTIR characterization disclosed that the surface concentration of a special Cu$^+$ site that can stably adsorb CO molecule reached the maximum on a Cu/SiO$_2$-3HZ-38 hybrid catalyst with the best catalytic activity. Therefore, Cu$^+$ sites might play important roles in the catalytic process of DMO hydrogenation to EG. Nonetheless, the presence of both Cu$^+$ and Cu^0 species on the functioning catalyst was confirmed because the synergetic effects of Cu^0 and Cu$^+$ surface species were involved in the catalytic cycle.

Appl. Catal. A, 2012, 445, 287-296

a Reaction conditions :catalyst weight = 0.5g, P = 2.5 M Pa, T = 488 K, H_2/MA = 80, and WLHSV$_{(MA)}$ = 2.0 g$g_{cat}^{-1}h^{-1}$.
b The selectivies were calculated excluded MeOH as a product since it was cycled back to the previous process to be a reactant.
c Others: CH_3CHO, $HCOOC_2H_5$, CH_3COOH, $HCOOCH_3$, C_3H_7OH and C_4H_9OH.

Catalytic performance

Catalyst	MA conversion (%)	Selectivity (%)[b] EtOH	EtOAc	Others[c]	$S_{Cu(0)}$ (m²/g)	$S_{Cu(I)}$ (m²/g)
5Cu/SiO$_2$-303-4h	12.3	44.3	55.2	0.5	5.0	8.9
5Cu/SiO$_2$-353-4h	22.1	43.0	56.8	0.2	11.2	6.9
5Cu/SiO$_2$-4h	19.8	44.4	55.3	0.3	7.6	6.7
10Cu/SiO$_2$-303-4h	26.4	43.3	56.5	0.2	12.0	8.6
10Cu/SiO$_2$-353-4h	51.2	52.4	47.5	0.1	19.9	8.0
10Cu/SiO$_2$-4h	42.3	50.0	49.9	0.1	18.6	8.5
20Cu/SiO$_2$-303-4h	79.7	76.4	23.5	0.1	23.1	17.0
20Cu/SiO$_2$-353-4h	59.9	59.4	40.6	[illegible]	[illegible]	[illegible]
20Cu/SiO$_2$-4h	61.2	62.2	37.7			
20Cu/SiO$_2$-16h	73.4	70.9	29.0			
20Cu/SiO$_2$-24h	81.7	77.8	22.2			
20Cu/SiO$_2$-40h	73.8	71.5	28.4			
30Cu/SiO$_2$-4h	62.7	65.3	34.6			
40Cu/SiO$_2$-4h	66.3	67.3	32.6			

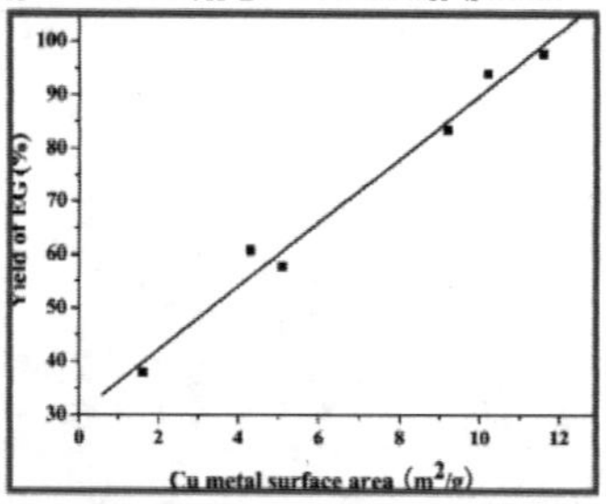

Figure 10. Yield of EG plotted against copper metal surface area for catalysts with different copper loadings.

a Reaction conditions :catalyst weight = 0.5g, P = 2.5 M Pa, T = 488 K, H_2/MA = 80, an
b The selectivies were calculated excluded MeOH as a product since it was cycled back
c Others: CH_3CHO, $HCOOC_2H_5$, CH_3COOH, $HCOOCH_3$, C_3H_7OH and C_4H_9OH.

J. Phys. Chem. C, 2009, 113, 11003-11013

Synergy of Cu^0 and Cu^+ on catalytic performance

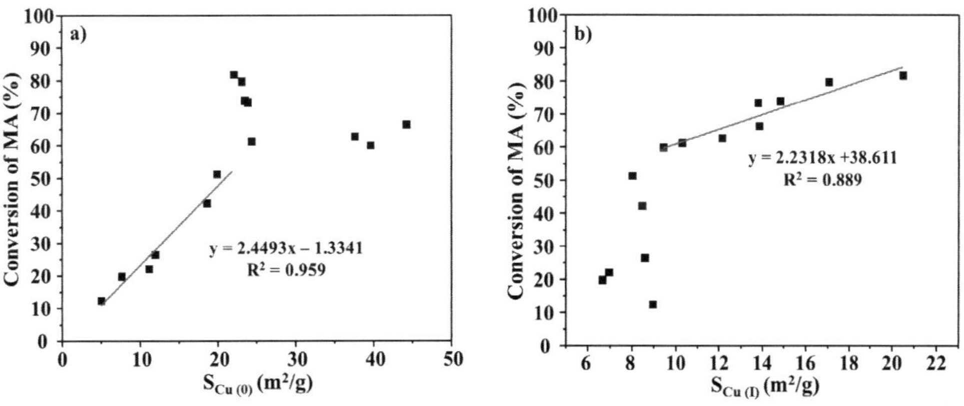

A synergy of active copper species on the hydrogenation of MA :

When there is insufficient of Cu (0) species, the catalytic activity of hydrogenation determined by the activity of H_2, whereas, it would be decided by the adsorption of methoxy and acyl species as the amount of surface Cu (0) achieved to a certain degree.

Conclusion

- **We successfully prepared a series of Cu/SiO_2 catalysts with remarkable differences on state of Cu species and exclued the other factors' influence.**
- **The conversion of MA grew linearly associated with the Cu (0) surface area increasement under a certain value. However, with more Cu (0) active sites, the samples catalytic activity increaed with enrichment of Cu (I) surface area.**
- **These relationships strongly proved there was a synergy of active copper species on the hydrogenation of MA and indicated that when there is insufficient of Cu (0) species, the catalytic activity of hydrogenation determined by the activity of H_2, whereas, it would be decided by the adsorption of methoxy and acyl species as the amount of surface Cu (0) achieved to a certain degree.**
- **The calculated results showed the MA could be adsorbed by $(Cu)_x^{\delta+}/SiO_2$ but not activated by Cu (111). And the H_2 is much easilier dissociated on Cu (111) than on $(Cu)_x^{\delta+}/SiO_2$. These calculatedresults are agreed with the experiments data.**

Progress on scale-up

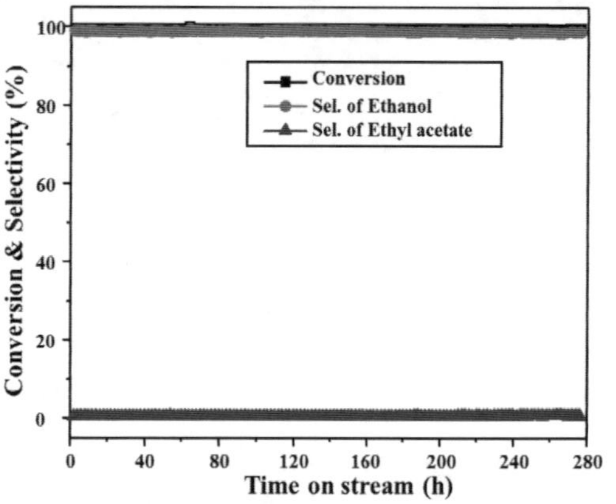

Reaction Conditions: P = 2.5 M Pa, T = 488 K, H_2/MA = 80, and WLHSV$_{(MA)}$ = 1.0 $gg_{cat}^{-1}h^{-1}$

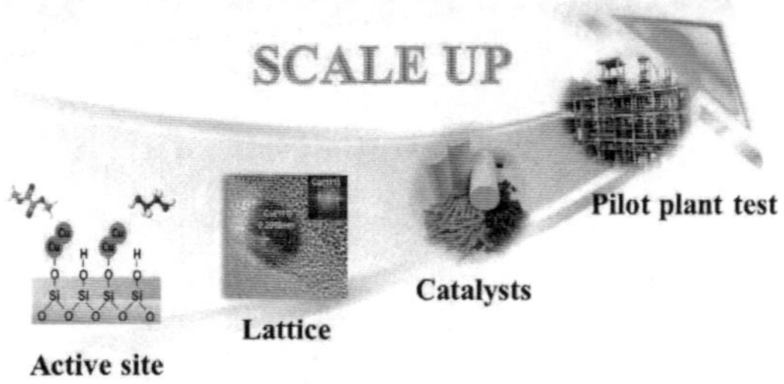

Cooperating with Wison Company for pilot plant test.

Acknowledgement

国家自然科学
基金委员会
National Natural Science
Foundation of China

Green Diesel Production by Hydrotreatment of Microalgae Oil using Pt Catalyst

Lin Zhou and Adeniyi Lawal

New Jersey Center for Microchemical Systems, Dept. of Chemical Engineering and Materials Science, Stevens Institute of Technology,
1 Castle Point Terrace,
Hoboken, NJ 07030

Hydrotreatment of algal-derived oil is considered a promising route to renewable liquid fuel production. The product comprises mostly straight chain alkanes, and is chemically similar to petroleum-derived diesel. With the advantage of fast growth rate and high acylglycerides content, microalgae is favored as a feedstock for green diesel production. Moreover, this process is CO_2-reducing and does not compete with food production for feedstock.

In this work, a 0.5wt% Pt commercial catalyst was used for hydrotreatment of Nannochloropsis salina algae oil provided by Valicor Renewables, LLC. The effects of reaction temperature, pressure, liquid hourly space velocity (LHSV) and gas hourly space velocity (GHSV) were investigated. Performance study results of carbon yield, hydrocarbon yield, product distribution and lipids conversion for this precious metal catalyst will be presented, and compared with the results of our previous work that used base metal catalyst NiMo. The optimal operating conditions for green diesel production for the Pt catalyst were determined.

Progress in Liquefaction of Biochar

Nikolaus Schwaiger

Technische Universität Graz
Institut für Chemische Verfahrenstechnik und Umwelttechnik

BDI-BioEnergy International AG

Outline

- Introduction
  - BPR Concept
- Liquid phase pyrolysis oil HDO
  - Results
- Direct pyrolysis char liquefaction
  - Thermal char liquefaction
  - Catalytic char liquefaction
- Process implementation
- Conclusion

Introduction | Pyrolysis Char | Liquefaction | Experimental | Summary

Chemical Engineering and Environmental Technology | www.icvt.tugraz.at 2

Introduction

bioCRACK

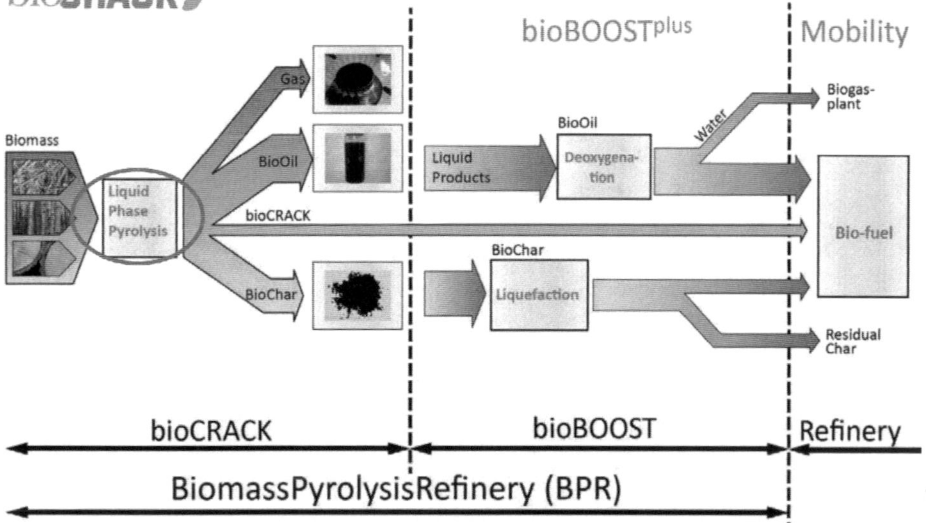

Liquid phase pyrolysis pilot plant

Feed: 100 kg/h Biomass
Budget: 7 Mio €

BDI
OMV
FFG

Liquid phase pyrolysis in the bioCRACK pilot plant

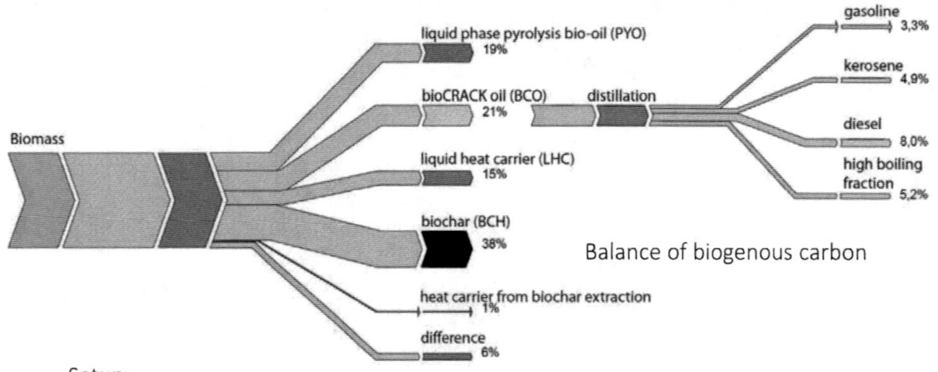

Balance of biogenous carbon

Setup:

- reaction temperature: 375°C
- duration of experiment: 38 h
- biomass feed rate 60 kg/h

CEET

Composition of bioCRACK oil and liquid phase pyrolysis oil

	Liquid phase pyrolysis oil	BCO
Watercontent [wt.%]	49,6	0,1
LHV [MJ/kg]	9,2	40,1
TAN [mg_{KOH}/g_{sample}]	82,2	-
Density [kg/m^3]	1,09	0,88
Viscosity [mPa s]	3,9	8,0
^{14}C-Content [wt.%]	100	9,8
	Elemental Composition	
Carbon content [wt.%]	25,6	83,6
Hydrogen content [wt.%]	9,2	11,8
Oxygen content [wt.%]	64,9	4,2
Nitrogen content [wt.%]	<1	<1

CEET

Introduction

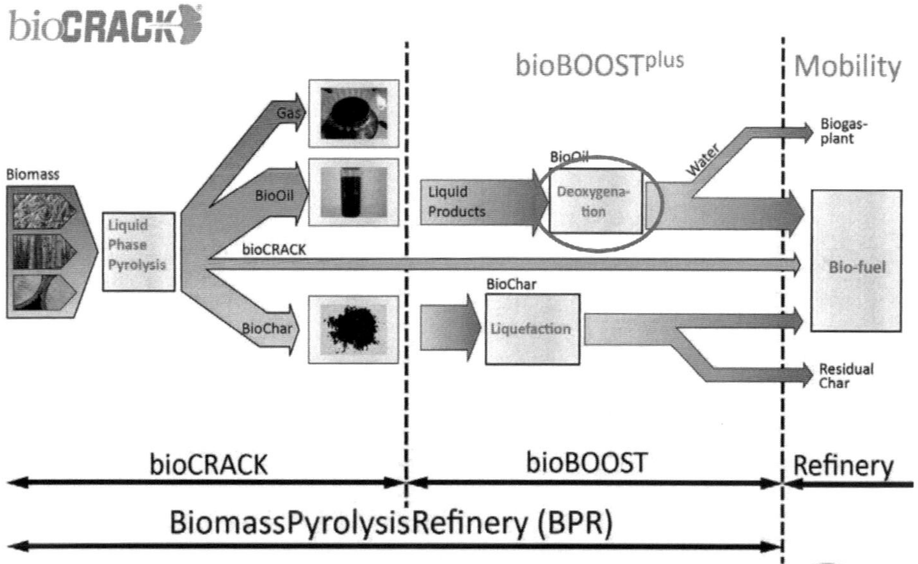

HDO of liquid phase pyrolysis oil

Two step HDO Process

	unit	Feed	HDO 1st step	HDO 2nd step	Diesel
Water content	[wt.%]	50	11	0.2	0.02
LHV	[MJ/kg]	8.7	26.4	41.2	42.5
Density	[kg/m^3]	1070	1100	870	845
Viscosity	[mPa s]	4	163	4.5	4.5
Carbon content	[wt.%]	25.6	62.5	85.5	85.9
Biogenous content	[wt.%]	100	100	28	<7
Hydrogen content	[wt.%]	9.2	8.3	12.1	13.3
Oxygen content	[wt.%]	64.9	28.7	1.9	<1
Nitrogen content	[wt.%]	<1	<1	<1	<1

Introduction

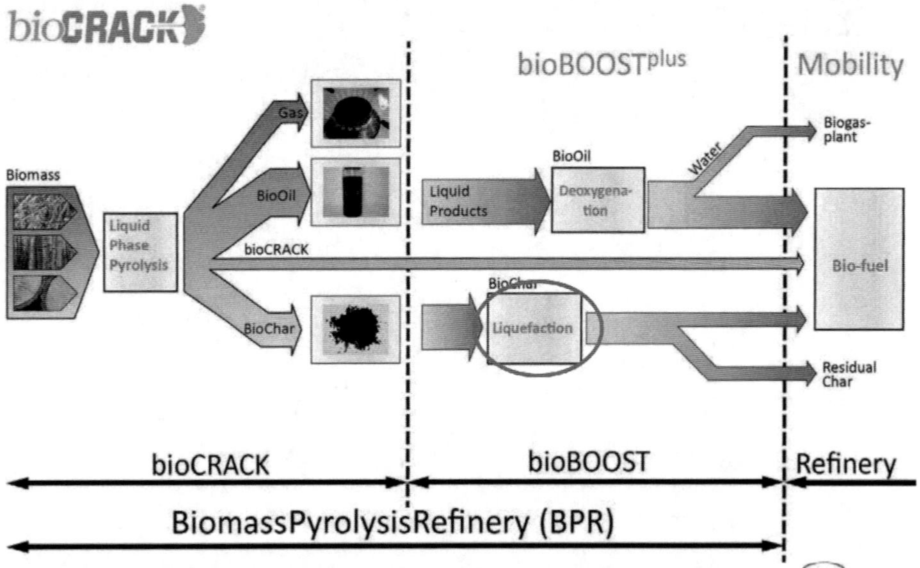

Introduction | Pyrolysis Char | Liquefaction | Experimental | Summary

Chemical Engineering and Environmental Technology | www.icvt.tugraz.at 9

CEET

Elemental composition

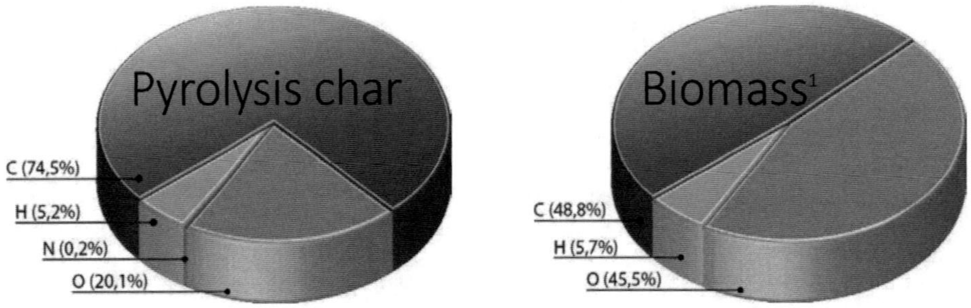

[1] **Behrendt, F. et al. (2008).** Direct Liquefaction of Biomass. Chemical Engineering & Technology, 31(5), 667–677.

Introduction | Pyrolysis Char | Liquefaction | Experimental | Summary

Chemical Engineering and Environmental Technology | www.icvt.tugraz.at 10

CEET

Pyrolysis char

Pyrolysis char: extracted and dried

Pyrolysis char: extracted, dried and milled <200 μm

Pyrolysis char SEM

Liquefaction of coal

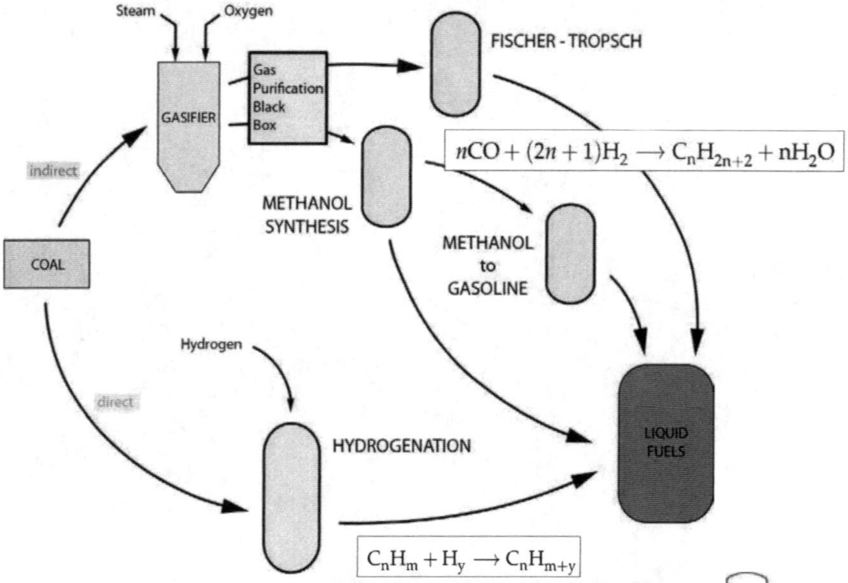

Introduction | Pyrolysis Char | Liquefaction | Experimental | Summary

Chemical Engineering and Environmental Technology | www.icvt.tugraz.at 13

CEET

Direct

Pyrolysis Char-liquefaction

Introduction | Pyrolysis Char | Liquefaction | Experimental | Summary

Chemical Engineering and Environmental Technology | www.icvt.tugraz.at 14

CEET

Product formation

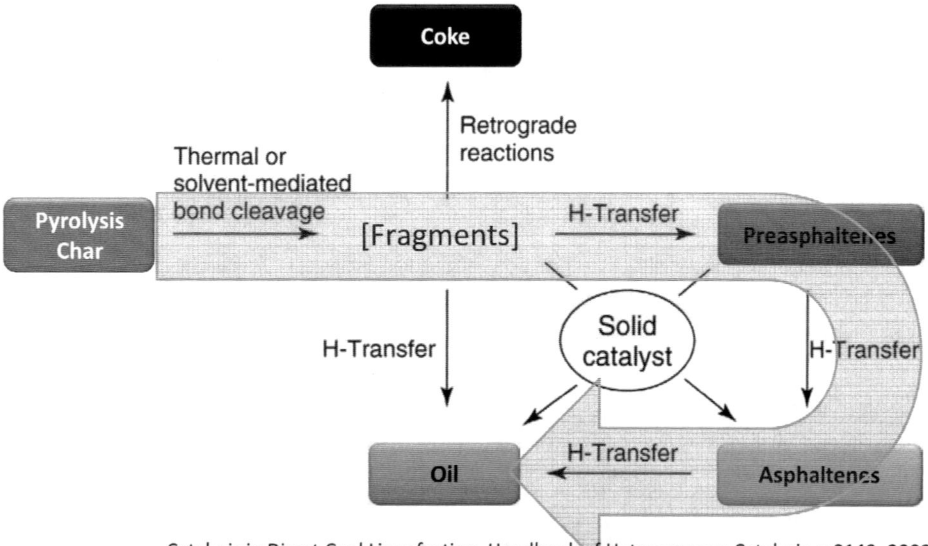

Catalysis in Direct Coal Liquefaction, Handbook of Heterogenous Catalysis, p 3149, 2008

Product formation

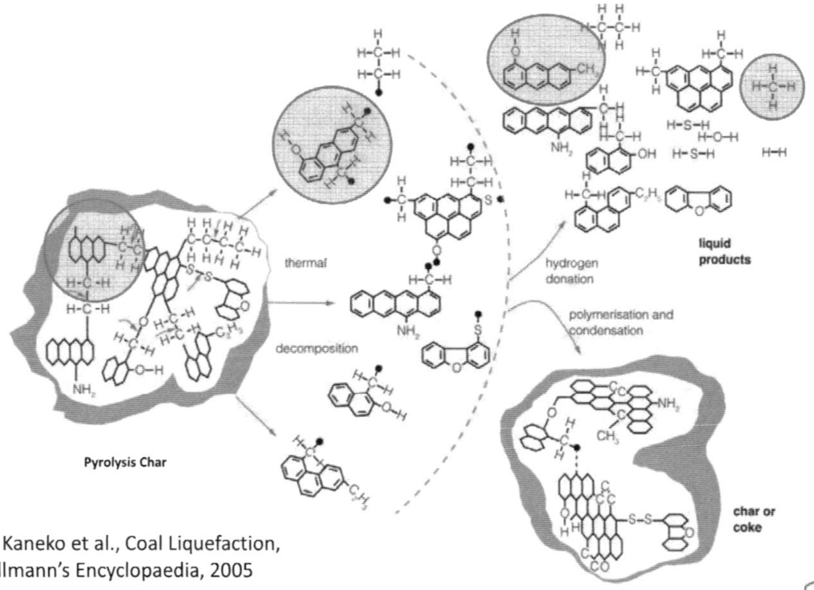

T. Kaneko et al., Coal Liquefaction,
Ullmann's Encyclopaedia, 2005

Hydrogen Donor-System

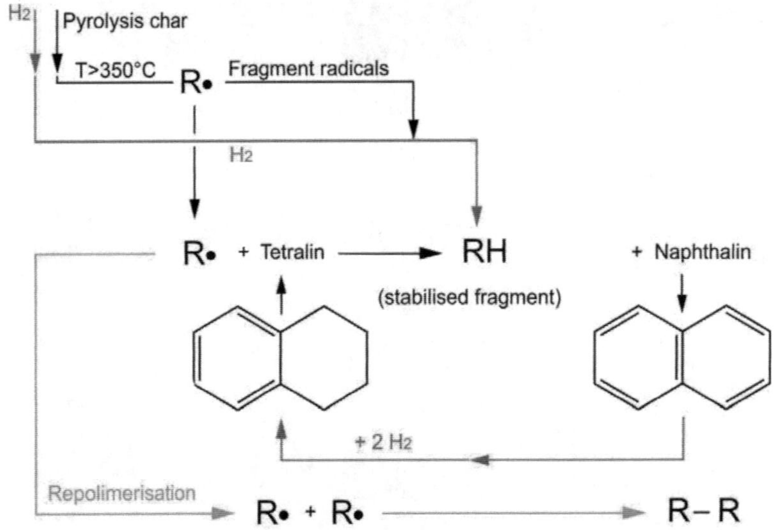

Scheme

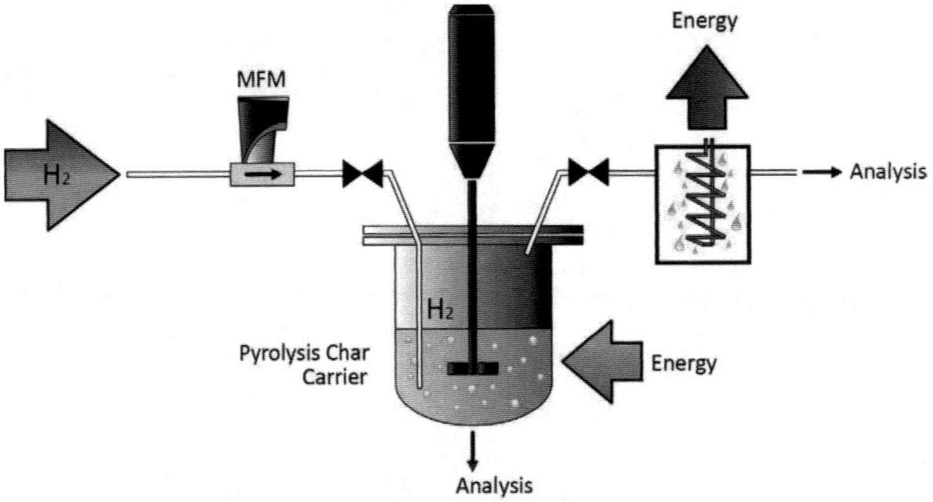

Product classification

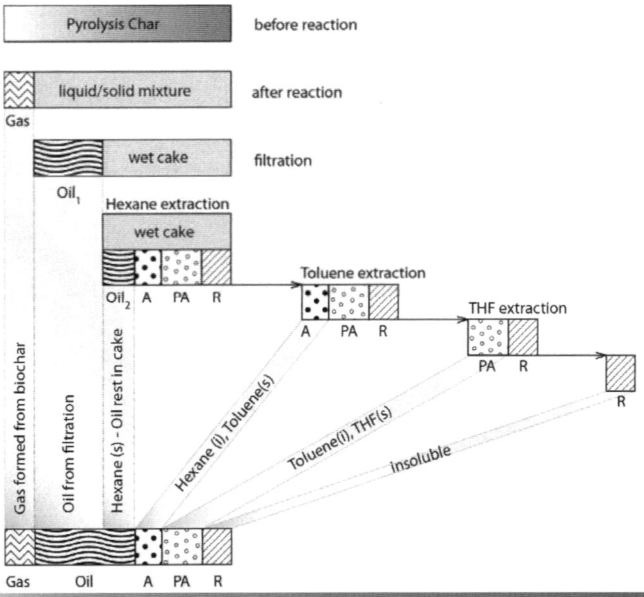

Extraction
Apparatus: fexIKA

Thermal char liquefaction

Experiments

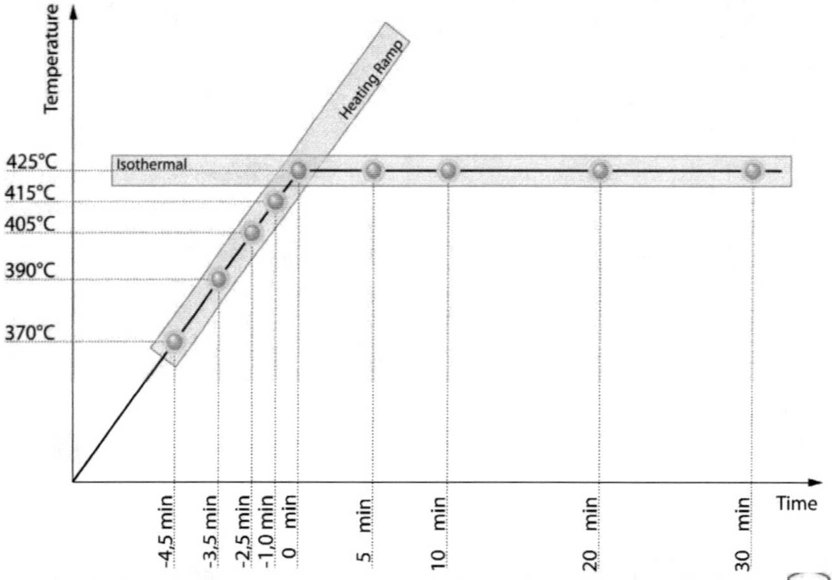

Results

Results (Heating Ramp) and (Isothermal stage)

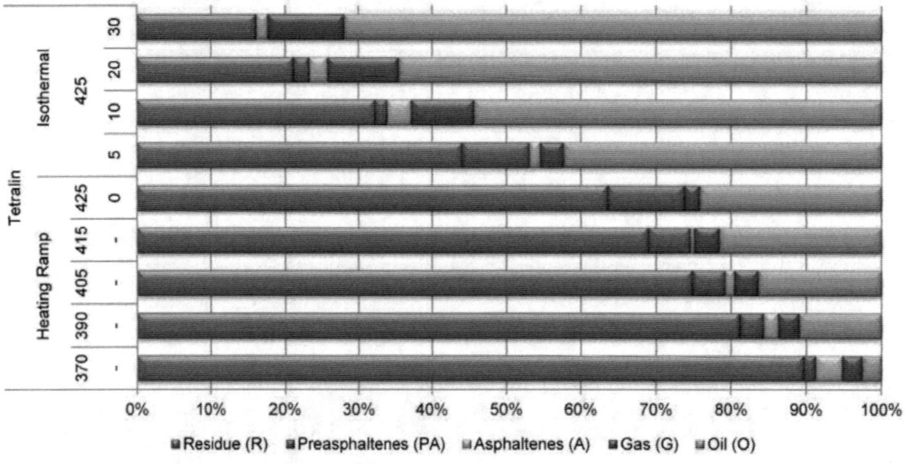

Resultate

Modelling

Li, Xian et al. (2008). *Fuel*, *87*, 508–513.

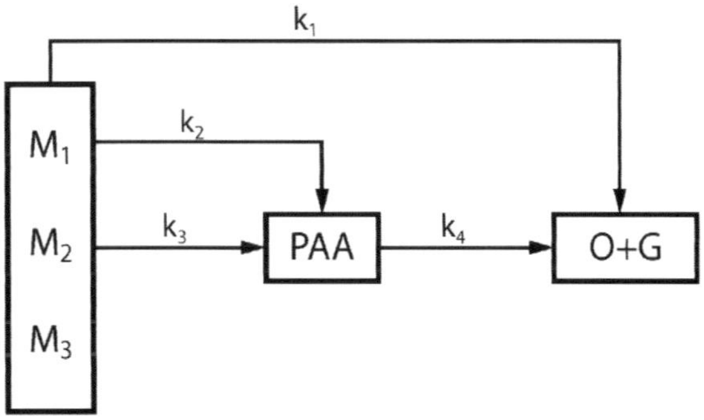

$$\frac{d\vec{\alpha}}{dt} = K\vec{\alpha} \qquad t = 0, \vec{\alpha} = \vec{\alpha}_0$$

$$\vec{\alpha} = \begin{Bmatrix} M_1 \\ M_2 \\ PAA \\ (O+G) \end{Bmatrix} \qquad K = \begin{Bmatrix} -(k_1+k_2) & 0 & 0 & 0 \\ 0 & -k_3 & 0 & 0 \\ k_2 & k_3 & -k_4 & 0 \\ k_1 & 0 & k_4 & 0 \end{Bmatrix}$$

<u>Assumptions:</u>

- Intermediates (PA and A) as well as products (O and G) are combined
- Kinetic equations: 1st order
- Coal is split into three parts:
  - M1 (easy reactive part)
  - M2 (hard reactive part)
  - M3 (none reactive part)
- (Reaction) water is included in oil
- Influence of hydrogen pressure, particle size, mass transfer etc. is ignored.

Results

Results Modelling (Isothermal)

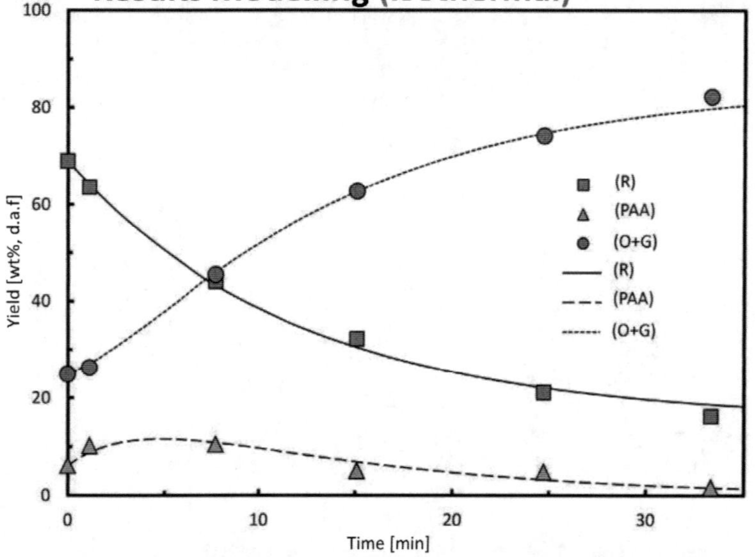

Introduction | Pyrolysis Char | Liquefaction | Experimental | Summary

Chemical Engineering and Environmental Technology | www.icvt.tugraz.at 25

CEET

Catalytic char liquefaction

Introduction | Pyrolysis Char | Liquefaction | Experimental | Summary

Chemical Engineering and Environmental Technology | www.icvt.tugraz.at 26

CEET

Catalysts

Catalysts:

- Raney-Nickel
  - Alumina-Nickel Alloy
- Red-Mud
  - Waste stream of alumina production
- Iron oxide

Results

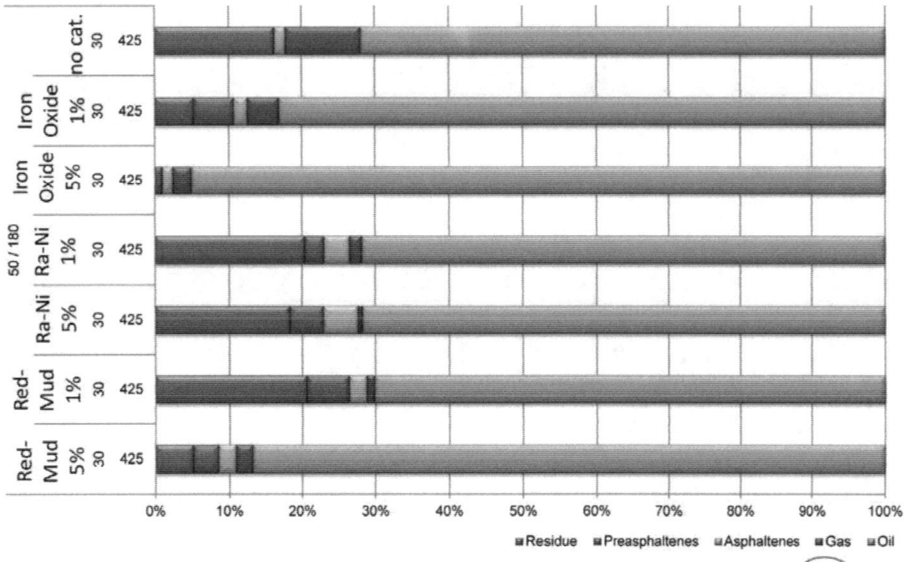

Implementation

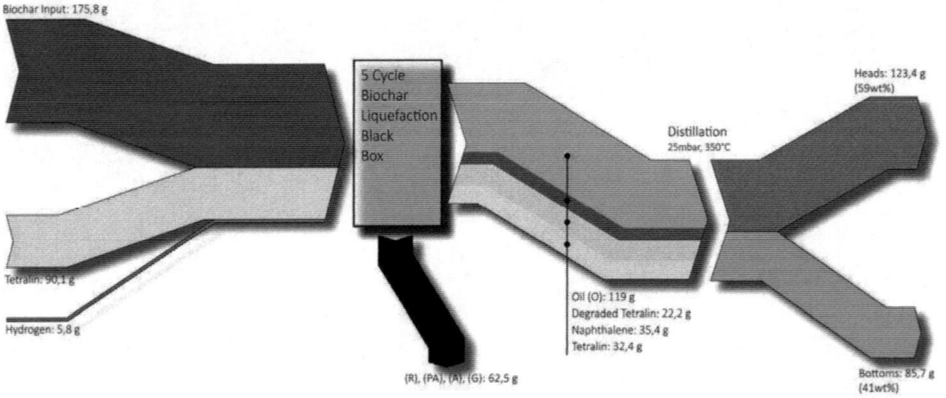

› Balance data:

- Biochar was liquefied (6 replicas)
- 90% conversion and 68-76% liquefaction rate
- Distillation: 59% of products in distillate

Conclusion

- 84% Conversion and 72% Oil yield
  for thermal char liquefaction conditions
  (425°C and 30 min)

- 99% Conversion and 95% Oil Yield
  for catalytic char liquefaction conditions
  (425°C und 30 min)

- For implementation less Tetralin consumption is favored->

  90% Conversion and 68-76% Oil Yield – with only 20% Tetralin consumption

Conclusion – Tetralin Recycling

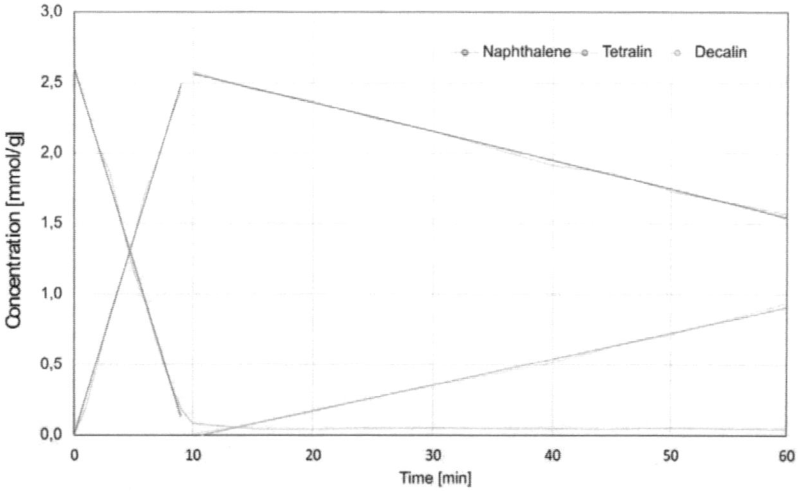

Naphthalene hydrogenation at 150°C with Ra-Ni

Acknowledgement

CEET Chemical Engineering and Environmental Technology

Prof. Dr. Matthäus Siebenhofer
Dr. Roland Feiner
Dipl.-Ing. Hannes Pucher

BDI

Dr. Peter Pucher
Dipl.-Ing. Lisa Ellmaier

FFG a3plus

Impression

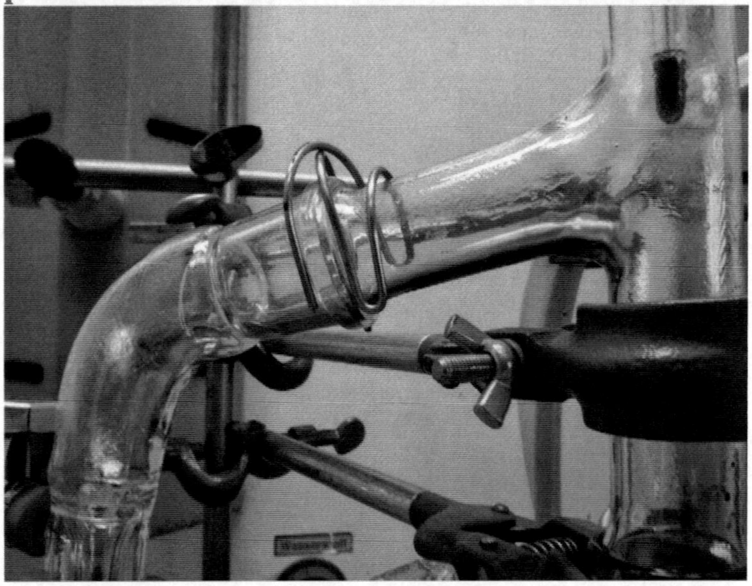

Introduction | Pyrolysis Char | Liquefaction | Experimental | Summary

Chemical Engineering and Environmental Technology | www.icvt.tugraz.at

33

CEET

CEET Chemical Engineering and Environmental Technology

Progress in Liquefaction of Biochar

TU Graz
Graz University of Technology

Nikolaus Schwaiger

Technische Universität Graz

Institut für Chemische Verfahrenstechnik und Umwelttechnik

Inffeldgasse 25 C, 8010 Graz

nikolaus.schwaiger@tugraz.at

bioboost@tugraz.at

Tel: +43(0)316/873-7965

CEET BDI FFG a3plus TU Graz Graz University of Technology

THERMAL CRACKING OF WASTE COOKING OIL: A PROPOSAL FOR KINETIC MODEL BASED ON GROUPS OF COMPOUNDS

FRAINER B L M[1], MEIER H F[1], WIGGERS V R[1], SIMIONATTO E L[2], ENDER L[1].
[1] Chemical Engineering Departament
[2] Chemistry Departament
Regional University of Blumenau, Blumenau – SC – Brazil
e-mail: brunafrainer@yahoo.com.br

ABSTRACT – In the thermal cracking reaction, the breakage of the triglyceride molecules leads to the formation of lower molecular weight compounds. When performed at temperatures around 500 °C this reaction produces a liquid fraction (bio-oil) with very similar composition and properties to those of fossil fuels. This study aimed to perform experiments in a pyrolysis reactor in order to obtain experimental data in order to construct a kinetic model based on chemical lumps. The chemical lumps evaluated were the yields of liquid, gas and solid obtained and five mechanisms based on these groups of compounds have been proposed. The dependence of the kinetic rates on the temperature and composition was subjected to regression and the kinetic parameters were determined.

Introduction

Many studies have been made over the years with greasy residues as raw material for thermal cracking. In this reaction, the breakdown of triglyceride molecules leads to formation of compounds of lower molecular weight. When conducted at temperatures of 500°C, this reaction produces a fraction of liquid (bio-oil) with properties very similar to those of fossil fuels ([1] [2] [3] [4] [5]).

The greatest difficulty in the fuel's production by thermal cracking process is the difficulty of scaling-up the process. This difficulty arises from limited knowledge about the kinetics of thermal cracking reaction, which consists of a series of complex reactions [6]. So, determine this kinetics is one of the keys to scaling up and application of this process. In this way, it is necessary to propose a mechanism and kinetic model for the reaction rates. Unfortunately, as can be noted in the literature, it is not possible to obtain a kinetic model based on chemical species conservation equations. One of the possibilities to reduce the number of reactions is to use the lumping of components where pseudo-components are defined. This procedure of mechanism reduction has been applied at petroleum processes ([7] [8] [9] [10] [11]).

Martinez and Ancheyta [12] applied a five lump model for hydrocracking of crude petroleum oil in a CSTR reactor involving fast deactivation of the catalyst. The model has 10 coefficients of reaction rate, and separates the lumps by boiling temperature range. The product's composition had provided good agreement with the experimental values, with less than 5% average absolute error.

Sadhukhan et al. [13] studied the pyrolysis of wood particles and developed a transient mathematical model for this process. Simulation studies were performed to analyze the effect of temperature and volume/size particle on the total pyrolysis time and final production of coal. The authors proposed two kinetic models composed of primary and secondary reactions, with two and three lumps, which showed a satisfactory outcome to this process.

So, the main objective of this work was to carry out an experimental study on the thermal cracking of waste cooking oil (WCO) using a continuous screw reactor in steady state and isothermal conditions. Based on the experimental results obtained, a kinetic model is presented, comprised of a mechanism and the expressions for the kinetic rates, based on the lumping strategy, considering the WCO as the raw-material and the gas (G), liquid (L) and solid (C) as lumps of the thermal cracking.

Experimental

The pyrolysis reactor consists of a cylindrical tube with an internal screw. The stainless steel tube has 920 mm of length and 25 mm of internal diameter and it is considered as a plug flow reactor. The heat source is provided by a set of electrical resistances divided into 4 regions of control, which provide an isothermal operation. A positive displacement pump is responsible for the constant mass flow. A schematic draw of the reactor used is presented in Figure 1.

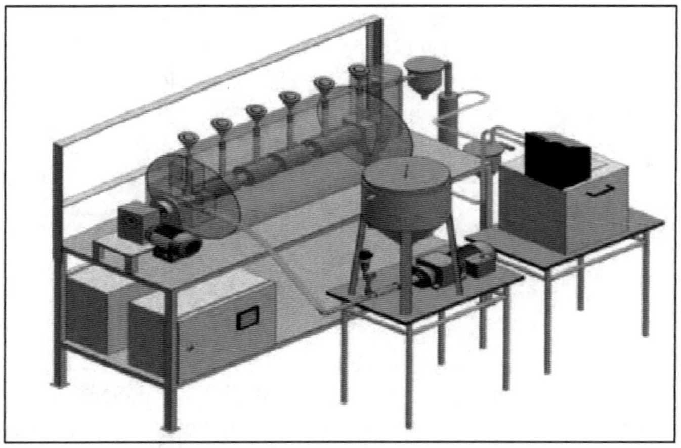

Figure 1. Pyrolysis screw reactor.

The material used for the thermal cracking consisted basically of waste cooking oil utilized as received. The experiments were conducted in three temperatures of thermal cracking (500, 525 and 550°C) and ten different flow rates of feed biomass have been defined, resulting in different spatial times. The mass flow range were between 120 and 500 g h^{-1}. So, ten experiments were performed at each temperature, totaling 30 experiments. They were performed in triplicate under the same conditions of the original experiments, totaling 90 experiments.

To close the mass balance, it is necessary to quantify the bio-gas fraction generated in the process. This measurement was made on the basis of volumetric flow rate and chemical composition of bio-gas obtained in the cracking of the WCO. In order to get these compositions gas chromatography assays were performed. Quantification of coke generated in thermal cracking is the result of the difference between the initial biomass, bio-oil and bio-gas (products). Using these data, we could then perform numerical simulations of the proposed kinetic models.

Kinetic Model

A macroscopic modeling was performed for the reaction system, taking into considering the conservation of chemical species for the determination of kinetic constants of the proposed mechanisms. The following simplifying assumptions were adopted: constant physical properties; perfect blend; negligible convective effects. Table 1 shows the equating corresponding to each proposed mechanism, where X_B is the mass fraction of WCO (unconverted biomass), X_L is the liquid fraction, X_G is the bio-gas fraction and coke fraction is represented by X_C.

Two proposals of kinetic models with 4 lumps were made for the thermal cracking reaction of the waste cooking oil in order to evaluate which one represents the process. The influence of two process variables (reaction temperature and the biomass flow) was analyzed, and with the solution of kinetic models, were determined which mechanism had a better approximation of the physical problem in question. Figure 2 shows the proposed mechanisms for this thermal cracking reaction where WCO is the biomass fraction, L is the liquid fraction, G is the gas, and C is the coke.

Table 1. Equating of proposed mechanisms

Reaction Rate			
$\frac{dX_i}{dt} = -r_i$			
Mechanism 1		**Mechanism 2**	
$r_{WCO} = -k_1 X_{WCO}$	(1.1)	$r_{WCO} = -(k_1 + k_2) X_{WCO}$	(2.1)
$r_L = k_1 X_{WCO} - k_2 X_L$	(1.2)	$r_L = k_1 X_{WCO} - k_3 X_L$	(2.2)
$r_G = k_2 X_L - k_3 X_G$	(1.3)	$r_G = k_3 X_L$	(2.3)
$r_C = k_3 X_G$	(1.4)	$r_C = k_2 X_{WCO}$	(2.4)

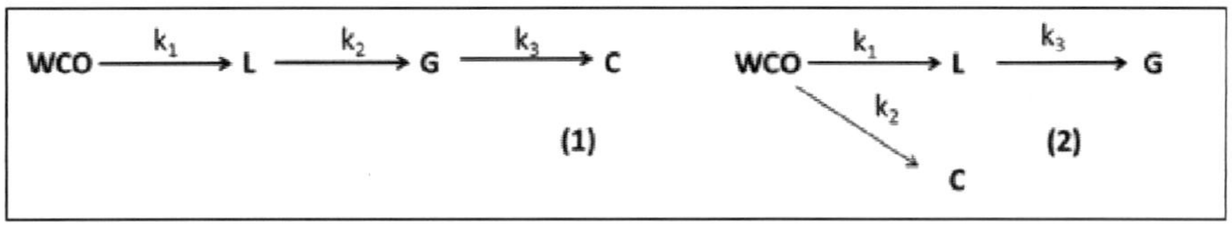

Figure 2. Proposed mechanisms for the thermal cracking reaction of waste cooking oil.

To compare the numerical and experimental data, a parameter that can be used to qualify the numeric setting is the value of the objective function, which is defined as the sum of squared errors between experimental values and simulated values. The difference between these values should be the lowest possible. The numerical and experimental data have been adjusted based on the mass fraction of the products of thermal cracking, using mass balance and chromatographic analyzes. The regression method used in this study was the Complex [13], which aims to direct search of the optimum (maximum or minimum) of an objective function with n variables. As a kinetic model relating the parameters of the chemical reaction kinetics.

Results and Discussion

Table 2 brings the average mass and space time yields of the three temperatures at which the experiments where τ is the spatial time, X_B is the mass fraction of oily sludge (unconverted biomass), X_L is the liquid fraction, X_G is the bio-gas fraction and coke fraction is represented by X_C.

Table 2. Mass yield of the products at 500, 525 e 550 °C

τ (min)	X_B	X_L	X_G	X_C	τ (min)	X_B	X_L	X_G	X_C	τ (min)	X_B	X_L	X_G	X_C
27.89	0.31	0.53	0.10	0.06	28.42	0.21	0.63	0.13	0.03	27.72	0.17	0.59	0.21	0.03
31.61	0.30	0.53	0.10	0.06	31.43	0.20	0.64	0.14	0.02	31.81	0.16	0.59	0.21	0.05
38.05	0.40	0.44	0.14	0.03	37.36	0.19	0.63	0.13	0.04	38.62	0.14	0.59	0.21	0.07
41.55	0.38	0.44	0.13	0.05	40.82	0.18	0.61	0.14	0.07	41.96	0.14	0.60	0.24	0.02
47.72	0.37	0.45	0.13	0.05	46.46	0.16	0.63	0.15	0.06	48.67	0.12	0.60	0.23	0.05
54.41	0.23	0.58	0.14	0.05	54.42	0.15	0.63	0.16	0.07	51.33	0.11	0.60	0.25	0.04
61.01	0.2.1	0.59	0.14	0.06	62.00	0.15	0.62	0.20	0.04	61.38	0.11	0.59	0.27	0.03
72.80	0.19	0.60	0.15	0.06	73.73	0.11	0.66	0.18	0.05	73.00	0.09	0.59	0.29	0.03
106.21	0.18	0.62	0.17	0.03	109.30	0.12	0.61	0.23	0.04	103.67	0.07	0.56	0.32	0.04

The regressions for the two proposed mechanisms (Figures 3, 4) were performed at the three temperatures worked experimentally. By comparison of the objective function, and also by the linearization of the velocity parameters obtained from the reaction, it was possible to determine which mechanism and what the working temperature are more appropriate to the thermal cracking of waste cooking oil process. The best results were for the mechanism 2, composed of four lumps and three parameters. It may notice some outliers, but all maintained the expected trend.

By comparing the numerical and experimental values, it can make the assessment of regression (Figure 5). The closer to the diagonal axis, the closer are the values calculated from experimental. It can be seen there through so that the dots were reasonably close to each other for all the fractions (lumps), which means that the regression was satisfactory, and that models represent positively the reaction studied, particularly the mechanism 2.

With the data obtained in the regression through the Arrhenius equation linearized gave the parameters that lead to the calculation of the activation energy.

$$\ln k = -\frac{Ea}{R}\left(\frac{1}{T}\right) + \ln k_o \quad (3)$$

From the regression were obtained data shown in Table 2. The objective function value was the lowest among all the studied mechanisms, especially in mechanism 2 to 525 °C. For all the studied mechanisms, there were points above and below the curve obtained numerically, but all the points kept the expected trend. Model 2 was best represented the process.

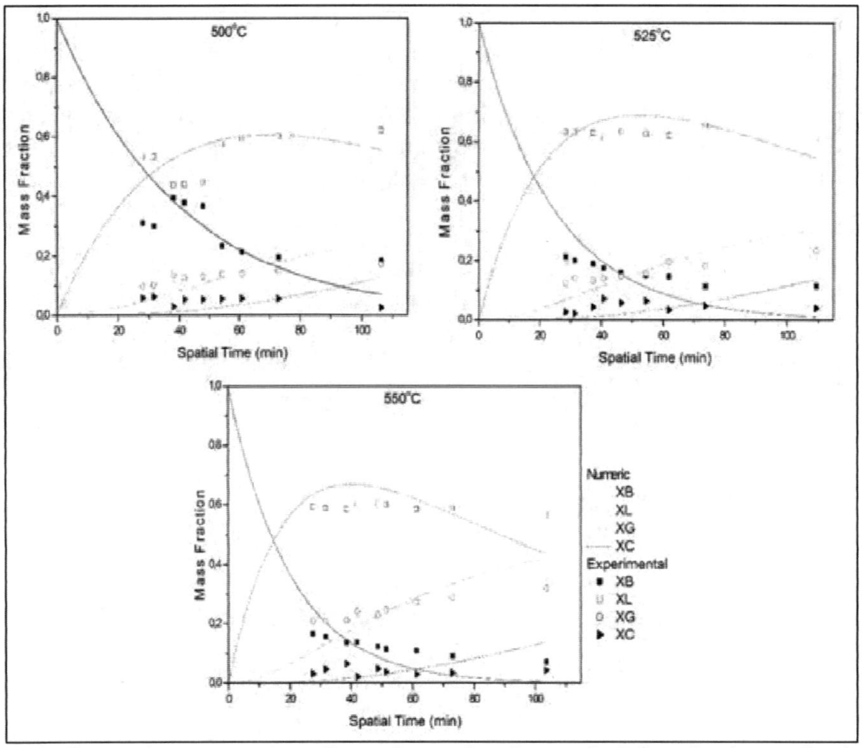

Figure 3. Comparison regression of experimental and numerical data for the mechanism 1.

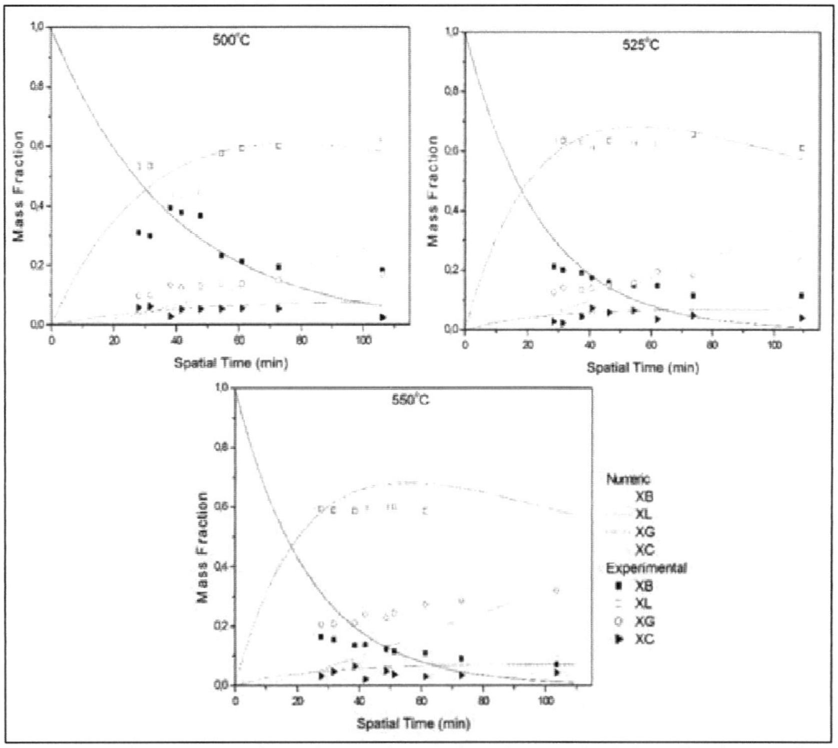

Figure 4. Comparison regression of experimental and numerical data for the mechanism 2.

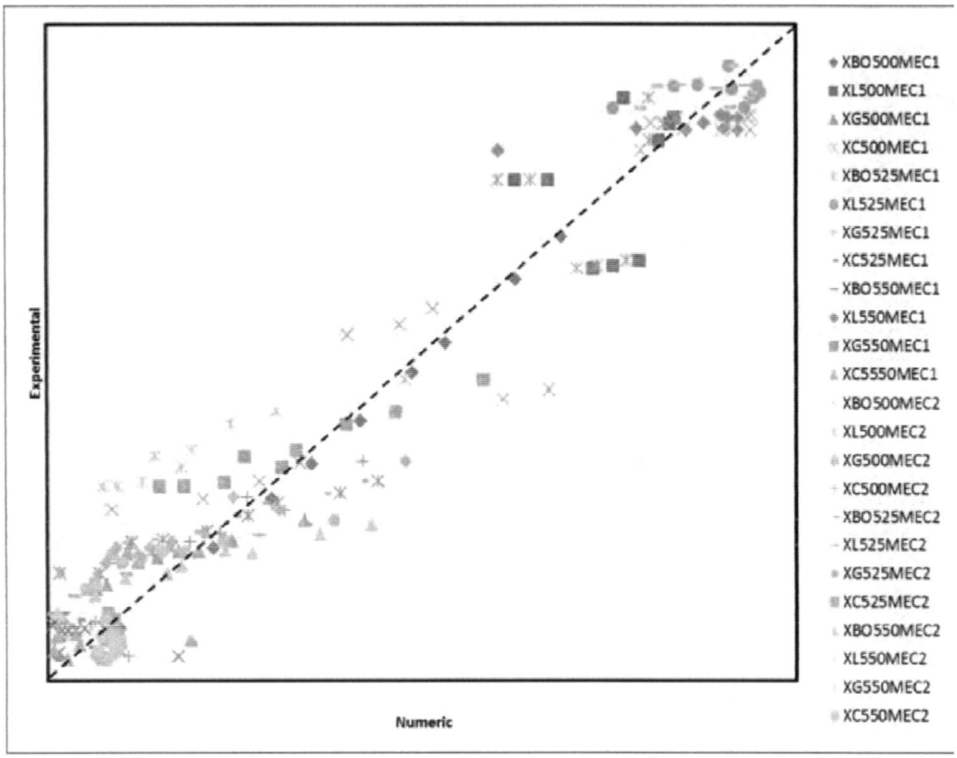

Figure 5. Comparison between experimental and numeric values.

Table 2. Data obtained from the Complex Method regression

Constant rate	Temperature (°C)					
	500		525		550	
	1	2	1	2	1	2
k_1	2,51E-02	2,37E-02	4,09E-02	3,89E-02	4,56E-02	4,48E-02
k_2	7,13E-03	2,19E-03	7,61E-03	3,01E-03	1,01E-02	3,16E-03
k_3	9,87E-03	5,29E-03	1,20E-02	5,49E-03	4,99E-02	8,40E-03
Objective Function	**1,63E-01**	**1,49E-01**	**3,26E-01**	**8,45E-02**	**6,37E-01**	**1,08E-01**
Iterations n°	170	207	119	181	124	177

With these data it was possible to apply Eq. 3 to obtain the values of the activation energies for the first order reaction mechanisms 1 and 2. Results obtained are shown in Table 3. The R^2 values show that linearization was satisfactory, thus confirming again that the proposed kinetic models represent well studied process, in particular mechanism 2.

Table 3. Parameters of the Arrhenius equation for the First order reaction order.

Pre-exponential factor (min^{-1})			R^2		AE (J/mol)	
	1	2	1	2	1	2
k_{01}	3,50E+05	2,90E+01	0,8985	0,9218	7,31E+04	2,93E+04
k_{02}	3,29E+00	8,00E-01	0,8664	0,8734	1,57E+04	1,48E+04
k_{03}	1,99E+01	7,47E-01	0,8183	0,8877	2,76E+04	2,08E+04

Conclusion

The comparison between the five proposed mechanisms identified that the mechanism 2 of 4 lumps and five parameters was best represented the reaction of thermal cracking of waste cooking oil, as generated the smallest value of the objective function, and also had the closest relationship between experimental and numerical values, showing that this type of kinetic approach satisfactorily represent this type of reaction.

REFERENCES

[1] V. R. Wiggers, H. F. Meier, A. Wisniewski Jr., A. A. Barros e M. R. Wolf Maciel, "Biofuels from continuous fast pyrolysis of soybean oil: A pilot plant study," *Bioresource Technology,* vol. 100, pp. 6570-6577, 2009.

[2] V. R. Wiggers, A. Wisniewski Jr., L. A. Madureira, A. A. Barros e H. F. Meier, "Biofuels from waste fish oil pyrolysis: Continuous production in a pilot plant," *Fuel,* vol. 88, pp. 2135-2141, 2009.

[3] A. Wisniewski Jr., V. R. Wiggers, E. L. Simionatto, H. F. Meier, A. A. Barros e L. A. Madureira, "Biofuels from waste fish oil pyrolysis: Chemical composition," *Fuel,* vol. 89, pp. 563-568, 2010.

[4] V. B. Camilis, D. Riva, E. L. Simionatto, V. R. Wiggers, L. Ender, H. F. Meier e A. A. Barros, "Thermocatalytic cracking of the mixture of used frying oil – textile stamping sludge for the production of oil with low acidity index," *Química Nova,* vol. 35(4), pp. 677-682, 2012.

[5] V. R. Wiggers, G. R. Zonta, A. P. França, D. R. Scharf, E. L. Simionatto, L. Ender e H. F. Meier, "Challenges associated with choosing operational conditions for triglyceride

thermal cracking aiming to improve biofuel quality," *Fuel,* vol. 107, pp. 601-608, 2013.

[6] K. D. Maher e D. C. Bressler, "Pyrolysis of triglyceride materials for the production of renewable fuels and chemicals," *Bioresource Technology,* vol. 98, pp. 2351-2368, 2007.

[7] X. Meng, C. Xu, J. Gao e L. Li, "Catalytic pyrolysis of heavy oils: 8-lump kinetic model," *Applied Catalysis A: General,* vol. 301, pp. 32-38, 2006.

[8] X. Meng, C. Xu, J. Gao e L. Li, "Seven-lump kinetic model for catalytic pyrolysis of heavy oil," *Catalysis Communications,* vol. 8, pp. 1197-1201, 2007.

[9] F. AlHumaidan, H. M. Lababidi e H. Al-Rabiah, "Thermal cracking kinetics of Kuwaiti vacuum residues in Eureka process," *Fuel,* vol. 3, pp. 923-931, 2013.

[10] J. R. Hernández-Barajas, R. Vázquez-Román e M. G. Félix-Flores, "A comprehensive estimation of kinetic parameters in lumped catalytic cracking reaction models," *Fuel,* vol. 88, pp. 169-178, 2009.

[11] N. Jia, R. G. Moore, S. A. Mehta e M. G. Ursenbach, "Kinetic modeling of thermal cracking reactions," *Fuel,* vol. 88, pp. 1376-1382, 2009.

[12] J. Martinez e J. Ancheyta, "Kinetic model for hydrocracking of heavy oil in a CSTR involving short term catalyst deactivation," *Fuel,* vol. 100, pp. 193-199, 2012.

[13] M. J. Box, "A new method of constrained optimization and a comparison with other methods.," *Comput J.,* vol. 8, pp. 42-52, 1695.

[14] A. K. Sadhukhan, G. Parthapratim e R. K. Saha, "Modelling of pyrolysis of large wood particles," *Bioresource Technology,* vol. 100(1), p. 2009, 3134-3139.

Weight loss and carbon loss kinetics in torrefaction of hardwood (yellow poplar, red oak) and softwood (loblolly pine)

Li Xiao, Sunkyu Park, Hasan Jameel, Steve Kelley

North Carolina State University, Raleigh, NC 27606

The goal of this study is to compare the kinetics of three woody biomass during torrefaction. Loblolly pine, yellow poplar, and red oak were torrefied at five different temperatures: 200oC, 225oC, 250oC, 275oC, 300oC in bench scale tube furnace in the atmosphere of nitrogen. Residence time was chose as 10min, 20min, and 30min. Carbon content of the fuel products were measured and carbon yield was calculated. Accordingly, activation energy, rate constant and exponential factor of mass loss and carbon loss kinetics were calculated and compared with existing kinetics data in literature. It was found that carbon loss rate is always less than mass loss rate, which demonstrated the fact of energy densification. Weight loss and carbon loss rate of hardwood is always slightly larger than that of softwood. Interestingly, there is a good linear relationship between mass yield and carbon yield, as well as between mass loss rate constant and carbon loss rate constant. This observation could be used to utilized to estimate mass yield and carbon yield, as well as the reaction rate during torrefaction from a specific torrefaction condition and wood species during torrefaction.

389254 Minimum Fill-up Time of Compressed Natural Gas Vehicle Tanks

Abstract:

The key challenge of gas fuel based fueling stations, such as those for CNG, is to provide a time efficient and safe fill-up for such vehicles. First, the fill-up process is simulated based on a self-consistent thermodynamic model and a dynamic model of the process. Then, the fill-up process is formulated as a minimum time optimal control problem with the aim to minimize the fill-up time while satisfying all safety constraints. The simulation and optimization predictions lead to fill-up strategies that can improve current practices.

384166 Optimization of Vanadium Redox Flow Battery: Ion-Exchange Membrane Screening and State of Charge Monitoring

Abstract:

Non stable and hardly predictable power of renewable sources demands an effective management on the power transmission level. Engineering development of new local energy storages forming a part of the modern transmission network and tailored to efficient and smart electricity is one of the major challenges.

Redox flow batteries are perspective devices mainly for the stationary storage of electric energy, due to their flexible installation and operation, versatile scalability, reduced self-discharge and high round-trip efficiencies. Among many possible combinations of redox couples, the so called all-vanadium redox flow battery (VRFB) offers the efficient energy storage of flexible capacity, predominantly for stationary applications as load levelling, smart-grids, peak shaving and back-up sources. The vanadium redox flow battery consist of two electrode spaces most often separated by ion-exchange membrane. Carbon felt is commonly used as electrodes / bipolar plates. The electrolytes are stored in external tanks and are pumped in to the battery stack where charging/discharging takes place. The usage of identical electro-active element on both sides of the cell (vanadium ions in four different oxidation states: 2+, 3+, 4+ and 5+) results in long-life operation without irreversible contamination of electrolytes. This concept was proposed by Skyllas-Kazacos [1] several decades ago. However up to this days there are relevant engineering problems preventing the wide utilization of the technology.

This work focus on the optimisation of VRFB in regards to ion-exchange membrane selection and online monitoring of the state of charge (SoC) of the battery.

The ion-exchange membrane (e.g. Nafion) is one of the key components of the VRFB battery stack, it has a direct influence on the power and the efficiency of the system. The reduction of the membrane cost should bring a significant economic advantage for the VRFB against other competitive technologies. Several polymeric ion-exchange membranes were characterized with respect to the conductivity and permeability for vanadium ions across the membrane. The impact of membrane composition, method of its fabrication, thickness and ion-exchange capacity on the measured characteristics was observed. The selected membranes were tested in our laboratory VRFB (efficiency 83% at 56 mA/cm^2 with Nafion N115).

Our state of the charge monitoring system is based on the combination of three techniques: conductometry, open cell voltage measurements and spectrophotometry. The combination of these methods enables for the on-line SoC monitoring of the individual VRFB electrolytes without the necessity of current load interruptions which is the desired property for the VRFB management. At the same time, the monitoring system enables a closer insight to the key phenomena as vanadium and proton diffusion through the membrane, hydrogen evolution etc.

[1] M. Skyllas-Kazacos, M. Rychcik, R. G. Robins, A. G. Fane, and M. A. Green, *Journal of The Electrochemical Society* **1986**, 133, 1057.

379291 A Zinc-Iron Redox Flow Battery for High-Performance and Low-Cost Energy Storage

Abstract:

Redox flow batteries are considered the most promising scalable technology for the storage of renewable wind and solar electricity. We have developed a new double-membrane triple-electrolyte concept that offers unprecedented freedom in choosing redox pairs and the pH of the electrolytes [1-3]. We have demonstrated a Zn-Ce flow battery that offers a 3.1 V cell voltage, the highest for an aqueous flow battery, and a S-Fe flow battery that is very inexpensive because S and Fe are the 5th and 1st most produced element, respectively. In this presentation, we show a new Zn-Fe flow battery that can deliver a peak power density of 647 mW/cm^2 and cycle at a current density of 100 mA/cm^2 with a round trip efficiency of 70%. Our state-of-the-art cost model shows that this Zn-Fe flow battery can meet the cost target of $100/kWh for commercialization.

[1] Shuang Gu, Ke Gong, Emily Z. Yan and Yushan Yan, A multiple ion-exchange membrane design for redox flow batteries, ***Energy Environ. Sci.***, 2014, **DOI:** 10.1039/C4EE00165F

[2] Y.S. Yan, S. Gu, K. Gong, Double-membrane triple-electrolyte redox flow battery design, Application No.: 13/918444, Publication date: 01/02/2014

[3] Y.S. Yan, S. Gu, K. Gong, Multiple-membrane multiple-electrolyte redox flow battery design, Application No.: 13/918452, Publication date: 01/02/2014

377740 Identification of Advanced Electrolytes for Non-Aqueous Redox Flow Batteries Via High-Throughput Screening Technique

Abstract:

2,5-di-tert-butyl-1,4-bis(2-methoxyethoxy)benzene (DBBB) has been identified as a promising active species for non-aqueous redox flow batteries. However, the development and optimization of suitable electrolytes is still required to realize this promise. Given the breadth and diversity of non-aqueous electrolytes, combinatorial screening using automated platforms offers an efficient means of mapping out the design spaces.

Employing an automated electrolyte synthesis and characterization system, we study electrolytes consisting of DBBB, various alkali ion salts ($LiBF_4$, $LiPF_6$, LiTFSI, LiTf, $NaBF_4$, and $NaPF_6$) in carbonate-based (EC, PC, DEC, and DMC) or ether-based (DME, DOL, TEGDME, and DEGDME) solvents, with an overarching goal of improving DBBB solubility and ionic conductivity. Via this high-throughput approach, binary and ternary mixtures of linear and cyclic carbonates were found to improve both DBBB solubility and electrolyte conductivity, a marked improvement over the previously base case scenario ($LiBF_4$/PC). A further improvement of the solubility of DBBB and the conductivity of the electrolyte was obtained using DME as the solvent. Moreover, in general, the addition of LiTFSI as a salt (or co-salt) was shown to improve DBBB solubility in electrolytes and the ionic conductivity of solutions.

Furthermore, based on the best combination in the corresponding carbonate- and ether- based electrolyte, the electrochemical behavior of DBBB at concentrations ranging from 5 mM (infinitely diluted solution) to the solubility limit (high-concentration solution) was systematically investigated using both 3 mm glassy carbon electrode and 11 μm carbon fiber microelectrode. The trend of key electrochemical parameters obtained here bridges the gap between electrochemical information of redox species at low concentration, which has been widely used to screen promising energy materials, and that at high concentration, which is directly relevant to the battery performance.

Aiming at the development of advanced electrolyte for non-aqueous redox flow batteries, this work demonstrates the archetype of the combination of the upstream screening work employing the automated, high throughput platform and the downstream validating work utilizing the electrochemical tools.

Metal Sulfides As Hydrogen Electrocatalysts for H_2-Br_2 Flow Batteries

Electrochemical energy storage will be useful for large scale applications only if cost-effective systems can be developed which match the efficiency and power requirements needed. For large-scale systems a flow battery using H_2 and Br_2 is one of the most efficient and potentially cost-effective technologies known.[1] Advantages include lower kinetic losses than other flow battery systems (leading to higher efficiencies), high power density, high energy density and high reliability.[2–4] The H_2-Br_2 cell consists of a hydrogen electrode for the half reaction:

$$H_2 \leftrightarrow 2H^+ + 2\,e^-$$

separated from the bromine electrode by a proton-exchange membrane. The bromine electrode half reaction is efficient and fast.:

$$Br_2 + 2e^- \leftrightarrow 2Br^-$$

The major present limitation of the H_2-Br_2 flow battery is the stability of the electrocatalysts in the corrosive HBr/Br_2 electrolyte. Corrosion may occur through formation of soluble compounds that dissolve into the electrolyte, as well as poisoning by bromide that blocks active surface sites. Although the hydrogen electrocatalyst is protected from bromide and bromine by the proton-exchange membrane in theory, in reality the bromide and bromine will crossover and affect the activity of the hydrogen electrocatalyst.

Here we will discuss our experimental methods and results of identifying and testing a wide variety of electrocatalysts and discuss the fundamentals of why certain materials are active or inactive for hydrogen evolution and oxidation. To test a wide range of materials, we have conducted high-throughput screening of metal sulfides,[5] using measurable experimental descriptors (gas evolution) to identify active electrocatalysts. The active materials were selected for further characterization, and testing in a H_2-Br_2 flow battery system. To understand the activity of certain materials, we performed gas-phase measurements of surface reactions, such as H_2-D_2 exchange and then coupled these with theoretical *ab initio* calculations of hydrogen's interaction with the catalyst surfaces[6] to identify the rate limiting steps for the reactions of interest (electrochemical hydrogen evolution and oxidation). We also measured the effect on activity following selective poisoning of certain sites, both experimentally and using Density Functional theory, to distinguish between different phases and sites of the catalyst. This led to understanding of what surface properties of materials result in an active electrocatalyst. For the best performing electrocatalysts, we also measured electrochemical conductivity and performed traditional spectroscopic and material characterization to identify limitations in the flow cell system, in hopes of improving the overall performance of the flow battery. The practical outlook based on these technological advancements will be discussed for an H_2-Br_2 flow battery system for grid-scale energy storage applications.

(1) Schuetz, G. H. *Int. J. Hydrogen Energy* **1977**, *1*, 379.

(2) Glass, W.; Boyle, G. H. In *Fuel Cell Systems*; 1969; pp. 203–220.

(3) Barno, G. G.; Frank, S. N.; Teherani, T. H.; Weedon, L. D. *J. Electrochem. Soc.* **1984**, *131*, 1973.

(4) Cho, K. T.; Ridgway, P.; Weber, A. Z.; Haussener, S.; Battaglia, V.; Srinivasan, V. *J. Electrochem. Soc.* **2012**, *159*, A1806.

(5) Ivanovskaya, A.; Singh, N.; Liu, R.-F.; Kreutzer, H.; Baltrusaitis, J.; Nguyen, T. Van; Metiu, H.; McFarland, E. *Langmuir* **2013**, *29*, 480.

(6) Singh, N.; Upham, D. C.; Metiu, H.; McFarland, E. W. *J. Electrochem. Soc.* **2013**, *160*, A1902.

Developing High Performance Zinc Alkaline Primary Battery Systems with Novel Anion Permeable Exchange Membrane Separators

Hooman Hosseini, Donald Dornbusch and Galen Suppes
Chemical Engineering, University of Missouri- Columbia, Columbia MO, USA

Thursday, November 20, 2014: 10:10 AM
M304 (Marriott Marquis Atlanta)

AIChE

1

Research Goal

- Developing an effective method for coating filters with ion exchange media to enhance the surface and solid phase migration mechanisms for use as separators in batteries.
- Obtaining experimental data on benefits of these coatings on separators.

AIChE

2

Materials:

Anode Composition

- Zinc Dust (< 10μm)
- Conductive Acetylene Black (30 to 40 nm)
- Inhibitor
- Binder

Separator

- Fast Flow Rate Whatman Filter Papers
- Pallflex Membrane Filters

Electrolyte

- Aqueous KOH 9 & 6 molar

Cathode Composition

- EMD
- Conductive Acetylene Black (30 to 40 nm)
- Binder

AIChE

3

Performance (low Drain- 17.5 mA/cm^2)

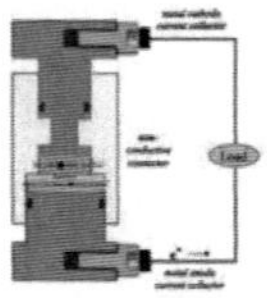

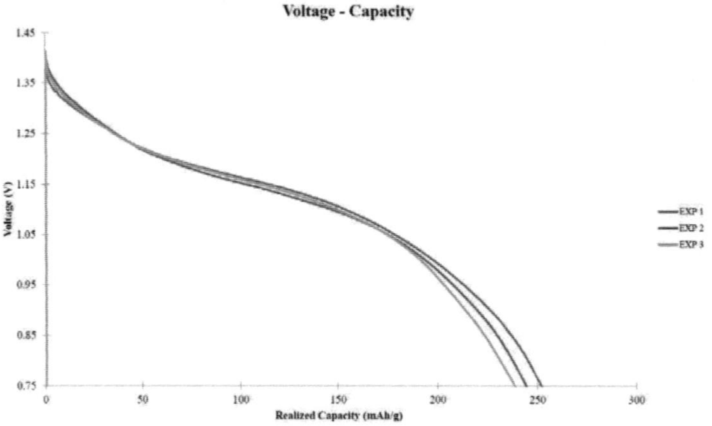

AIChE

4

Diffusion Overpotential Phenomena

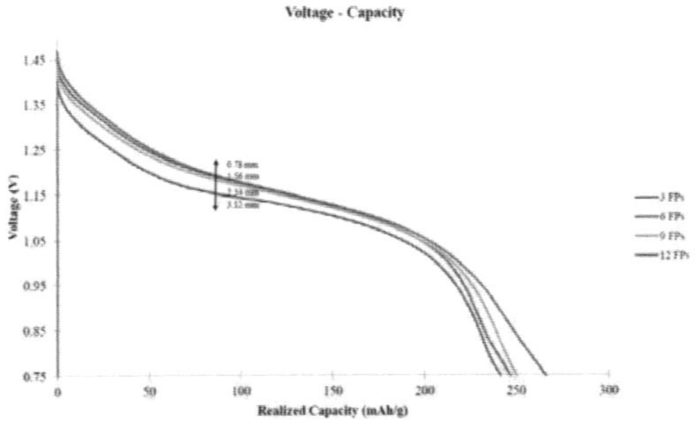

low Drain - 17.5 mA/cm²

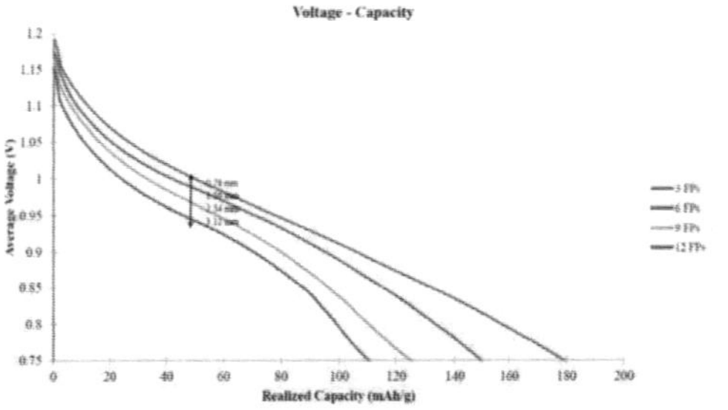

High Drain – 93.5 mA/cm²

AIChE

5

Surface Modification – PE Synthesis

- ❑ **Physical Grafting**
  Radiation, generating the radicals, grafting the quaternary ammonia
  (Varco, Slade and coworkers APE based on perfluorocarbon membranes)
  Unable to produce ionomer solution

- ❑ **Chemical Grafting**
  Chemical Reaction, Directly grafting the quaternary ammonia groups
  (Pan et al, Polysulfone based backbones as a linear chain)
  Strong and flexible membrane, efficient method, high mechanical strength

- ❑ **Polymerization**
  Polymerization of QA containing monomers (Coates and Co-workers)
  Very high IEC, No testing on Mechanical and Thermal Stability

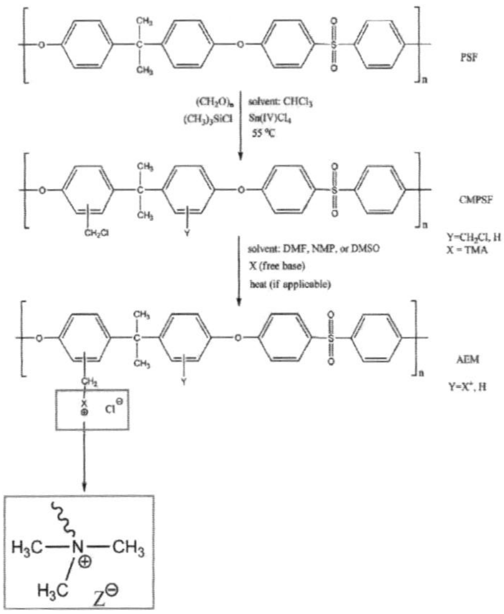

AIChE

ACCOUNTS OF CHEMICAL RESEARCH ,Vol. 45, No. 3 2012 473–481

6

CMPSF

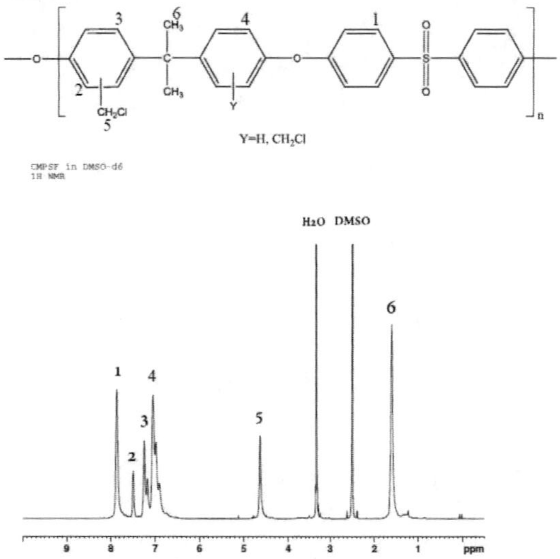

QAPSF

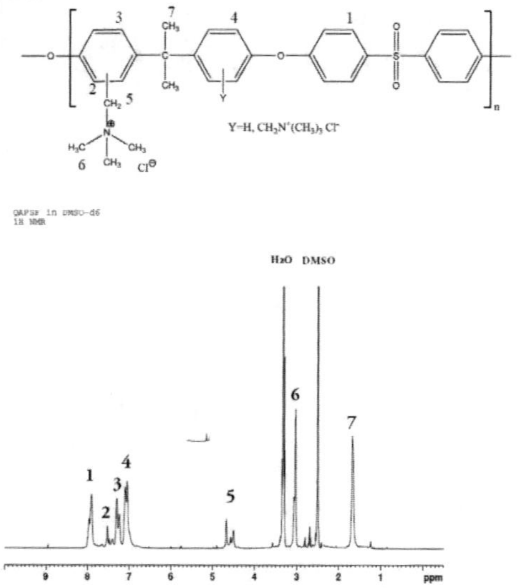

AIChE

7

Process

1. Polyelectrolyte Synthesis

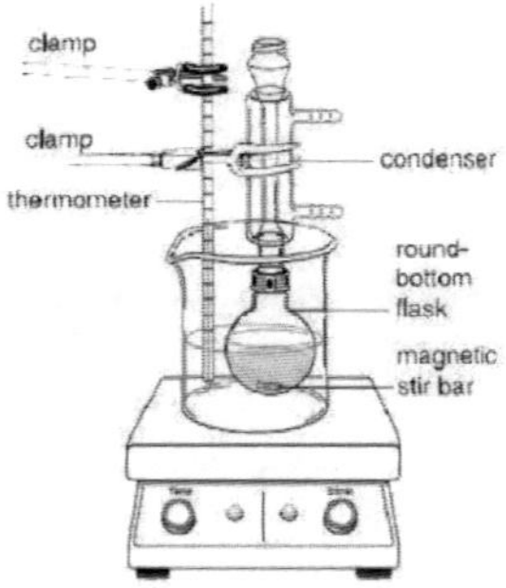

2. Dilutions

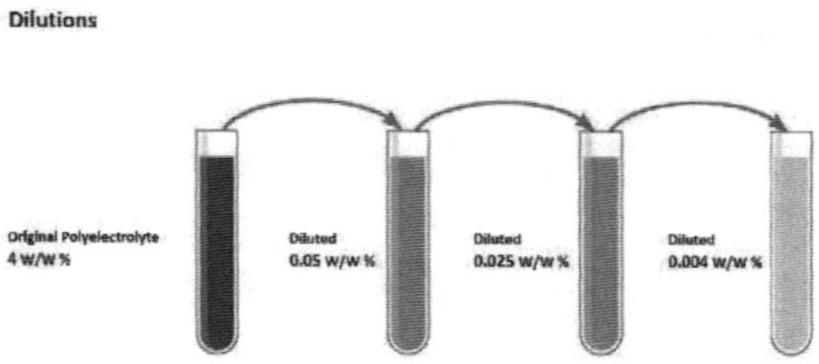

AIChE

9

Process

3. Dipping the Filter Papers in Diluted Polyelectrolytes

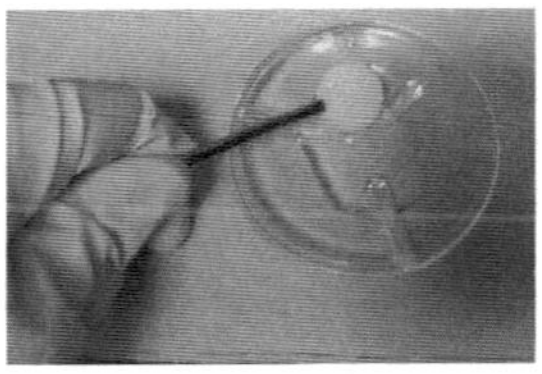

4.Vaccum Oven for Solvent Vaporizing

5. Dipping the Coated Filter Papers in Electrolytes

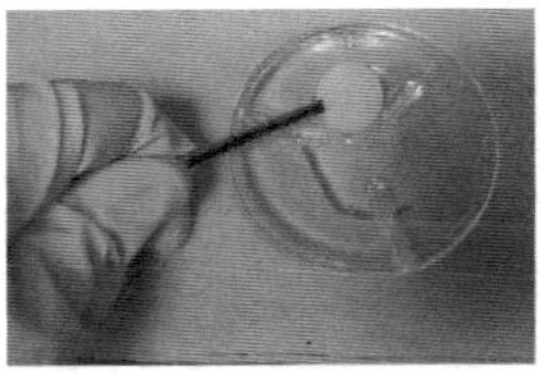

6. Testing the Coated Filter Papers in Diffusion Based & Convetive Flow Batteries

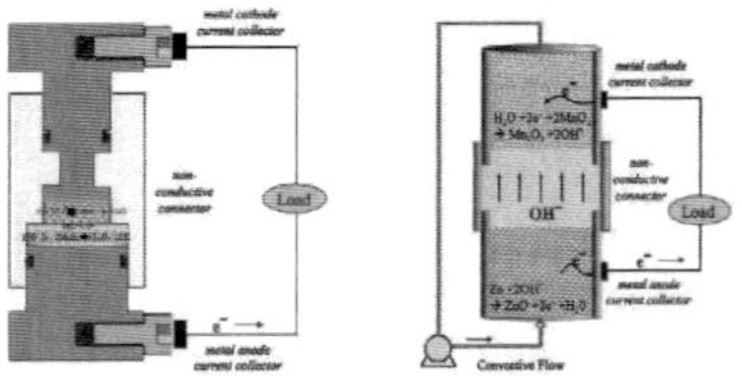

AIChE

10

Mass Gain Calibrations on filter paper fibers

Substrate	Concentration of PE	Mass ratio of Absorbed QAPSF on Substrate
Pallflex Membrane Filters	4 % QAPSF	79 %
Pallflex Membrane Filter	0.05 % QAPSF	3.9 %
Pallflex Membrane Filter	0.004 % QAPSF	0.3 %
Whatman Filter Paper GF-D	4 % QAPSF	16 %
Whatman Filter Paper GF-D	0.05 % QAPSF	0.8 %
Whatmsn Filter Paper GF-D	0.004 % QAPSF	0.064 %

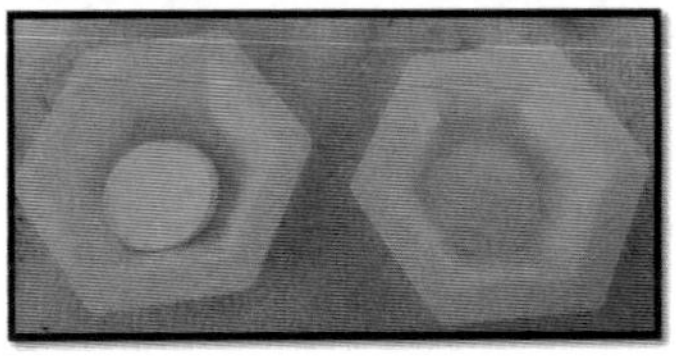

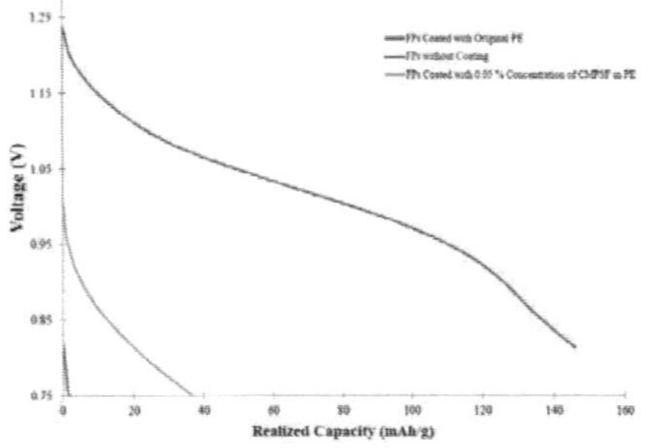

High Drain – 80 mA/cm²

11

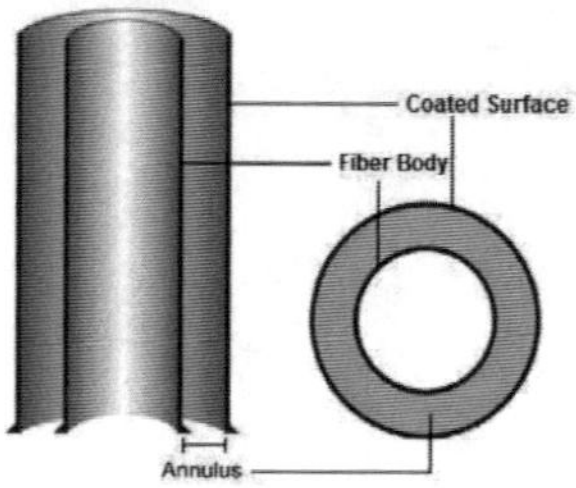

$$V_{cm}/V_{Fiber} = 0.2\ \%$$
$$\boldsymbol{\rho}_{QMPSF} = 1.5$$
Mass Gain= 0.3 %
10 nm layer of QAPSF on Pallfelx

AIChE

Improvements: Controls (Uncoated FPs)

Normalizing

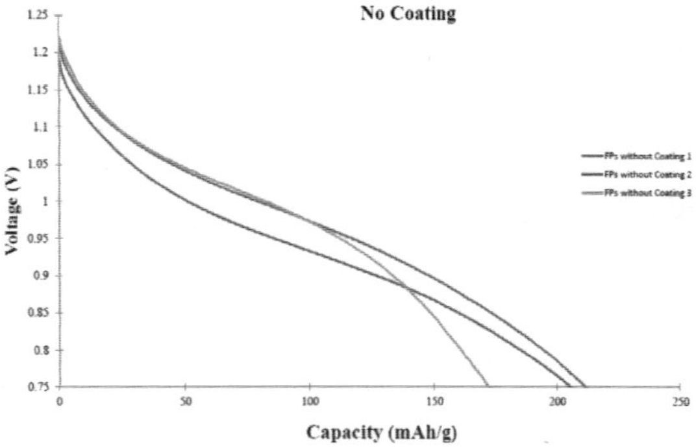

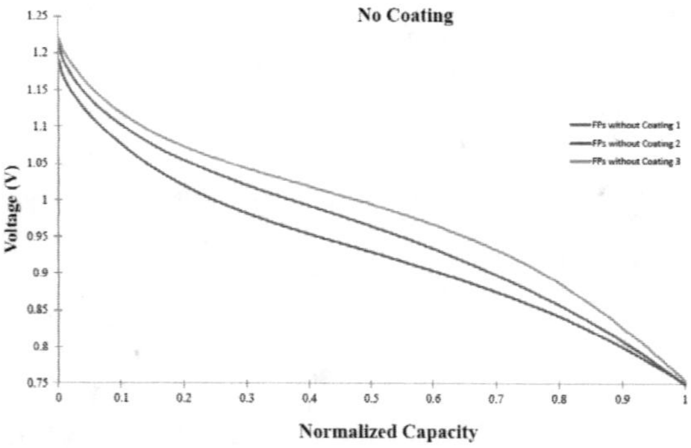

AIChE

High Drainage – 80 mA/cm²

Improvements: Coated Filter Papers with QAPSF-PE

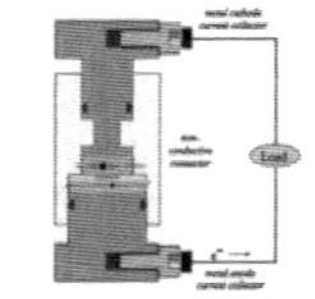

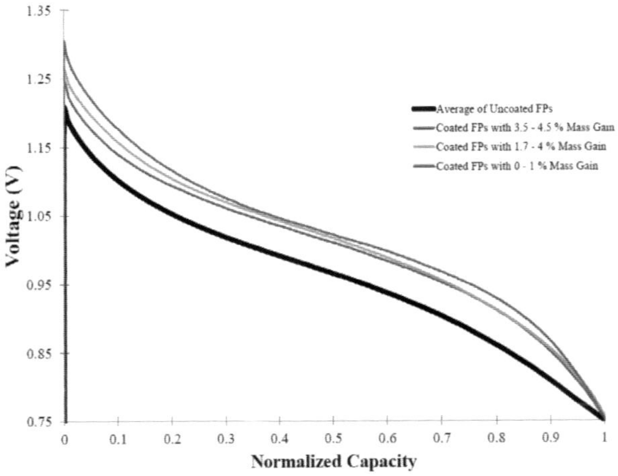

Percentage of Mass Gain on Substrate (%)	Capacity (mAh/g)	Utilization (%)
0	210	69
< 1	260	85
1 - 3	203	66
3 - 5	194	63

AIChE

14

Convective Flow

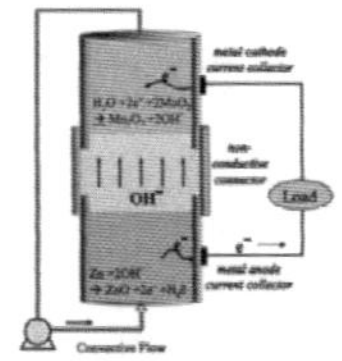

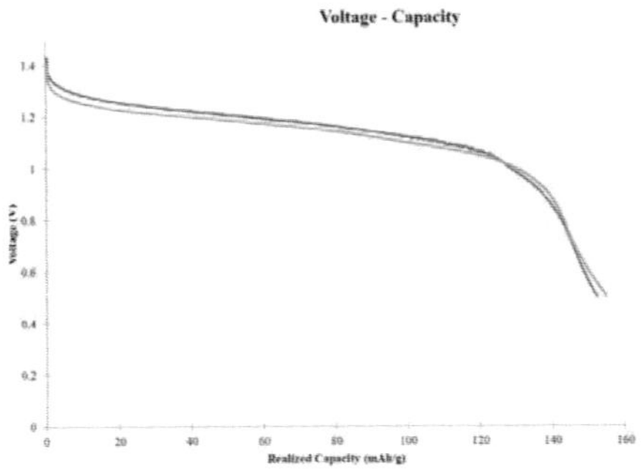

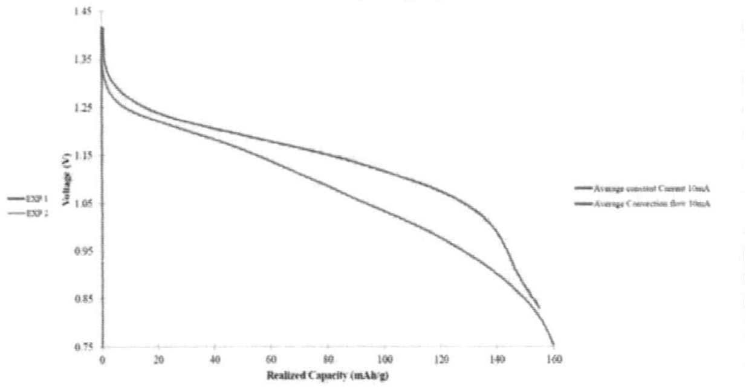

low Drain- 7.5 mA/cm^2

AIChE

15

Mechanisms of Ion Conduction Improvements

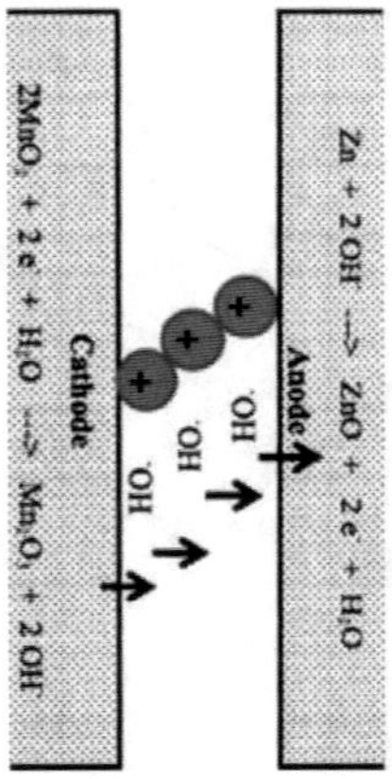

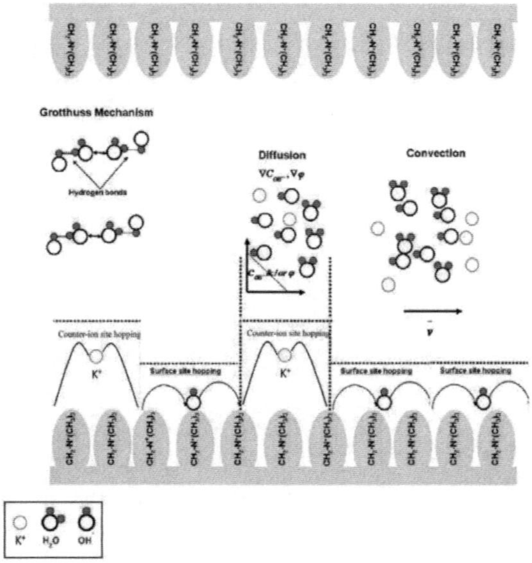

16

Surface Conduction Modeling

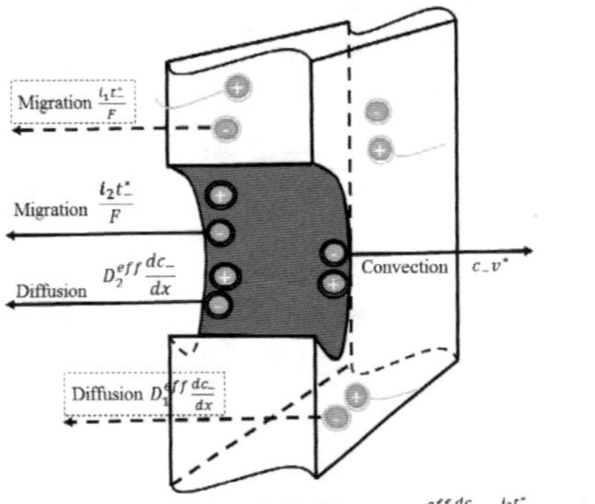

Mass Flux(liquid): $N_- = -D_2^{eff}\frac{dc_-}{dx} - \frac{i_2 t_-^*}{F} + c_- v^*$

Mass Flux(solid): $N_- = -D_1^{eff}\frac{dc_-}{dx} - \frac{i_1 t_-^*}{F}$

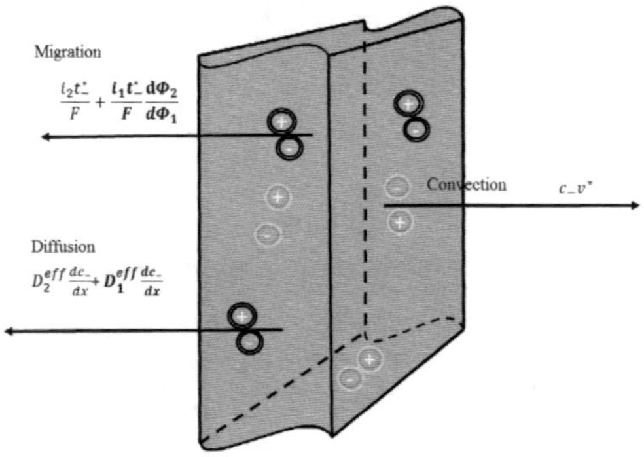

$N_- = -(D_2^{eff}\frac{dc_-}{dx} + \boldsymbol{D_1^{eff}\frac{dc_-}{dx}}) - (\frac{i_2 t_-^*}{F} + \frac{\boldsymbol{i_1 t_-^* d\Phi_2}}{\boldsymbol{F\ d\Phi_1}}) + c_- v^*$

Impact of Separator's Solid-Phase Ion Conductivity Parameter
on Convection Battery Performance and Modeling, Ramsey Hilton et all, AIChE Journal, Vol. 60, No. 11

17

Surface Conduction Modeling

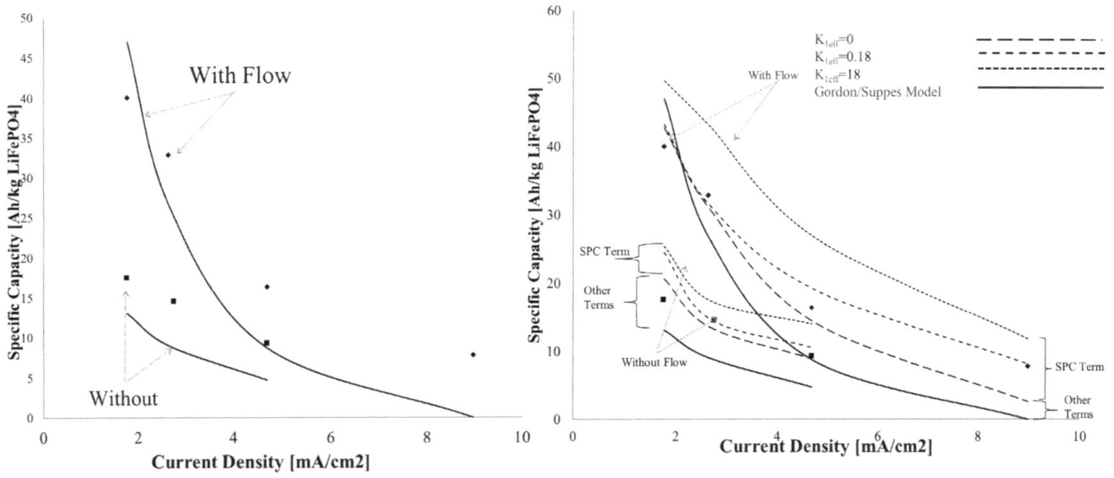

AIChE

18

Impact of Separator's Solid-Phase Ion Conductivity Parameter on Convection Battery Performance and Modeling, Ramsey Hilton et all, AIChE Journal, Vol. 60, No. 11

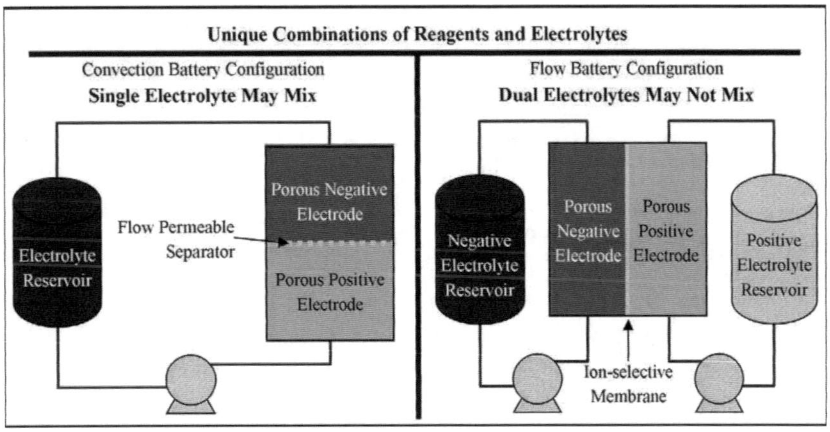

AIChE

Gordon, M. and G. Suppes, Convection battery—modeling, insight, and review. AIChE Journal, 2013. 59(8): p. 2833-2842.

19

Summary

- Filter papers have been successfully coated with AEPE to examine the surface ion conduction transport.
- Process required much lower loadings than initially anticipated.
- Preliminary data appears to validate both the method of applying coating, a method to prepare separators with ionic exchange media coatings for use with convection batteries, and hypothesized improvements in performance.
- Approach should be applicable to a range of materials and may provide effective screening tool in evaluating different materials for use as separators.
- These types of separators create new and improved opportunities for separator development since mechanical properties (filter) of separator can be decoupled from ionic properties (coating).

AIChE

20

References

- Highly Flexible, Printed Alkaline Batteries Based on Mesh-Embedded Electrodes,. Abhinav M. Gaikwad , Gregory L. Whiting , Daniel A. Steingart , and Ana Claudia Arias, Adv. Mater 2011, 23, 3251–3255
- Fabrication of High-Performance Flexible Alkaline Batteries by Implementing Multiwalled Carbon Nanotubes and Copolymer Separator , Zhiqian Wang , Zheqiong Wu , Natalia Bramnik , and Somenath Mitra, Adv. Mater. 2014, 26, 970–976
- Impact of electrode separator on performance of a zinc/alkaline/ manganese dioxide packed-bed electrode flow battery, Bryan D. Sawyer, Galen J. Suppes , Michael J. Gordon, Michael G. Heidlage, J Appl Electrochem (2011) 41:543–550
- Ion Transport by Nanochannels in Ion-Containing Aromatic Copolymers, Nanwen Li and Michael D. Guiver, Macromolecules 2014, 47, 2175−2198
- High-Energy Density Flow Battery Validation, Galen J. Suppes et all, AIChE Journal, Vol. 57, No. 7
- Impact of Separator's Solid-Phase Ion Conductivity Parameter on Convection Battery Performance and Modeling, Ramsey Hilton et all, AIChE Journal, Vol. 60, No. 11
- Designing Advanced Alkaline Polymer Electrolytes for Fuel Cell Applications, Accounts of Chemical, Jing Pan, Chen Chen, Lin Zhuang and Juntao Luresearch ,Vol. 45, No.3 2012 473–481

AIChE

21

Acknowledgment

- Professor Galen J Suppes, University of Missouri Columbia
- "G.J.Suppes" Research Group Members: Donald Dornbusch & Sam Lohman
- NSF:Project funding provided by National Science Foundation award #123350
- TRONOX Company

AIChE

22

Thanks for Attention...

AIChE

22

383788 Characterization of Glycerol-Diesel Emulsions Fuels

Abstract:

Diesel fuel emulsions have demonstrated reductions of unwanted combustion emissions. Glycerol, a renewable and abundant resource, is an attractive fuel component since it may lower energy costs and reduce unwanted engine emissions. In this presentation, various glycerol-diesel emulsions are prepared using an ultrasonic processing technique and resultant fuel properties are presented. The fuels typically have higher viscosity (5-20 cps) and density (0.9-0.98 g/mL) and lower fuel energy density (36-42 MJ/kg). The high glycerol oxygen content (52 wt.%) reduces stoichiometric A/F ratio and flame temperatures. Surface analytical techniques and long-term stability evaluations are used to identify optimal surfactant composition for extended shelf-life. A narrow droplet size distribution with mean of approximately 3.3 microns is achieved, which is found to be broadened with the addition of water in the glycerol up to 50/50 vol.%. Glycerol-diesel emulsions prepared at 10 and 20 vol.% glycerol-phase resul in reduced NOx and PM emissions of 5-15% and 25-50%, respectively in a naturally-asiprated signle-cylinder diesel engine. Thermal efficiency improvements are observed at high loads .

384836 Surrogates for Fuel Property Modeling in Alternative Fuel Blends

Abstract:

Alternatives to petroleum based fuels are essential for the future transportation industry. For a fuel to be a viable replacement for a petroleum fuel it must meet a number of specifications. The prediction of fuel properties for blends of alternative and petroleum fuels is becoming increasingly important. This work considers the compatibility of potential alternative fuels with petroleum diesel and jet fuels, with a focus on property prediction and the development of surrogates for the prediction of multiple properties. Important cold flow properties addressed in this presentation include cloud point and pour point temperatures. Important engine performance-related properties addressed include the cetane number, boiling curve, and liquid phase compressibility. Modeling is performed using surrogates which represent the overall classes and distribution of components found in these fuels.

The temperature at which solids first form in a fuel mixture is referred to as the cloud point temperature. Cloud point measurements are presented for additives in two test diesel blends and a jet fuel. The behavior of potential fuel additives is found to relate more to the oxygen containing functional group rather than extent of branching or size of the alkane groups. Additive functionalities studied include, diesters, esters, ketones, alcohols, ethers, and alkanes. Long chain ester components were also tested. Experimental results are interpreted with thermodynamic models using fuel surrogates.

The cetane number of a fuel is a measure of the ignition delay after injection in a compression-ignition engine. Engines are tuned to have peak performance using fuels within a certain range of cetane numbers. Measured cetane numbers of blended fuels are modeled using composition-specific blending numbers based on the functionality class of the fuel constituents. This presentation will include brief comments on the compressibility of fuels and fuel blends.

ANALYSIS OF PYROLYSIS OIL DURING COAL-BIOMASS CO-PYROLYSIS

Gifty Osei-Prempeh, Asad Davari, James Ingles
West Virginia University Institute of Technology, Montgomery, WV 25136

Bio-oil from biomass pyrolysis shows a higher percentage of carboxylic acids. The potential exist to utilize coal and catalysts to enhance the bio-oil by blending coal and biomass before pyrolysis. A micro-pyrolyzer connected to a GCMS was used to collect data on the evolved pyrolysis oil during the coal-biomass co-pyrolysis process. Two types of biomass (hardwood and softwood sawdust) were blended with West Virginia low sulfur bituminous coal and pyrolyzed at different temperatures in the micro-pyrolyzer. The organics evolved during the pyrolysis is analyzed by the GCMS and the data provided the necessary information for evaluation of the possible chemical processes taking place during the pyrolysis. The analysis of the data by looking at the effect of the pyrolysis conditions and the effect of the ratio of coal on the organic species evolved will be presented.

ANALYSIS OF THE ASHES OBTAINED IN COMBUSTION PROCESSES OF LIGNOCELLULOSIC WASTES AS RAW MATERIALS FOR COMPOSITES

R. A Tolosa[1, 2, 6], N.P Arias[1, 5], O.Giraldo[3, 5], C.A. Cardona[4, 6*]

[1]Departamento de Ingeniería Eléctrica, Electrónica y Computación.
[2]Escuela de Arquitectura y Urbanismo.
[3]Departamento de Física y Química.
[4]Departamento de Ingeniería Química
[5]Laboratorio de Materiales Nanoestructurados y Funcionales.
[6]Instituto de Biotecnología y Agroindustria.
Universidad Nacional de Colombia Sede Manizales.
Cra 27 N° 64-60
*corresponding author cacardonaal@unal.edu.co

Plants contribute in an important way to food sustainability; production of medicaments, oils, biofuels, and textiles. Moreover, it must be considered that once the plant has been used, a number of residues are generated, which although mostly are biodegradable, they cause contamination and produce harmful effects to the environment. Furthermore the agroindustrial production maximizes the plant productivity, but at the same time it generates a larger amount of residues. In this context, not only it must be think which usages give to the plant, but also to the residues, especially when they are produced in large volumes. Residual biomass of the agroindustrial process is a highly polluting source, due its degradation contributes the output of greenhouse gases, pollution of air, in the case of being disposed to open sky, it occupies lands agriculturally active, and when it is deposited in landfills, it degrades the soil as well as the hydric sources. In this line, generate proposals that allow its use, implementation, and insertion to the productive chain is really attractive, and the utilization of these residues as raw material that conforms a new construction material results a viable alternative. One of the processes that can overcome this disadvantage is produce ashes from agroindustrial wastes to use them in the construction material as additive in order to improve the mechanical properties especially the compressive strength and durability of cement matrix.

The biomass calcination offers as added value the generation of energy, furthermore, this process can achieve the considerate reduction of biomass volume, and if the technique applied to generate energy is suitable, the resulting ash will have space as feedstock for the production of materials in the construction sector. The combustion generates two types of ashes, flying and bottom boiler It is worth to indicate that the ASTM C-618 framed the fly ash as the one coming from coal combustion, and only making clear what is consider as fly ash does not include residues of incineration of municipal or industrial waste, ASTM C618-08a, furthermore, it does not consider other biomasses. This is why; this definition has been revalued by several authors on the understanding that the fine-grained ash or fly ash, is not necessarily exclusive of coal calcination, and this type of ash is produced by different types of biomass, therefore the denomination given to the term should be more extensive. On the other hand, is has been talking about the term “addition”, which applies to pozzolanic materials that are used as raw material aggregated to the matrix, with the aim to improve it.

A pozzolan, as it is defined by the ASTM-618, is a siliceous material or alumino-siliceous that by itself possesses little or no cementitious value, but when it is finely divided and in the presence of water, it chemically reacts with calcium hydroxide at room temperature forming compounds with cementitious properties. Similarly, an addition may be active if it has pozzolanic capacity, if it has not, in which case it acts as filler within the matrix

In the present work different agroindustrial Colombian wastes from rice, sugar cane and palm oil industries were systematically submitted to combustion process between 400-800°C for 6h at heat velocities of 1°C/min and 10°C/ min, with the aim to produce silica as pozzolanic material for use it as addition in concrete. The ashes were characterized by X ray diffraction (XRD) and the amount of these ashes (wt%) were measured. The XRD pattern shown in Figures 1-6 show mainly crystallographic phases associated with amorphous silica, cristoballite and quartz. Amorphous silica from combustion process of rice husk were produced in all temperature range studied meanwhile the production of amorphous silica is limited to 600°C for palm oil raquis and 700°C for sugar cane bagasse. When the ashes were washed with DDW (Distilled deionized water). Soluble compounds were eliminated from de surface, in most of the case studied, and amorphous silica was detected. The silica change its crystallographic phase when temperature was raised as is reported in the literature. The table 1 exhibit the amount of ashes produced in each operational conditions. As it was observed in the table mentioned above, the rice husk produce more ashes than the other lignocellulosic sources studied. In most of the cases, the amount of ashes produced was independent of heat velocity.

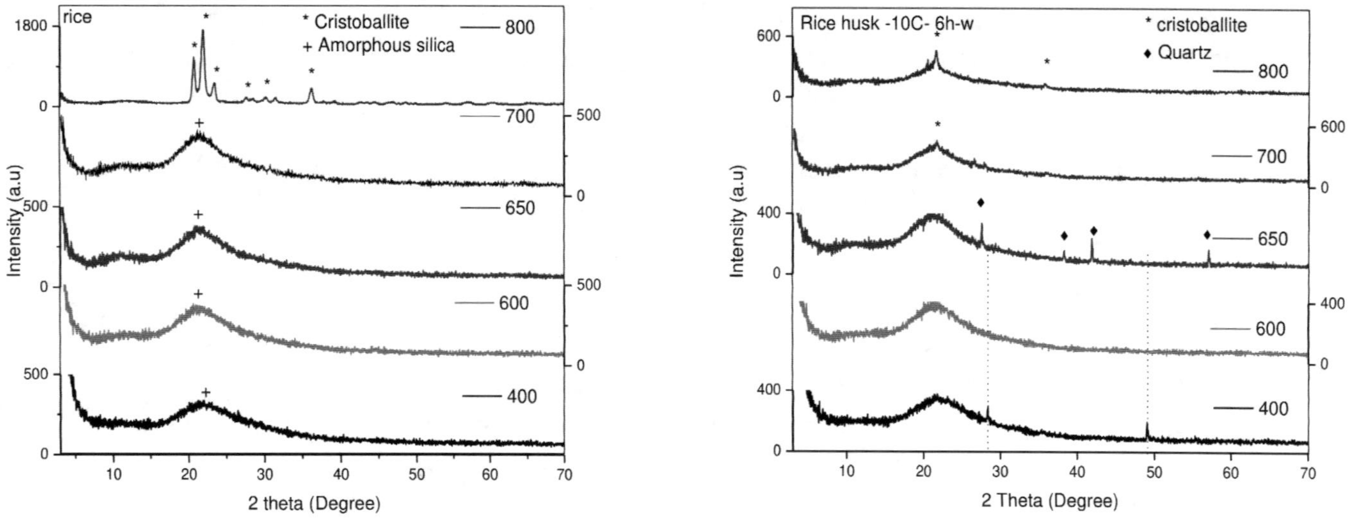

(a) (b)

Figure 1Ashes from rice husk, 10°C/min, 6h.
(a) calcinated , (b) calcinated and washed

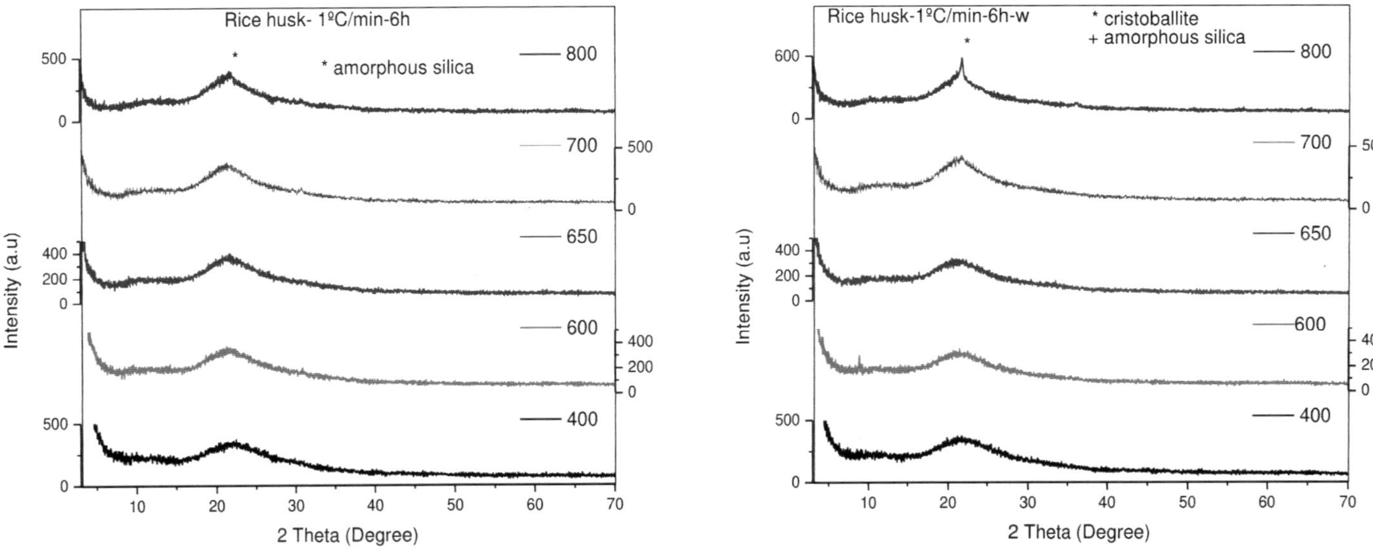

(a) (b)

Figure 2 Ashes from rice husk, 1°C/min, 6h.
(a) calcinated , (b) calcinated and washed

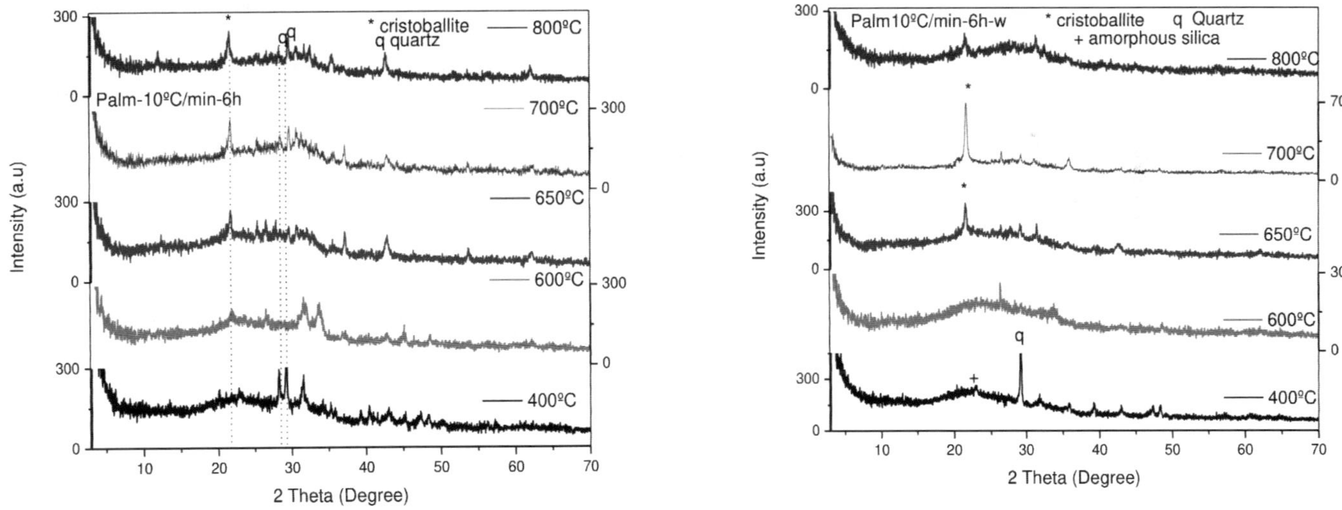

(a) (b)

Figure 3 Ashes from rice palm, 10°C/min, 6h.
(a) calcinated , (b) calcinated and washed

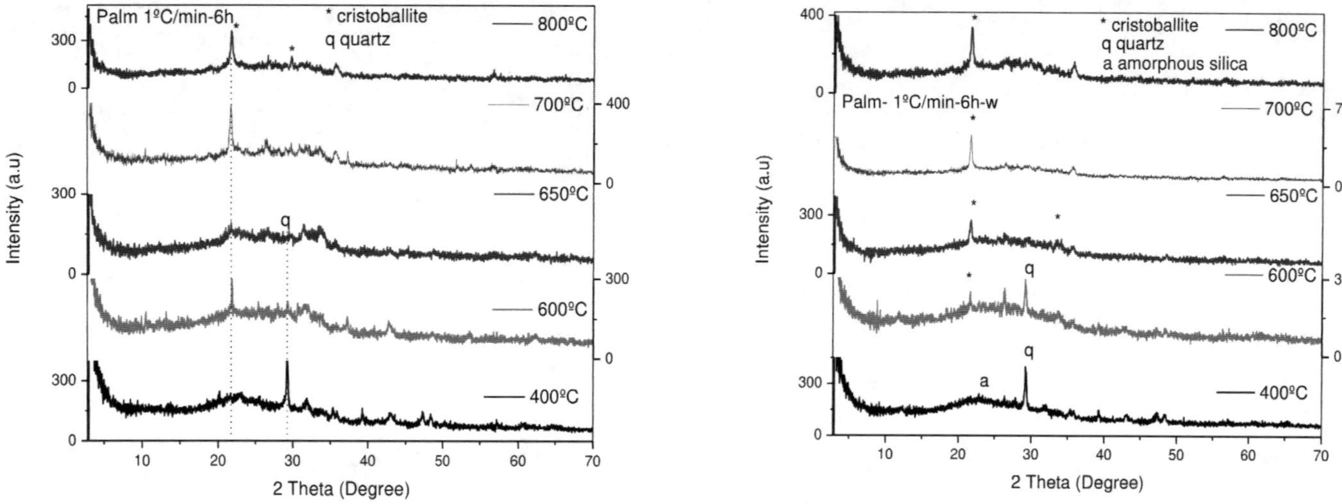

(a) (b)

Figure 4 Ashes from rice palm, 1°C/min, 6h.
(a) calcinated , (b) calcinated and washed

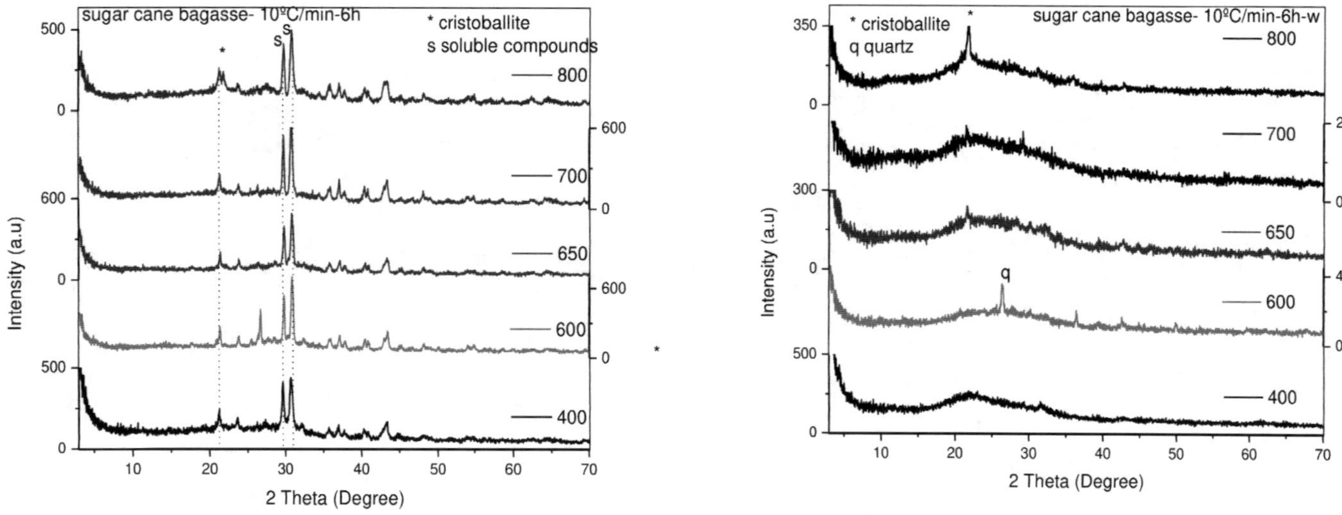

(a) (b)

Figure 5 Ashes from sugar cane bagasse, 10°C/min, 6h.
(a) calcinated , (b) calcinated and washed

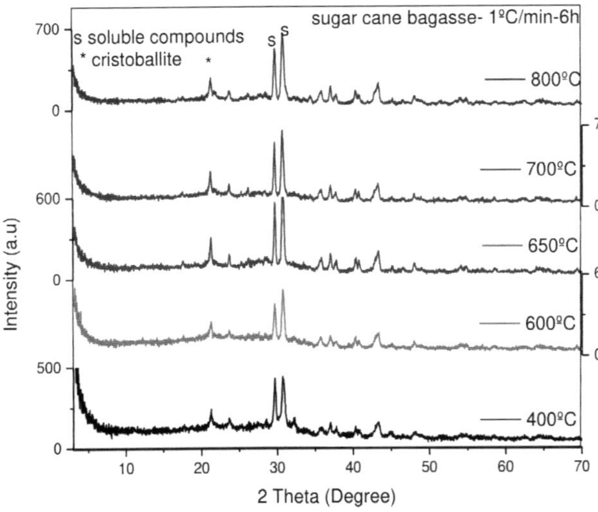

(a)

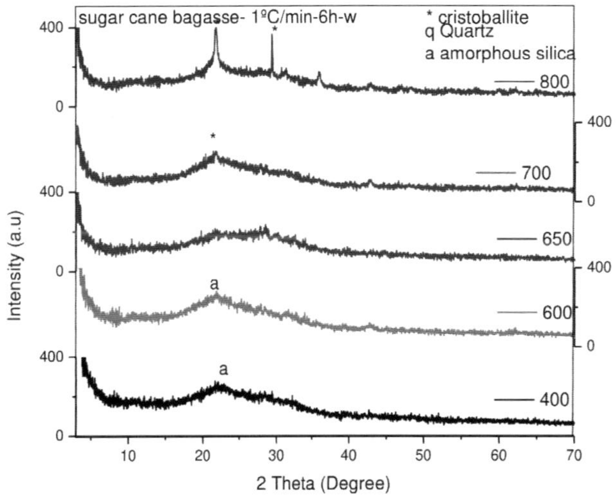

(b)

Figure 6 Ashes from sugar cane bagasse, 1°C/min, 6h. (a) calcinated , (b) calcinated and washed

Table 1. Ashes (wt%) from combustion process

Material	Temperature (°C)	10º C/min, 6h	1º/min, 6h
Palm oil raquis	400	2.06	2.48
	600	0.76	2.43
	650	2.22	2.20
	700	2.88	1.92
	800	3.31	1.55
Rice husk	400	19.92	19.75
	600	14.20	18.48
	650	18.40	10.13
	700	17.64	17.93
	800	16.01	18.17
Sugar cane bagasse	400	2.47	2.85
	600	2.09	2.36
	650	1.48	0.88
	700	1.71	2.49
	800	1.95	1.95

Influence of electrochemical cell configurations and surface engineering on microbial electrocatalysis

Amit Kumar[a&b], Piet Lens[c], Dónal Leech[a]

[a]Biomolecular Electronics Research Laboratory, School of Chemistry & Ryan Institute, National University of Ireland Galway, Ireland.
[b]Current address: Department of Chemical Engineering, Massachusetts Institute of Technology, 77 Massachusetts Ave, Cambridge, MA, USA.
[c]Department of Environmental Engineering and Water Technology, Unesco-IHE, Delft, The Netherlands.

Outline:
There is an increasing interest in microbial electrochemical cell technologies because of their potential application in bioelectricity and chemicals production. In these cells the anode potential, electrochemical cell configuration and the anode surface structure and chemistry plays a distinct role. We report here on the voltammetric behaviour of electroactive bacterial (EAB) films, induced to grow on carbon-based electrodes under different growth conditions. The approach initially examines the role of an applied controlled-potential to affect biofilm electrochemical catalytic activity using mixed-culture studies on model fuel/wastes in two different configurations. Increased biofilm current production is observed to occur as a function of increasing the applied anode potential (up to +0.2 V vs Ag/AgCl), in single-chamber electrochemical cell configuration. In contrast highest currents are observed at lower applied anode potential (−0.2 V vs Ag/AgCl), in dual-chamber, membrane-separated, electrochemical cell configuration. In addition, we also attempted surface modification of the carbon surface, using aryldiazonium salt reduction, for the improved adhesion, growth and electrical power production. Arylamine modified anodes achieve a current density of 3.4 A m^{-2}, while unmodified anodes achieve only 1.3 A m^{-2} during the first batch feed cycle.
Over all, this insight into differences in optimal applied potentials based on cell configuration can play an important role in selection of parameters required for microbial fuel cells and bio-electrochemical systems. The surface functionalization strategy provided a route to enhancing microbial bioelectrochemical systems process performance and for studying the complex mechanisms involved in such systems.

Funding: Support for this research was provided by a Charles Parsons Energy Research Award through Science Foundation Ireland and an EU FP 7 Marie Curie Fellowship to AK for Career Development.

369871 Immobilizing Sulfur for Lis Batteries

Abstract:

Sulfur cathodes are anticipated to be the next step in cathode chemistries, providing a unique combination of low-cost active material, high energy density, and high specific capacity. With theoretical specific capacities of 1672 mAh/g, sulfur cathodes have the potential of obtaining near tenfold capacity increase over current conventional lithium-ion cathodes.

However, the loss of soluble polysulfide intermediates results in a significant capacity fade that must be overcome. The immobilization of polysulfide intermediates to inhibit loss of active cathode material through modification of the cathode would allow for vigorous use in diffusion and convection batteries. The convection battery can be used to prevent dendritic growth on lithium anodes, which is another potential problem with lithium-sulfur cells on extended cycling. This presentation is on research to immobilize sulfur and its intermediates including analysis of the impact on cycling and performance in the convection battery.

362271 Importance of Conductivity and Structure in the Capacity Retention of NiO Anodes in Li Batteries

Abstract:

Over the past several years, there has been a considerable amount of effort put forth to increase the energy density of Li-ion batteries. At the anode, transition metal oxides have emerged as one of the most promising classes of materials, offering multiple electron transfer steps per mol of reactant, acceptable electronic conductivity, and relatively low cost (though higher than traditional graphite materials). Several mechanisms are active in this class of materials, with the most common being the conversion of the metal oxide to its base metal:

Charge: $M_xO + 2Li^+ + 2e^- \rightarrow xM + Li_2O$

where M is the transition metal (Ni, Fe, etc). Unfortunately, the reversibility of the conversion reaciton is not very good, leading to precipitous losses in the reversible capacity; sometimes almost all of the capacity can be lost in less than 20 cycles as material is trapped in the charged state.

One of the most pressing questions in the literature for metal oxides is root cause for the poor reversibility. In this talk, we will discuss the roles of conductivity and initial metal oxide nanostructure on capacitiy retention, with NiO being a probing compound. We will also summarize results that have allowed our group to produce high loading NiO-based electrodes that have reversible capacities around 700 mA/g, even after 100 cycles.

383973 Synthesis and Electrochemical Performance of LiMnPO4 Nano Materials As a Cathode Material for Lithium-Ion Batteries

Abstract:

Exploring novel cathode materials is crucial to meet the demand of high-performance lithium ion battery. The increasing research interests have been focused on the new active materials to replace the typical LiCoO2 regarding to the costs and the toxicity of cobalt. The potential alternatives are lithium transition metal phosphates, such as LiFePO4 or LiMnPO4, thanks to their thermal and electrochemical stability, low toxicity, low costs and their environmental friendly. The low operating potential of LiFePO4 indicates the low energy density which limits its application for electric vehicles. The successful commercialization of LiFePO4 has stimulated the search for other olivine's such as LiMnPO4. LiMnPO4 exhibits an operating voltage of about 4.1 V vs. Li+/Li, making the theoretical energy density (171 mAh·g-1× 4.1 V= 701 Wh·kg-1) 1.2 times larger than that of LiFePO4 (170 mAh·g-1× 3.45 V= 586.5 Wh·kg-1). However, the electrochemical performance of LiMnPO4 is poor due to the slow lithium diffusion kinetics within the grains and the low intrinsic electronic conductivity, even much worse than those of LiFePO4. Excellent electrochemical performance, especially the rate capability of the material can be obtained by particle-size minimization, carbon coating, cation substitution and morphology control.

LiMnPO4 is prepared by a hydrothermal method with different solvents such as PEG 400, NMP and EG, different surfactants such as SDBS, CTAB and PVP, different kinds of carbon coating to find the higher performance.

Electrochemical impedance spectroscopy (EIS) has been used extensively as an in-situ experimental tool for understanding kinetic and transport processes within electrochemical systems. Equivalent circuits, composed of passive electrical elements, are the most common approach used to determine parameters from impedance spectra. However, this fitting technique can be complicated by multiple equivalent circuits corresponding to the same impedance spectra. A more robust approach to understanding experimental spectra is through the development of macroscopic models, which account for the underlying physics and chemistry of the system. In this work, we develop a pseudo-2D macro homogeneous dual lithium ion insertion cell model integrated with the ability to simulate electrochemical impedance spectroscopy EIS to investigate the cause of time constants in lithium ion battery impedance spectra. Our model accounts for physics based transport and growth of the solid electrolyte interphase (SEI), the surface film that forms on the negative electrode particles which consumes active lithium.

FIRE SAFETY WITH HIGH ENERGY DENSITY BATTERIES

Michael C. Stern,[1] *William Braff,*[1] *Hubert Biteau,*[2] *Kevin Marr,*[1] *Vijay Somandepalli*[1]
[1] *Exponent, Inc., Natick, MA* [2] *Exponent, Inc., Bowie, MD*

Abstract

As high energy density battery technology develops to meet the rapidly increasing demands of consumer electronics, electric vehicles, and grid level storage, corresponding battery safety issues need to be understood and risk management strategies must be developed to mitigate potential failures. In this paper, battery fire safety will be discussed in terms of the underlying chemistry with a specific focus on lithium-ion battery cathodes. The ability of certain lithium-ion cathode materials to release oxygen during thermal runaway events can increase the severity of the failure. In this paper, we propose two potential metrics for battery fire safety: the oxygen balance and the anaerobic heating potential. The oxygen balance is calculated for a $LiCoO_2$ battery to compare the oxygen content to the available fuel. In addition, the anaerobic heating potential, or available heat without the need for external oxygen, is calculated based on the oxygen balance of the active materials contained within a $LiCoO_2$ battery. The anaerobic heating potential is important because it represents the heat that can be released spontaneously during a failure. It is shown that the heat generated from the combustion of the internally generated oxygen in a $LiCoO_2$ battery is comparable to the resistive heating generated by an internal short-circuit. This framework can be applied to alternative cathode materials, including iron phosphate, aluminum, nickel, or manganese. This framework delivers a readily calculable property that may provide a useful screening tool for evaluating new battery materials. Further work is needed to refine the methodology and quantify its accuracy.

Introduction

The widespread adoption of mobile electronics has fueled the need for batteries with continually improving power and energy densities to supply power-hungry devices for longer periods of time. Lithium-ion batteries, which have been at the forefront of meeting these demands, have fire safety challenges when compared to batteries with aqueous electrolytes. Concurrently, larger packs of lithium-ion batteries are also being deployed in electric vehicles and are even being considered for considerably larger energy storage applications.[1] As the energy and power demands increase, the fire safety concerns also increase. Safety concerns regarding the widely adopted lithium cobalt oxide (LiCoO2) cells have attracted significant attention.[2,3] In response to these concerns, industry has shown considerable interest in lithium iron phosphate ($LiFePO_4$) cells that are believed to be more stable.[4] New cathode materials based on manganese, nickel, and aluminum oxides are also being developed.[5]

New technologies such as lithium metal anodes and magnesium-ion batteries offer the potential for even higher energy densities, which will likely further increase the challenges of battery fire safety management.[6,7] As these new technologies and materials are investigated, screening methods to estimate the potential hazards could help researchers focus on new materials that will likely lead to safer batteries.

One potential method to evaluate the hazards posed by severe battery failures is to apply well-established concepts from the discipline of fire science. Somandepalli and Biteau demonstrated the effectiveness of cone calorimetry, a common experimental technique in fire science, to evaluate the fuel load and heat release rate of a lithium cobalt oxide battery at different states of charge.[8] In this paper, we evaluate how the calculation of a battery's oxygen balance can be used as a quantitative predictive tool to estimate the internal temperature rise during an internal short-circuit. Lithium cobalt oxide cathodes, which undergo thermal runaway events with faster heat rates, have higher oxygen balances when compared to other cathode materials. When considering a complete battery, the heat released by the generated oxygen is comparable to the resistive heating generated by a short-circuit. Therefore, a battery with the capacity to generate oxygen could undergo greater self-heating and become a more competent source of ignition during a short-circuit failure.

Background

Lithium-ion battery technologies

Lithium-ion batteries offer higher voltages and higher energy densities than many other common types of secondary batteries, such as alkaline, lead-acid, or nickel-metal hydride. The high voltage is directly related to lithium's highly negative electromotive force (i.e., strong reducing properties). Because of the strong reducing capacity of lithium, it is unstable when exposed to either air or water. To make a stable lithium, or lithium-ion, battery, the battery must be sealed from ambient oxygen and humidity and use a non-aqueous electrolyte. Lithium-ion batteries often use carbonate solvents (e.g., dimethyl carbonate and ethylene carbonate) with lithium hexafluorophosphate salt ($LiPF_6$).

Lithium-ion batteries have a range of cathode materials including lithium cobalt oxide ($LiCoO_2$), lithium iron phosphate ($LiFePO_4$), as well as several nickel and manganese cobalt oxide mixtures (NMC). Lithium-ion anodes are most commonly made with graphite that can be intercalated by the lithium metal atoms.

Existing testing methods

A variety of testing methodologies can be used to evaluate the thermal stability of batteries including several forms of calorimetry (ARC, DSC, Cone Calorimetry) and induced failures in sealed environments.[9] Accelerating Rate Calorimetry (ARC) and Differential Scanning Calorimetry (DSC) are common analytical tools used to assess the thermal stability of materials. Cone calorimetry, conversely, is used to evaluate the heat release rate of burning materials in fire scenarios; this provides an effective measure of the severity of a battery fire.

Cone calorimetry, however, has traditionally been applied to measure hydrocarbon based fires and needs modification in order to handle batteries that are primarily of inorganic composition. Somandepalli and Biteau incorporated oxygen release from $Li_{0.5}CoO_2$ in their measurements of a $LiCoO_2$ battery to calculate the heat release rate.[8] More work to incorporate the contributions of metal oxidation and reduction reactions are needed to further develop cone calorimetry for battery failure purposes.

To quantitatively assess the composition of the vent gas from a failed battery, sealed chamber experiments have been performed in which thermal failures are induced and the quantity of composition of the released gases are measured. Testing with $LiCoO_2$ has shown that a significant quantity of gas is released during a thermal runaway event. Although tests were performed in the absence of oxygen, around 21% to 32% of the generated vent gas was the fully oxidized species, CO_2, indicating reaction with oxygen from within the battery.[9]

Application of Fire and Explosion Science Methodologies to Batteries

Calorimetric techniques like DSC and cone calorimetry measure both the total heat release and heat release rate of material under different scenarios. Materials can have very similar total heat releases, but drastically different heat release rates, creating significantly different hazards in some scenarios. Often the heat release rate, or reaction rate of decomposition for condensed materials is strongly impacted by the oxygen balance of the material. The oxygen balance, however, is not sufficient to define a material as hazardous. It only compares the ratio of oxidizer to fuel, not an energy or power density. The oxygen balance is defined as the mass ratio of oxygen produced (if positive) or oxygen consumed (if negative) per mass of the initial molecule during a combustion of the material. We can mathematically define the oxygen balance (*OB*) for a molecule:[10]

$$\mathrm{OB} = -\frac{8}{MW}\left(\sum_i v_i N_i\right) \tag{1}$$

Where:
MW = Molecular weight of molecule
N_i = Number of atoms of element *'i'* per molecule
v_i = Oxidation state of element *'i'* after combustion (e.g., hydrogen is +1, oxygen is -2)

For transition metals, the heuristic of a +2 post combustion oxidation state is often applied. However, at representative combustion temperatures (i.e., 1000 °C), Fe_2O_3 is more stable than FeO and therefore, a +3 value is more appropriate. Table 1 shows the most stable oxide and oxidation state of several common metals based on formation free energies. Further study is required to determine if a more appropriate temperature or methodology is required to calculate the dominant oxidation state. One limitation of this method is that it assumes the end state is an oxide material. For phosphate salts that are stable to high temperatures, this may not characterize the final state correctly.

Table 1. Stable oxides and metal oxidation states for several metals

	0 °C		500 °C		1000 °C	
Metal	**Most Stable Oxide**	**Metal Oxidation State**	**Most Stable Oxide**	**Metal Oxidation State**	**Most Stable Oxide**	**Metal Oxidation State**
Fe	Fe2O3	+3	Fe2O3	+3	Fe2O3	+3
Co	Co3O4	+8/3	Co3O4	+8/3	CoO	+2
Mn	MnO2	+4	Mn2O3	+3	Mn3O4	+8/3
Al	Al2O3	+3	Al2O3	+3	Al2O3	+3
Ni	NiO	+2	NiO	+2	NiO	+2

Table 1 shows a trend of Ni > Co > Mn > Fe = Al in oxidizing ability if initially starting at the same oxidation state. Typically, the oxidation state of one of these metals would be +3 or +7/2 when in a charged cathode. Lithium nickel oxide ($LiNiO_2$) cathodes are known to be thermally unstable, which is consistent with the above table showing they could potentially act as an oxidizers even at low temperatures.

Previous decomposition studies of lithium cobalt oxide cathode have shown that the cobalt transitions from a +7/2 oxidation state ($Li_{0.5}CoO_2$) to a +8/3 oxidation state (Co_3O_4) and then finally to the +2 state (CoO).[3] The higher stable oxidation states of manganese, aluminum, and iron (and therefore lower oxygen balances) explain why cathode materials with those components are often found to be more stable.

Phosphorous in the form of phosphate, will maintain a +5 oxidation state even at elevated temperatures. Alkali Metals (e.g., lithium) will have a combusted oxidation state of +1 while Alkaline Earth Metals (e.g., magnesium) will have a combusted oxidation state of +2. The oxygen balance of a material composed of multiple molecular types can be calculated as the mass weighted average. The oxygen balance of a lithium-ion $LiCoO_2$ 2100 mAh pouch cell is calculated in Table 2 based on oxidation states at 1000 °C.

Table 2. Composition and oxygen balance (OB%) of $LiCoO_2$ battery

Component	Chemical Compound	Molecular Formula	Mass %	OB %	Mass Weighted OB%
Tape	Polyester Terephthalate	$(C_{10}H_8O_4)_n$	0.2	-166.7	-0.3
Label	Polypropylene	$(C_3H_6)_n$	0.5	-342.9	-1.7
Separator	Polypropylene	$(C_3H_6)_n$	6.4	-342.9	-21.9
Electrolyte (No Salt)	Carbonates	$C_5H_{10}O_3$	9.5	-149.1	-14.2
Outer Packaging (Top)	Nylon 6	$(C_6H_{11}NO)_n$	2.2	-233.6	-5.1
Outer Packaging (Bottom)	Polypropylene	$(C_3H_6)_n$	2.1	-342.9	-7.2
Positive Tab	Polyester Terephthalate	$(C_{10}H_8O_4)_n$	0.9	-166.7	-1.5
Negative Tab	Polyester Terephthalate	$(C_{10}H_8O_4)_n$	0.9	-166.7	-1.5
$LiCoO_2$ coated, positive electrode (aluminum)	Lithium Cobalt Oxide (Charged)	$Li_{0.5}CoO_2$	42.4	+12.7	+5.4
Graphite-coated negative electrode (copper)	(Charged)	LiC_6	34.9	-253.2	-88.4
Total:					**-137.2**

Composition from Ref. 8

Electrolyte formula shown as Diethyl Carbonate (DEC) as reported solvent was 76% DEC. OB% based on full composition excluding salts.

The calculations in Table 2 show that the oxygen balance for the $LiCoO_2$ cell is still significantly negative despite the positive contribution of the $LiCoO_2$ electrode. Significantly more oxygen than can be produced in the cathode is required to combust the fuel load of the anode, separator, and solvent. It is possible that the anode does not significantly contribute to the fuel load, however, because of the high ignition temperature of graphite. Based on the results in this table, $Li_{0.5}CoO_2$ produces 12.7 grams of oxygen per 100 grams of material or 5.4 grams of oxygen per 100 grams of battery. Lithium perchlorate, a strong oxidizer sometimes used in battery electrolytes has an oxygen balance of 60%, and therefore, produces five times the oxygen per gram as $Li_{0.5}CoO_2$.

For hydrocarbon fuels, combustion generates approximately 13.1 kJ/g of O_2.[8] Therefore, the oxygen capacity of the charged cathode could contribute 707 J/g (0.2 Wh/g) of anaerobic heating. The anaerobic heating potential, which is the amount of heating possible without external air, represents the heat release possible during a rapid internal short-circuit before material is exposed to atmospheric air. During an internal short-circuit, the electrical energy is dissipated causing localized hotspots in the battery. If the cathode of the battery is capable of releasing oxygen, chemical reactions may occur and lead to more anaerobic heating. The anaerobic heating potential could be a useful parameter when evaluating the initial stages of failure in vented batteries, which are sealed from external air until prior to venting.

For the 38 gram 7.8 Wh battery described in Table 2, the anaerobic resistive heating generated by the dissipation of the stored electrical energy through a short circuit is approximately equal to the anaerobic heating due to the combustion of internally generated oxygen. Aerobic heating can also be important when material is ejected from a cell and mixes with atmospheric oxygen.

Table 3 calculates the anaerobic and aerobic heating potentials for the $LiCoO_2$ cell. The potential anaerobic and aerobic heating are both calculated using the approximation that the battery materials produce 13.1 kJ of heat per gram of oxygen consumed.

Table 3. Heating modes and capacities of $LiCoO_2$ battery

Mode of Heating	**Heat (Wh)**	**Temperature Rise (°C)**
Short Circuit	7.8	742
Combustion with Generated Oxygen (Anaerobic)	7.6	720
Combustion with External Oxygen (Aerobic)	191	2410
Combustion with External Oxygen (Inert Anode, Aerobic)	69	1950

Potential temperature rises calculated with a battery heat capacity of 1 J/g/K. Temperature rises for aerobic processes based on the heat capacity of the battery and the air required for combustion.

The results in Table 3 highlight the similarity in anaerobic heating capacity for the combustion of generated oxygen and the resistive heating through a short-circuit. It also shows that the anaerobic processes represent a relatively small amount of energy when compared to the aerobic combustion of the battery regardless of the inclusion the anode. Evaporation, melting, radiative heat losses, and heat loss to the surroundings are not considered in the temperature rises and would reduce the potential temperature rises if included.

Accelerating rate calorimetry experiments performed by Exponent on $LiCoO_2$ (0.21 Wh/g) and $LiFePO_4$ (0.12 Wh/g) pouch cells demonstrate the difference in performance of the two cathode materials. The oxygen balance of the $LiFePO_4$ cathode material is zero from 0 °C to 1000 °C, based on the oxidation states in Table 1. Therefore, it is expected that the thermal runaway of the $LiFePO_4$ battery will be less severe since there is no potential for anaerobic heating due to oxygen generation from the cathode.

The ARC experiments were performed in air, which may have allowed for some combustion of ambient oxygen. The inner chamber of system, however, is expected to contain less than a gram of oxygen based on its volume. Based on the oxygen balance calculated in Table 2, this should only be sufficient to combust a few percent of the total fuel available fuel in the $LiCoO_2$ battery. $LiCoO_2$ accelerates more rapidly than $LiFePO_4$ cells once the runaway occurs. The $LiCoO_2$ battery at 50% state of charge (SOC) heats more quickly than the fully charged $LiFePO_4$ cell at temperatures over 200 °C despite having slightly less stored electrical energy than the $LiFePO_4$ cell. This suggests that a source of heat other than electrical energy, may be participating the thermal runaway events.

The low self-heating rates of the fully discharged battery indicates that without the localized resistive heating that occurs due to a short-circuit, the battery temperature does not reach localized temperatures high enough to induce the additional self-heating phenomena. This is not surprising considering that temperatures beyond 500 °C are required to cause the decomposition of Co_3O_4 to CoO, and without localized short-circuits, those temperatures are not reached during the ARC experiments.

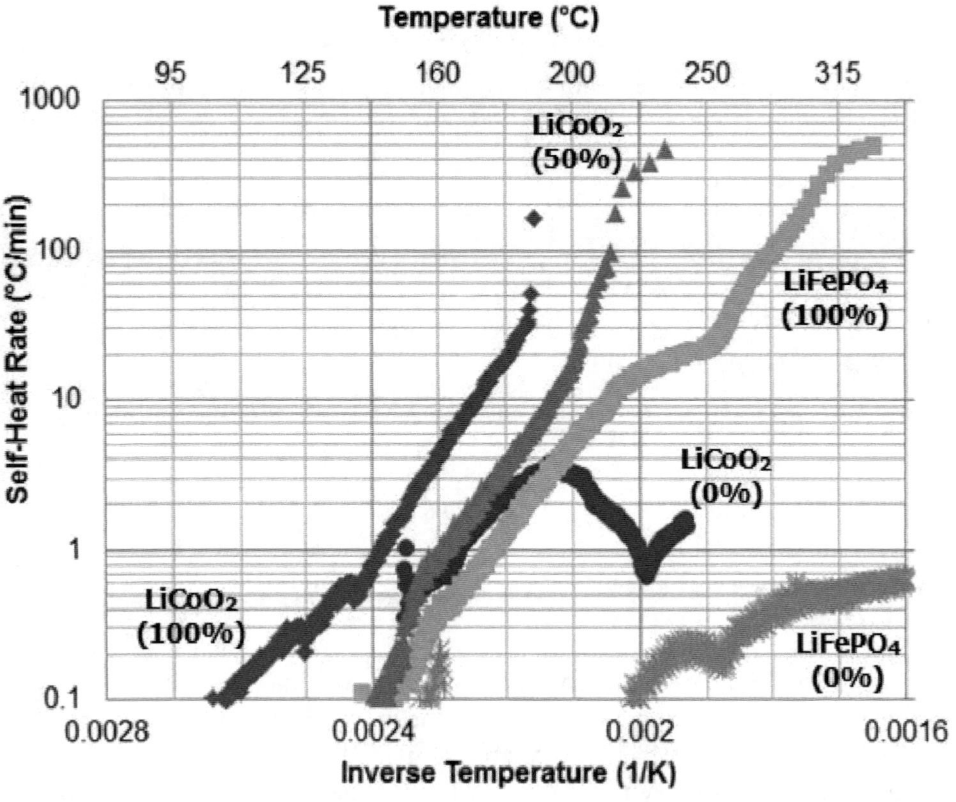

Figure 1. Self-heat rates for $LiCoO_2$ and $LiFePO_4$ cells and different states of charge.

Discussion and Conclusions

The oxygen balance and the associated anaerobic heating capacity of a battery have been proposed as metrics framework for battery safety. An internal short-circuit could initiate the decomposition of the cathode leading the generation and subsequent reaction of the oxygen, which could double the temperature rise of the battery. This additional temperature could increase the severity of the failure if it led to the external ignition of the much larger fuel loads in the battery.

Further work is needed to compare calculated oxygen balances to experimental results to determine the effectiveness of this method. Additional work may also indicate a more appropriate methodology for selecting the appropriate oxidation states to use for different metals.

Characterizing the anaerobic self-heating potential provides a quick estimate of the potential temperature rise of the battery during a severe failure. This can be used to compare the potential hazard new cathode materials and new battery technologies to existing materials. The potential temperature rise is a relevant property because the competency of a shorted battery to act as an ignition source for its own combustible contents or nearby combustible materials is dependent on the temperatures that could be obtained during a short-circuit induced anaerobic heating event.

References

1. Dunn, B., Kamath, H., and Tarascon, J-M (2011), "Electrical Energy Storage for the Grid," Science 334:928-935.
2. Doughty, D.H. and Pesaran, A.A. (2012), "Vehicle Battery Safety Roadmap Guidance", NREL Technical Report NREL/SR-5400-54404.
3. Leising, R.A.,, Palazzo, M.J., Takeuchi, E.S., Takeuchi, K.J. (2001) "Abuse Testing of Lithium-Ion Batteries: Characterization of the Overcharge Reaction of $LiCo_{O2}$/Graphite Cells," J. Electrochem. Soc. 148(8):A838-A844.
4. Etacheri, V., Marom, R., Elazari, R., Salitra, G., and Aurbach, D. (2011), "Challenges in the development of advanced Li-ion batteries: a review," Eng. Env. Sci. 4:3243-3262.
5. MacNeil, D.D., Lu, Z., Chen, Z., and Dahn, J.R. (2002), "A comparison of the electrode/electrolyte reaction at elevated temperatures for various Li-ion battery cathodes," J. Power Sci. 108:8-14.
6. Chusid, O., Gofer, Y., Gizbar, H., Vestfrid, Y., Levi, E., Aurbach, D., and Riech, I., (2003), "Solid-State Rechargeable Magnesium Batteries," Adv. Mat. 15:627-630.
7. Takehara, Z., (1997), "Future prospects of the lithium metal anode," J. Power Sci. 68(1):82-86.
8. Somandepalli, V. and Biteau, H. (2014), "Cone Calorimetry as a Tool for Thermal Hazard Assessment of Li-Ion Cells," SAE Int. J. Alt. Power. 3(2):222-233.
9. Somandepalli, V., Marr, K., and Horn, Q. (2014), "Quantification of Combustion Hazards of Thermal Runaway Failures in Lithium-Ion Batteries," SAE Int. J. Alt. Power. 3(1):98-104.
10. Shanley, E.S., and Melhem, G.A. (2004) "The oxygen balance criterion for thermal hazards assessment," Proc. Safety Prog. 14(1):29-31.

AUTHOR INDEX

AUTHOR INDEX

AUTHOR INDEX

AUTHOR INDEX

AIChE
120 Wall Street, FL 23
New York, NY 10005-4020

ISBN 978-1-5108-1261-1
9 781510 812611

CURRAN ASSOCIATES INC.
proceedings.com